D1329647

Inorganic Biochemistry

Inorganic Biochemistry

An Introduction

Second Edition

J. A. Cowan

 **WILEY-VCH**

NEW YORK · CHICHESTER · WEINHEIM · BRISBANE · SINGAPORE · TORONTO

J. A. Cowan
Evans Laboratory of Chemistry
The Ohio State University
100 West 18th Avenue
Columbus, OH 43210-1173

This book was printed on acid-free paper ∞

Library of Congress Cataloging-in-Publication Data

Cowan, James A.
  Inorganic biochemistry : an introduction / James A. Cowan.
    p.   cm.
  Includes bibliographical references and index.
  ISBN 0-471-18895-6 (hc : alk. paper).
  1. Bioinorganic chemistry.   2. Bioinorganic chemistry.   I. Title.
  QP531.C68   1996
  574.19'214--dc20
                                                                96-14100
                                                                    CIP

Printed in the United States of America

ISBN 0-471-18895-6   Wiley-VCH, Inc.

10 9 8 7 6 5

Dedicated to my parents and brother

(Helen, Jim, and Robert)

Preface

This new edition maintains the approach adopted in the earlier version, by focusing on the key concepts and underlying principles of inorganic biochemistry. Major additions to the text include a substantive review section on reaction kinetics and thermodynamics. Although these topics are perhaps among the most daunting for the beginning student, they also provide great insight into, and an understanding of, the dynamics and equilibria of biochemical reactions. For this reason, a close study of this section will pay rich dividends in later chapters. A second major addition is a comprehensive set of problems and study questions. Many of these are based on examples from the original literature. Complete literature citations are provided to facilitate and encourage the reading of the original papers. This will allow the reader to check the answers to problems and also serve as a tool to encourage and develop the habit of reading the primary literature. Such problems should also illustrate the value of library work in solving real scientific problems. In several chapters, some illustrative problems have been worked through, with complete answers provided in the text. Review questions have been added after most of the summary sections in each chapter, and some of these are appropriate for group discussions.

In keeping with the earlier edition, the text does not aim to be comprehensive but to illustrate the use of basic principles to tackle important problems in inorganic biochemistry. Several topics have been updated where recent progress has been made. The most important updates and revisions include, as noted earlier, a new section on basic kinetic and thermodynamic principles in Chapter 1. Also, a discussion of transient kinetic methods has been added to Chapter 2. The molecular mechanisms of two redox enzymes, methane monooxygenase and sulfite reductase, are now sufficiently developed to merit inclusion in Chapter 5. Chapter 6 includes a new section on ribozyme chemistry. The coverage of iron regulation in Chapter 7 has been expanded to include a discussion of iron-response elements (IREs) and the iron regulatory protein (IRP). A new section on radiopharmaceuticals has been added to Chapter 8, and the discussion of peroxidase enzymes has been expanded to highlight recent progress in that area. Finally, a new case study on the chemistry of the DNA-cleaving antibiotic bleomycin has been added to Chapter 10. General references have been updated for each chapter.

The first edition of this text prompted much constructive criticism and valuable suggestions. I have tried to accommodate many of these in this revision. Further comments and suggestions resulting from this edition would certainly be most welcome.

J. A. Cowan
Columbus, Ohio

Preface to the
First Edition

Inorganic biochemistry is an expanding field of modern chemistry. It lies at the interface of biochemistry and classical coordination chemistry and draws on a large number of subdisciplines. Synthetic chemists, spectroscopists, electrochemists, theoreticians, biochemists, and molecular biologists meet at the frontiers of chemistry and biology and are challenged by exciting problems. Unfortunately, the rapid progress of the last two decades has met little in the way of nonspecialist reviews to allow the general reader access to the field. I have tried to develop a text that adopts a pedagogical style to teach the concepts and underlying principles of modern metallobiochemistry. Essential background material in physical methods and biochemistry is covered. Building from a firm basis in inorganic coordination and reaction chemistry, the biological chemistry of metal ions is developed. Emphasis is placed on chemical principles rather than on the techniques employed. No effort has been made to provide detailed coverage of every area that might deserve attention. Rather, specific examples have been chosen that illustrate important general principles. General references (normally other texts or review articles) are detailed at the end of each chapter. Original literature references are cited in figure legends and table footnotes to allow interested readers, especially those utilizing the book as an adopted course text, to review and develop the material more thoroughly.

In preparing the content of each chapter I felt strongly that the chemistry should evolve from consideration of biochemical topics (e.g., transport and storage, regulation, redox pathways, etc.), rather than a description of the chemistry of specific metals, cofactors, or enzymes. Chapters 1 and 2 will allow the student to develop a background in inorganic solution chemistry, structural and mech-

ix

anistic biochemistry, and physical inorganic methods that forms a necessary foundation for subsequent chapters. I do not consider Chapter 2 to be essential reading, at least in the initial stages. For some readers it may best serve as a reference source for the methodology utilized in later chapters. In my own opinion it is more important for the student to appreciate the chemistry underlying biology than it is to know the spectroscopic nuts and bolts that are often required to develop this chemistry. Taken out of context, physical methodology can often leave the beginning student frustrated and uninterested. Spectroscopy can be tough, especially when the pay-off is nowhere in sight. Of course, instructors will have their own views on the best way of introducing the subject. It is hoped that the book will prove flexible enough to accommodate all styles. I would certainly welcome all suggestions and comments on any aspect of the text.

The ubiquity of metal ions in biology has led to widespread interest in this area among workers from other disciplines. Although this book is written primarily as a course text for senior undergraduates and beginning graduate students in chemistry and biochemistry, it should be of general value to the biological community and to those researchers requiring an introduction to this field of study. It should arm the student with the basic knowledge required to tackle more advanced reviews. In and of itself this text is not designed to cover all the details of inorganic biochemsitry. If it generates interest and enthusiasm for this exciting area of chemistry and stimulates the reader to learn more, it will have served its purpose well.

Special thanks go to Lee Stout for typing the body of the text and Jeanne Jaros, who prepared the graphic art. Thanks are also due to B. A. Averill and J. T. Bolin, who each provided copies of manuscripts prior to publication. I am especially grateful to a large number of graduate students (A. Agarwal, C. B. Black, H-W. Huang, K. R. Kneten, S. Kim, W. Liang, S. M. Lui, A. Soriano, A. Tevelev, and B. M. Wolfe) for their valuable comments while the text was in preparation. I am also indebted to colleagues (J. K. M. Sanders, B. A. Averill, and L. Que) for constructive criticism and suggestions for improvement. I would welcome further comments or criticisms from readers. Ultimately, it is the students who must decide the value of this book, since it is for them that it is primarily written.

J. A. Cowan
Columbus, Ohio

Contents

Chapter 5. Metalloproteins and Metalloenzymes: (II) Redox Chemistry   203

Chapter 6. Alkali and Alkaline Earth Metals 257

Chapter 7. Metals in the Regulation of Biochemical Events   291

Chapter 8. Cell Toxicity and Chemotherapeutics   319

CHAPTER

1

Fundamentals of Inorganic Biochemistry

Inorganic biochemistry can be defined most simply as the investigation of the chemical reactivity of metal ions in biological environments. It is a broad and relatively young science that has attracted researchers from many diverse and apparently unrelated backgrounds. The multidisciplinary nature of this field may well be what differentiates it best from other areas of biochemical science. Synthetic chemists, spectroscopists, electrochemists, theoreticians, biochemists, and molecular biologists meet at the frontiers of chemistry and biology and are challenged by exciting problems. These problems are investigated by use of a wide array of spectroscopic, chemical, and biochemical techniques. However, it is toward the elucidation of the underlying chemical principles that such studies are directed. The apparent diversity of the field, in fact, does hide a simple truth: The rules governing the chemistry of biological molecules are the same as those that define the chemical and physical properties of typical organic and inorganic molecules and materials. For this reason a thorough understanding of the fundamental kinetics, thermodynamic, and structural principles underlying organic and inorganic chemistry will prove extremely useful in unraveling the reactivity of apparently complex biological systems. This chapter provides a crash-course introduction to solution properties of coordination complexes and attempts to outline some underlying principles of inorganic chemistry and biochemistry that provide a general reference point for the entire text. Part A covers important aspects of inorganic solution chemistry; Part B summarizes important ideas in chemical kinetics and thermodynamics; and Part C outlines key concepts in molecular and cell biology. Inorganic biochemistry is built around such a framework.

1

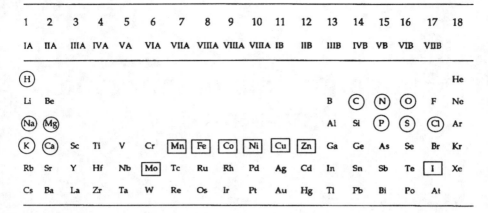

Figure 1.1   Elements known to be essential for life are highlighted ($\bigcirc$ bulk elements, $\square$ trace elements). Labels 1–18 (new system) or IA–VIIB (old system) at the heads of columns represent groups of electronically and chemically related elements. The important valence electrons for each category of element are noted in parentheses as follows: group 1, alkali (ns); group 2, alkaline earth (ns); groups 3–12, transition metals ($[n-1]d$); groups 13–17, main group (np); group 18, noble gases. Essential ultratrace elements include Si, V, Cr, Se, Br, Sn, F.

PART A.   FUNDAMENTALS OF INORGANIC SOLUTION CHEMISTRY

1.1   The Elements

The bulk and trace elements that form the core of inorganic biochemistry are noted in Figure 1.1 and are the essential building blocks for all life forms. Table 1.1 illustrates how each class of element serves a distinctive biochemical role (structural, catalytic, regulatory). The field of inorganic biochemistry is directed toward the understanding of this chemistry. The biochemistry of carbon, hydrogen, nitrogen, oxygen, and phosphorus, which comprise the largest fraction of the material found in living organisms, will not be considered directly, although the chemistry of these main group elements forms a structural core around which our considerations of the biochemistry of metal ions and their relationship with proteins, enzymes, nucleic acids, lipids, sugars, vitamins, and hormones will be formed.

Table 1.1   Overview of the Biochemical Roles of the Elements

Alkali metals	Charge neutralization, voltage gating, structure stabilization
Alkaline earths	Messengers, enzyme activators, structure regulation
Transition metals	Electron transfer, redox catalysis
Main group	Structural elements of biological materials, conformational triggering

1.2　Formal Oxidation States and Coordination Geometries of Biologically Important Metal Ions

The alkali, alkaline earth, and main group elements tend to adopt an oxidation state that corresponds to a stable noble gas configuration (Figure 1.1). The noble elements (or inert elements) in group 18 have filled outer valence orbitals. Such an electronic configuration is extremely stable, and so these elements have little reason to engage in further chemistry. Other elements also tend to react so that they lose electrons to form positively charged species (cations) or accept electrons to form negatively charged species (anions) in order to attain such a configuration. In the early periods this results in cations of oxidation state equal to the group number (e.g., Na^+, Mg^{2+}, Al^{3+}), whereas the later nonmetals more readily accept electrons to give, in a formal sense, anions (e.g., O^{2-}, Cl^-). The oxidation state of an ion is an extremely useful formalism that aids in electron counting, although the effective charge at the atomic center may be different as a result of electron delocalization from neighboring atoms.

The transition metals offer the most diverse and interesting redox chemistry in the periodic table. Multiple stable oxidation states are common, reflecting the relative ease of removing d-electrons. The stability of a given oxidation state is a trade-off between the ionization energy (IE) required to remove electrons from the valence orbitals, and the solvation ($\Delta H_{\text{solvation}}$) or ligation energy ($\Delta H_{\text{ligation}}$) obtained by surrounding the metal cation with solvent or ligand atoms (Figure 1.2). In the early part of the first transition series there are few d-electrons to be

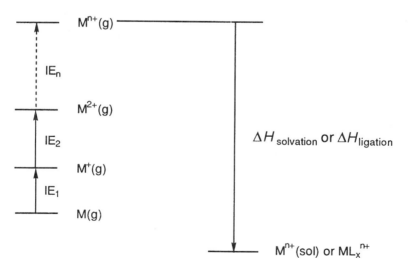

Figure 1.2 The formation of a solvated (sol) or ligated (L) metal ion involves several steps. Both the ionization energies (IE) and the solvation/ligation energies (ΔH) increase with the charge on the metal. For convenience, all abbreviations used in the text are summarized in Appendix 2.

lost, and so only lower oxidation states are available. At the other extreme it becomes progressively more difficult to remove electrons as we move across the series, since the d-orbitals are stabilized and their energies are lower. Also, the elements toward the right of the transition series have fewer unfilled d-orbitals to participate in bonding with electron-donating ligands. Table 1.2 lists the biologically important oxidation levels for the essential bulk and trace metals noted in Figure 1.1. Although several transition metals may form complexes with higher oxidation states, the ligand environments required to stabilize such valencies, which are highly oxidizing, are not commonly found in biology. For similar reasons, low oxidation states (M^{n+}, $n \leq 1$) are also uncommon.[1] Other than the organic frameworks of biological macromolecules, the main group elements tend to be found as small anionic (e.g., SO_4^{2-}, NO_3^-, HCO_3^-, Cl^-) or gaseous species (e.g., O_2, N_2, CO_2, SO_2, CH_4).

The aqueous solution chemistry of first-row transition metal ions is dominated by low or moderate oxidation states. Higher oxidation states tend to be the rule for second- and third-row transition metals, which consequently require electronegative ligands for stability to avoid oxidation by the high valent metal center. For example, anionic oxo and halo ions are formed (ReO_4^-, $PtCl_4^{2-}$, $PtCl_6^{2-}$). Since O^{2-} and Cl^- do not result in biologically functional complexes of these metals, second- and third-row transition metals are generally unsuitable for use in biological chemistry.[2] A second, evolutionary more pragmatic reason for the use of first-row transition elements stems from the larger abundance of these metals in nature. This provides a strong driving force in the world of natural selection.

Table 1.2 Oxidation States Commonly Available to Essential Bulk and Trace Metals

Metal	Available Oxidation States[a]						
Na	1						
K	1						
Mg		2					
Ca		2					
V		2	(3)	(4)	(5)		
Cr		2	(3)	(4)	(5)	(6)	
Mn		2	3	4	(5)	(6)	(7)
Fe		2	3	4	(5)		
Co	1	2	3				
Ni	1	2	3				
Cu	1	2					
Zn		2					
Mo		(2)	(3)	4	5	6	

[a] Numerals indicate positive oxidation levels. Those in parentheses are not generally found in biological molecules.

Table 1.3 Coordination Numbers (C.N.) and Preferred Geometries for Selected Metal Ions

Cation	C.N.	Geometry	Biological Ligands
Na^+	6	Octahedral	O, ether, hydroxyl, carboxylate
K^+	6–8	Flexible	O, ether, hydroxyl, carboxylate
Mg^{2+}	6	Octahedral	O, carboxylate, phosphate
Ca^{2+}	6–8	Flexible	O, carboxylate, carbonyl, (phosphate)
Mn^{2+} (d^5)	6	Octahedral	O, carboxylate, phosphate
			N, imidazole N
Mn^{3+} (d^4)	6	Tetragonal	O, carboxylate, phosphate, hydroxide
Fe^{2+} (d^6)	4	Tetrahedral	S, thiolate
	6	Octahedral	O, carboxylate, alkoxide, oxide, phenolate
			N, imidazole N, porphyrin
Fe^{3+} (d^5)	4	Tetrahedral	S, thiolate
	6	Octahedral	O, carboxylate, alkoxide, oxide, phenolate
			N, imidazole N, porphyrin
Co^{2+} (d^7)	4	Tetrahedral	S, thiolate
			N, imidazole N
	6	Octahedral	O, carboxylate
			N, imidazole N
Ni^{2+} (d^8)	4	Square planar	S, thiolate
			N, imidazole N, polypyrrole (F-430)
	6	Octahedral	Uncommon
Cu^{1+} (d^{10})	4	Tetrahedral	S, thiolate, thioether
			N, imidazole N
Cu^{2+} (d^9)	4	Tetrahedral	S, thiolate, thioether
			N, imidazole N
Cu^{2+} (d^9)	4	Square planar	O, carboxylate
			N, imidazole N
	6	Tetragonal	O, carboxylate
			N, imidazole N
Zn^{2+} (d^{10})	4	Tetrahedral	O, carboxylate, carbonyl
			S, thiolate
			N, imidazole N
	5	Square pyramidal	O, carboxylate, carbonyl
			N, imidazole N

Coordination data for common biological ions are noted in Table 1.3 and Appendix 3, and common coordination geometries are illustrated in Figure 1.3. Note that the coordination numbers and geometries of the alkali and alkaline earth metals depend on the relative sizes of the cation and ligand. Sodium and magnesium readily accommodate six ligands, whereas the larger potassium and calcium ions may expand their coordination shells to accommodate seven or eight ligands and display a flexible coordination geometry. Coordination preferences for the transition metals again depend, in part, on the steric bulk of the ligands surrounding the metal ion. Small cations can accommodate fewer li-

octahedral (O$_h$)   square pyramidal (spy)   trigonal bipyramidal (tbp)

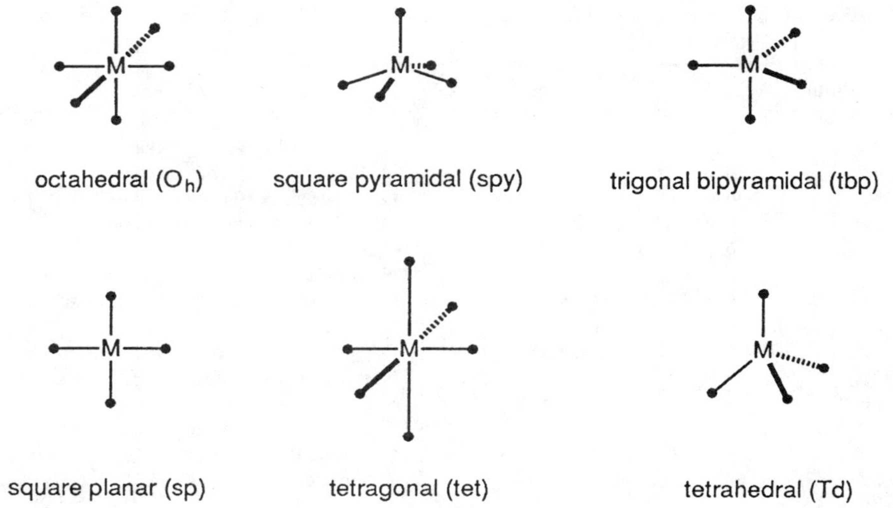

square planar (sp)   tetragonal (tet)   tetrahedral (Td)

Figure 1.3 Common coordination geometries around a metal ion. In biological macromolecules the ligand atoms (represented by ●) may each be distinct, and structural distortions from the regular geometries shown are common.

gands in their inner coordination spheres and tend to adopt a tetrahedral geometry that minimizes steric and electrostatic repulsion relative to square planar.[3] Unlike the alkali or alkaline earth metals, where the bonding is predominantly electrostatic in origin, the coordination complexes of transition metals contain a substantial degree of covalency. The d-orbital and ligand orbital energies are comparable and may interact favorably in a bonding fashion. Section 1.6 will illustrate these points in greater detail. Distortions from the ideal symmetries shown in Figure 1.3 are common in biology and provide important mechanisms for fine-tuning the physicochemical properties of the metal center.

Summary of Sections 1.1–1.2

1. Elements in the same group of the periodic table (alkali, alkaline earth, transition metals, main group, noble gases) have similar chemical properties. Elements in different groups have different chemistries, and this is reflected in their distinct biological roles.
2. Transition metals have variable oxidation states. There are two main reasons for the dominance of first-row metals: (1) they commonly adopt low valence states (M^{n+}, $n = 1, 2, 3$) that are compatible with biological ligands, and (2) they are the most abundant metals. Consequently, they are the most accessible for evolutionary development.

1.3 Classifications of Metal Ions and Ligands: Hard–Soft Acid–Base (HSAB) Theory

In 1963 Pearson coined the terms *hard* and *soft* to describe metal ions and ligands.[4] As a general rule, hard cations form their most stable complexes with hard ligands and soft cations with soft ligands. Table 1.4 classifies specific examples. The idea underlying this scheme is fairly simple. Hard metal ions (characterized by small ionic radii or high charge) have little electron density to share with a ligand. Hard ligands do not readily give up their electron density, and so a combination of the two results in a complex that is stabilized by simple electrostatics (e.g., Mg^{2+} and HPO_3^{2-}, or Na^+ and $CH_3CO_2^-$). On the other hand, soft metal ions and ligands (typically possessing polarizable electron clouds) are more prone to sharing electron density with a greater degree of covalency in the bonding, and so form a mutually stable complex (e.g., Cu^+ and I^-, or Hg^{2+} and CH_3S^-). If, for example, we combine a hard metal with a soft ligand, the metal does not readily accept the electron density being offered by the ligand, and so the resulting complex is less stable, since both partners are incompatible. Although rather simple-minded, the preceding illustration paints a clear picture of the origin of the effect. Certain metal ions and ligands display intermediate behavior and can interact reasonably well with both hard and soft species. As a general rule, the affinity of a donor atom in a ligand for a hard metal ion varies as $F > O > N > Cl > Br > I > C \sim S$. This order is inverted for a soft metal. In practice the HSAB concept has proved remarkably useful in indicating favorable combinations of ligands and metal ions. A close inspection of Table 1.3 will provide many useful examples that demonstrate the general validity of this approach.

Table 1.4 Classification of Lewis Acids and Bases by Hard/Soft Criteria

	Acids	Bases
Hard	H^+, Li^+, Na^+, K^+ Mg^{2+}, Ca^{2+}, Mn^{2+} Al^{3+}, Ln^{3+} Cr^{3+}, Co^{3+}, Fe^{3+}, VO^{2+}, MoO^{3+} SO_3, CO_2	H_2O, ROH, NH_3, RNH_2 RCO_2^-, Cl^-, F^-, PO_4^{3-}, HPO_4^{2-}, $H_2PO_4^-$, SO_4^{2-}
Intermediate	Fe^{2+}, Co^{2+}, Ni^{2+}, Cu^{2+}, Zn^{2+}, Pb^{2+}, Sn^{2+}, SO_2, NO^+, Ru^{2+}	Imidazole, pyridine
Soft	Cu^+, Ag^+, Au^+, Tl^+, Hg^+ Cd^{2+}, Pd^{2+}, Pt^{2+}, Hg^{2+}	RSH, R_2S, CN^-, I^- $S_2O_3^{2-}$

R denotes a general organic side chain. Intermediate acids and bases bind moderately strongly either to soft or hard bases and acids, respectively.

1.4   Stability Constants[5]

The selectivity of biological ligands for particular metal ions depends on both
the nature of the ligands (hard or soft) and the coordination geometries commonly adopted by the metal center. Table 1.3 notes the ligand and coordination
preferences for most of the important functional metal ions in biology. The
thermodynamic stability of a metal complex may be denoted by stepwise formation constants (or stability constants) K_n or by overall stability constants β_n.
For example, the stepwise replacement of water molecules by ligand L at a metal
ion $M(H_2O)_n$ in aqueous solution may be written as

$$M + L \rightleftharpoons ML \qquad K_1 = [ML]/[M][L]$$
$$ML + L \rightleftharpoons ML_2 \qquad K_2 = [ML_2]/[ML][L]$$
$$ML_2 + L \rightleftharpoons ML_3 \qquad K_3 = [ML_3]/[ML_2][L]$$

where the K_n represent individual stability constants for the reactions noted.[6]
Water molecules are typically not included in the equilibrium equation, since
$[H_2O] = 55.5$ M is a constant for aqueous solutions. A more correct repre-
sentation is given by $[M(H_2O)_n] + L \rightleftharpoons [M(H_2O)_{n-1}L] + H_2O$, where $K = [M(H_2O)_{n-1}L][H_2O]/[M(H_2O)_n]([L])$. Hidden solvent factors are frequently significant and must be accounted for in detailed analyses of the thermodynamics
of complex formation.[7,8] As an alternative to the preceding representation, the
reactions may be represented as

$$M + L \rightleftharpoons ML \qquad \beta_1 = [ML]/[M][L]$$
$$M + 2L \rightleftharpoons ML_2 \qquad \beta_2 = [ML_2]/[M][L]^2$$
$$M + 3L \rightleftharpoons ML_3 \qquad \beta_3 = [ML_3]/[M][L]^3$$

where the β_n represent overall stability constants. There is a simple relationship
connecting β_n and K_n:

$$\beta_n = K_1 K_2 \ldots K_n$$

We have already noted that the selectivity of a ligand for a metal ion is
sensitive to such factors as oxidation state, geometry, and HSAB criteria. Inspection of the β_1 values for metal–ligand complexes in Table 1.5 shows not
only a clear dependence of β_1 on the charge of the metal ion and ligand, but
also the hard/soft character of the metal and ligand. Systematic trends noted in
the literature relate to conditions where each metal ion is equally available.
However, in biology the availability of specific metal ions is rigorously controlled in both intra- and extracellular compartments; consequently, the competition of metal ions for specific ligands is also regulated by their local con-

Table 1.5　Stability Constants $(\beta_n)^a$ and Uptake Factors $(\beta_n \cdot [M^{n+}])$ for a Variety of Intracellular Metal Ions[b] and Biological Ligands

	Na^+	K^+	Mg^{2+}	Ca^{2+}	Cu^{2+}	Zn^{2+}
Glycine[c]	—	—	3.44	1.38	8.62	5.52
Glycine[d]	—	—	0.44	−5.62	−5.38	−4.48
Cysteine[c]	—	—	—	—	19.2	9.86
Cysteine[d]	—	—	—	—	5.2	−0.14
Aspartate[c]	—	—	2.43	1.6	8.57	5.84
Aspartate[d]	—	—	−0.57	−5.4	−5.43	−4.16
Histidine[c]	—	—	—	—	18.3	12.9
Histidine[d]	—	—	—	—	4.3	2.9
Citrate[c]	—	—	3.6	3.3	5.9	5.0
Citrate[d]	—	—	0.6	−3.7	−8.1	−5.0
Valinomycin[c]	0.7	4.9	—	2.7	—	—
Valinomycin[d]	−0.3	2.9	—	−4.3	—	—
NTA[c]	2.1	1.0	7.0	6.4	12.8	10.5
NTA[d]	1.1	−1.0	4.0	−0.6	−1.2	0.5

[a]　Taken from L. G. Sillen and A. E. Martell, *Stability Constants*, Special Publication 25, Royal Society of Chemistry, 1971.
[b]　Assuming, $[Na^+]_i \sim 0.1$ M, $[K^+]_i \sim 0.01$ M, $[Mg^{2+}]_i \sim 10^{-3}$ M, $[Ca^{2+}]_i \sim 10^{-7}$ M, $[Cu^{2+}]_i \sim 10^{-14}$ M, $[Zn^{2+}]_i \sim 10^{-10}$ M.
[c]　The first row of data for each ligand gives the log of the stability constant (log β_1).
[d]　The second row of data for each ligand gives the log of the uptake factor (log($\beta_1 \cdot [M^{n+}]$)). NTA = nitriloacetic acid (Appendix 1). The amino acids are illustrated in Figure 1.12. Valinomycin is related to enniatin, which is shown in Figure 1.4. Note that $\beta_1 = K_1 = K_{ML} =$ formation constant for the metal–ligand complex.

centrations (Table 1.5). To illustrate the point, consider a stable complex (ML) with a large formation constant K_{ML}. In the extreme limit, if there is no metal ion M in a solution of L, then clearly no ML can be formed! The local concentration of a complex ML depends, therefore, on both the magnitude of the formation constant and the availability of the metal ion. To account for this fact, an uptake factor ($K_{ML} \cdot [M]$) can be defined that explicitly considers the local concentration of the metal ion and gives a better measure of the probability of complex formation.[9] We shall briefly examine the effect of each factor that might influence metal binding constants. However, in any realistic case it is likely that several of these will be involved.

Ligand Preference

The preference of metal ions for ligands is summarized in Table 1.3. Clearly, there is a correlation on the basis of hard/soft criteria. Hard metal ions (e.g., Fe^{3+}, V^{3+}, Mn^{3+}) can be selectively ligated by small hard anions, of which alkoxide (RO^-) derivatives [tyrosinates, hydroxamates, and catecholates (see siderophores in Section 3.2)] are the only reasonable candidates at biological pH. Alternatively, softer metal ions are preferred by thiolate ligands.

Coordination Geometry

Table 1.3 summarizes the preferred ligand geometries for a variety of metals in different oxidation states. Since the architecture of ligand geometry is inherent in the design of a protein or a natural metal-binding antibiotic, this can form the basis for selective uptake of metal ions.

Ion Size

Many proteins fold in a manner that defines a cavity that selectively binds metal ions of a particular size. This mechanism is used in calcium-binding proteins. Similarly, natural metal-carrier ligands are synthesized that possess a central hole or channel that again closely matches the size of a specific metal ion. Such ligands are often cyclic and are the subject of discussion in Chapter 3. Figure 1.4 illustrates some examples.

Influence of pH

The competition between metal ions for coordination sites has been alluded to earlier. However, a more predominant competitor is the proton (H^+), particularly when the ligands derive from ionized functionality. In the case of transition metals the formation constants at neutral pH are normally much greater than the acid ionization constants (pK_a) that correspond to the affinity of a ligand for a proton, as shown. However, when the pH is lowered and $[H^+]$ increases, the proton may compete effectively with the metal ion.

$$[HL]^+ \rightleftharpoons L + H^+ \qquad K_a$$

In some intracellular compartments and vesicles the pH can be lowered from the normal physiologic level of 7.4 to a pH of 5.5. In Chapter 3 we shall see

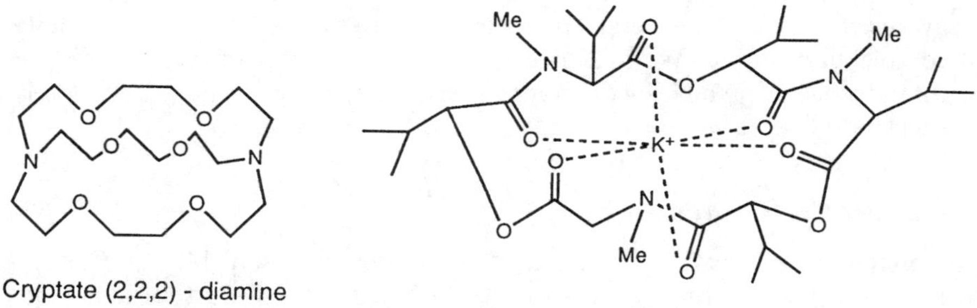

Cryptate (2,2,2) - diamine

Enniatin

Figure 1.4 A cryptand ligand and a K^+ binding antibiotic (enniatin), respectively, illustrate synthetic and biological ring chelates. Selectivity for metal ion binding depends on both the size of the ion and the preferred coordination geometry. For the (2, 2, 2) cryptand shown, the ordering is $K^+ > Rb^+ > Na^+ > Cs^+ > Li^+$.

that this can be used to effect the release of iron from the iron transport protein transferrin. Competitive binding by H^+ is more common for monovalent ions, which tend to form weak complexes with ligands.

$$[HL]^+ + M^{n+} \rightleftharpoons [ML]^{n+} + H^+$$

1.5 Stabilization of Oxidation States

The most common oxidation states for the transition metals are noted in Table 1.2. The stabilization of high or low oxidation levels requires specific ligands. Metals in high oxidation states coordinate hard electronegative ligands, and lower oxidation levels are stabilized by softer ligands of lower electronegativity. The stability of a given oxidation level for a metal ion, reflected in the redox potential $(E°)$,[10] may also be controlled by using the preferences for particular coordination geometries or ligand donor atoms by that ion. Clearly, negatively charged ligands tend to stabilize a higher oxidation state. Compare the reduction potentials for the Fe^{3+}/Fe^{2+} couples listed in Table 1.6. Relative to the hard ligand H_2O, negatively charged CN^- stabilizes the oxidized state (less positive $E°$), whereas the softer phenanthroline (phen)[11] ligand stabilizes the reduced state (more positive $E°$).

Summary of Sections 1.3–1.5

1. Hard ligands bind to hard metal ions (high charge, small radii). Soft ligands bind to soft metal ions (low charge, large and polarizable).
2. Biological molecules bind metal ions optimally when there is a match between the ligand atoms and metal ion on the basis of HSAB criteria, the geometric arrangement of the ligand atoms, the size of the binding pocket, and the absence of competing metals or ligands.
3. There is a large variation in metal ion concentrations in intracellular and extracellular environments. Less abundant transition metals require ligands with large stability constants (relative to the more abundant alkali and alkaline earth metals) to promote effective uptake. Stability constants for the

Table 1.6 Comparison of Reduction Potentials for Some Iron Complexes

Reaction	Reduction Potential $E°$ (V vs. NHE)
$Fe(H_2O)_6^{3+} + e^- \rightleftharpoons Fe(H_2O)_6^{2+}$	+0.77
$Fe(CN)_6^{3-} + e^- \rightleftharpoons Fe(CN)_6^{4-}$	+0.36
$Fe(phen)_3^{3+} + e^- \rightleftharpoons Fe(phen)_3^{2+}$	+1.14

Reduction potentials are discussed in more detail in Chapter 2. A positive potential implies the equilibrium lies to the right of the equation. The larger positive potentials correlate with a more stable reduced complex.

most abundant monovalent ions (K^+, Na^+) are negligible for most biological ligands, and so these ions exist predominantly as aquated species [M^+(aq)].
4. Specific oxidation states may also be stabilized by the criteria summarized in 2.

Review Questions

- Rationalize the relative magnitudes of the binding constants for Cu^{2+} and the following ligands: $Cu(NH_3)_4^{2+}$, $\log_{10}\beta_4 \sim 13.1$; $Cu(en)_2^{2+}$, $\log_{10}\beta_2 \sim 19.9$; $Cu(trien)^{2+}$, $\log_{10}\beta_1 \sim 20.5$; $Cu(cyclen)^{2+}$, $\log_{10}\beta_1 \sim 24.2$, where en = ethylenediamine, trien = triethylenetetramine, and cyclen is a cyclic ligand with four nitrogens linked through ethylene chains.

[Burgess, J. *Ions in Solution* pp. 80–92, Halsted Press, John Wiley & Sons, 1988]

- Dynamic cation hydration numbers, the average number of H_2O molecules associated with a metal ion as it moves through solution, are noted below. Explain these relative values. Li^+, 13-22; Na^+, 7-13; K^+, 4-6; Mg^{2+}, 12-14; Ca^{2+}, 8-12.

[Burgess, J. *Ions in Solution* pp. 32-35, Halsted Press, John Wiley & Sons, 1988]

1.6 Ligand Field Stabilization Energy

As a result of their distinct spatial orientations, d-orbitals lose their degeneracy in the presence of a ligand set.[12] As illustrated in Figure 1.5, σ-bonds are char-

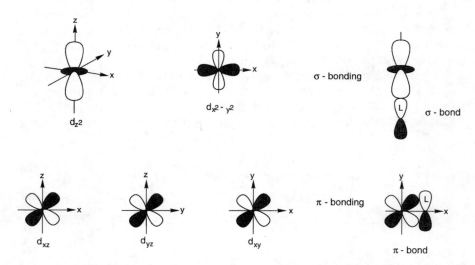

Figure 1.5 Orientation of d-orbitals relative to an octahedral ligand field. The $d_{x^2-y^2}$ and d_{z^2} orbitals have the correct geometry for σ-bond formation, whereas the remaining orbitals participate in π-bonding.

acterized by orbital overlap (electron density) along the axis joining the two nuclei that are connected by the bond. For π-bonds, orbital overlap lies off the internuclear axis. In a formal sense there is no bonding electron density along the axis. The $d_{x^2-y^2}$ and d_{z^2} orbitals are directed along the Cartesian axes and favor σ-bonding, whereas the d_{xy}, d_{xz}, and d_{yz} lie between the axes and favor π-bonding. The magnitude of the orbital splitting energy (Δ) depends on the identity of the ligands and the geometry of the resulting complex. The influence of geometry is illustrated for octahedral (O_h), tetragonal (tet), square planar (sp), trigonal bipyramidal (tbp), tetrahedral (T_d), and square pyramidal (spy) coordination in Figure 1.6. Consider the octahedral ligand geometry shown in Figure 1.3. Six ligands are directed along the Cartesian axes. It is clear that the ligands point directly at the $d_{x^2-y^2}$ and d_{z^2} orbitals but lie between the d_{xy}, d_{yz}, d_{xz} orbitals, and so these two sets of orbitals have different energies (Figure 1.6). The difference in energy is termed the ligand-field splitting energy (Δ). In biological systems metal ions are seldom bound in a symmetric ligand environment. However, the assumption of an idealized symmetry is often sufficiently valid to provide good insight on the spectroscopic and chemical behavior of the metal center. When these d-orbitals are populated by electrons ($\uparrow$), the filling order will vary according to the relative magnitudes of the ligand-field splitting energy (Δ) and the electron-pairing energy (P) (Figure 1.7A). Figure 1.7B illustrates the

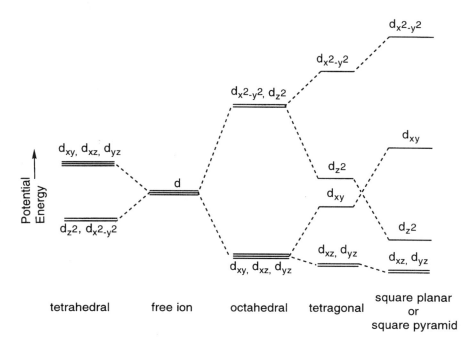

Figure 1.6 Loss of d-orbital degeneracy in a variety of ligand field geometries. There must be no net change in the overall energy of all the levels.

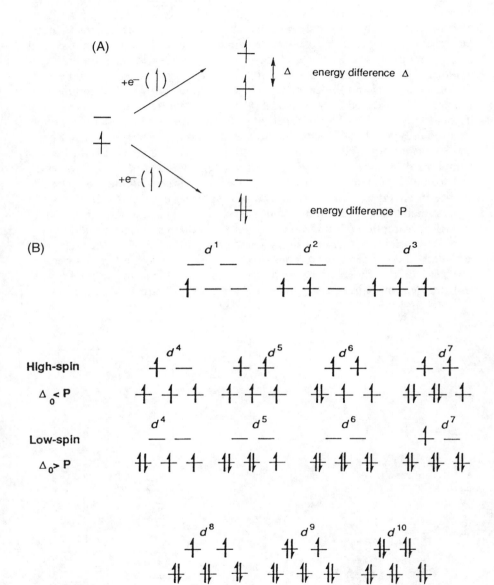

Figure 1.7 (A) Population of d-orbitals depends on the relative magnitudes of the ligand-field splitting energy and electron-pairing energy. (B) Electron configurations for d^1 to d^{10} ions in an octahedral ligand field.

possible electron configurations for an octahedral geometry. Note that, although there is only one way to fill the available orbitals for d^1, d^2, d^3, d^8, d^9, and d^{10} configurations, two possibilities exist for d^4–d^7. For strong-field ligands (large Δ), the energy gained by populating the lower-energy orbitals makes up for the interelectron repulsion (P) between electrons occupying the same orbital. For weak-field ligands (small Δ), the stabilization energy is not large enough to overcome the spin-pairing energy P; the electrons populate the empty orbitals of higher energy. The total number of unpaired spins is lower in the former case; thus, these are termed *low-spin configurations.* Similarly, *high-spin configurations* arise when the number of unpaired spins is greatest.[13] The energy separation Δ may be estimated by monitoring (d–d) bands in the absorption spectra of $[Co(III)(NH_3)_5X]^+$ complexes. Note that there is no correlation between the magnitude of Δ and the concept of hard and soft ligands. For octahedral aquo complexes $M(H_2O)_6^{n+}$, $\Delta \sim 10{,}000$ cm^{-1} for divalent metals and $\sim 20{,}000$ cm^{-1} for trivalent ions. The relative splitting energies for a variety of ligands have been determined: $I^- < Br^- < Cl^- \sim SCN^- \sim N_3^- < NO_3^- < F^- < (NH_2)_2CO < OH^- < C_2O_4^{2-} \sim H_2O < NCS^- \sim H^- < CH_3CN < H_2NCH_2CO_2^- < NH_3 \sim$ pyridine $<$ en $\sim SO_3^{2-} < NH_2OH <$ bipy $<$ phen $< NO_2^- < PPh_3 < CH_3^- < CN^- < CO$.[14] This ordering of ligands is termed the *spectrochemical series.* In biological systems oxygen and sulfur donors tend to be weak-field ligands, and nitrogen donors are strong-field ligands. Ligand environments such as these give rise to high- and low-spin metal ions, respectively. Since Δ(tetrahedral) $\sim$ 4/9 Δ(octahedral), metal ions in a tetrahedral ligand environment are typically high spin. The value of Δ also depends on the metal. Δ generally increases with oxidation state, across a row, and on descending a group. For example, $Mn^{2+} < Co^{2+} < Ni^{2+} < V^{2+} < Fe^{3+} < Cr^{3+} < Co^{3+}$. Relative to the first transition series, Δ increases by approximately 50 percent and 100 percent for aquo complexes of the second and third transition series, respectively.

Many kinetic and thermodynamic properties of metal ions depend on the magnitude of Δ. Metal ions may derive extra stability when bound by ligands as a result of the orbital splitting depicted in Figures 1.6 and 1.7. The sum total of the contributions from all the d-electrons is termed the *ligand field stabilization energy* (LFSE). For reference, Appendix 4 notes the relative energies of d-orbitals in a variety of coordination geometries. The LFSE's for the various electron configurations of the octahedral coordination geometry shown in Figure 1.7 are noted in Table 1.7. Since the increased energy of the two upper levels

Table 1.7 Ligand Field Stabilization Energies for an Octahedral Complex

d^1	$-\frac{2}{5}\Delta$	d^4(HS)	$-\frac{3}{5}\Delta$	d^4(LS)	$-\frac{8}{5}\Delta$	d^8	$-\frac{6}{5}\Delta$
d^2	$-\frac{4}{5}\Delta$	d^5(HS)	—	d^5(LS)	$-\frac{10}{5}\Delta$	d^9	$-\frac{3}{5}\Delta$
d^3	$-\frac{6}{5}\Delta$	d^6(HS)	$-\frac{2}{5}\Delta$	d^6(LS)	$-\frac{12}{5}\Delta$	d^{10}	—
		d^7(HS)	$-\frac{4}{5}\Delta$	d^7(LS)	$-\frac{9}{5}\Delta$		

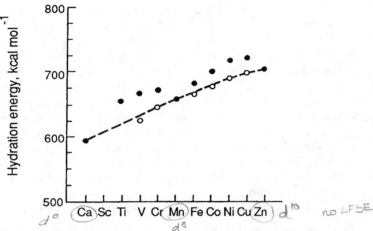

Figure 1.8 Hydration energies for the divalent metal ions of the first transition series plotted against the elements as a function of increasing atomic number. Solid circles are experimental points and open circles correspond to values from which spectrally evaluated ligand–field stabilization energies have been subtracted. [Adapted from F. A. Cotton, *J. Chem. Educ.*, *41*, 466 (1964).]

must match the energy decrease of the two lower levels, we assign a contribution of $-2/5\Delta$ to each electron in the lower levels and $+3/5\Delta$ to each electron in upper levels. LFSE's for other geometries may be deduced in a similar fashion. These data are invaluable for understanding the kinetic and thermodynamic properties of metal complexes.

The hydration energies of the divalent ions of the first transition series provide a standard example of the influence of LFSE (Figure 1.8). For the reaction

$$M^{2+}(g) + H_2O(l) \rightarrow M^{2+}(aq) \qquad \Delta H^{\circ}_{hyd}$$

the hydration energy is greatest for those metals that derive additional stability when coordinated by six water molecules. The additional stabilization arises from the LFSE. Subtraction of the LFSE yields a straight-line relationship, where the increase in ΔH°_{hyd}, going from left to right, results from the decrease in ionic radius (and increased electrostatic attraction for H_2O) across the transition series. As a result of LFSE, many physical parameters (e.g., ionization energy, lattice energies, formation constants) show a similar deviation from the expected uniform change with increasing atomic number.

1.7   Kinetics and Mechanisms of Reactions Involving Metal Complexes

In biology the ligand environments of metal ions are often in a state of change. For example, the activation of the protein calmodulin by $Ca^{2+}(aq)$ requires the replacement of calcium-bound water molecules by protein ligands, whereas the activation of phosphate ester bond cleavage by Zn^{2+} in alkaline phosphatase (Chapter 4) requires coordination of the phosphate to the metal site. The loss, addition, or exchange of ligands at metal centers plays a key role in the regulation of cellular metabolism. For example, the rapid ligand exchange rates of Ca^{2+} ($k_{ex} \sim 10^9$ s^{-1}) relative to Mg^{2+} ($k_{ex} \sim 10^5$ s^{-1}) explains the evolutionary selection of the former as a secondary messenger system (see Section 7.1). Clearly, a knowledge of the reaction mechanisms for substitution of metal-bound ligands is essential for a proper understanding of inorganic biochemistry.

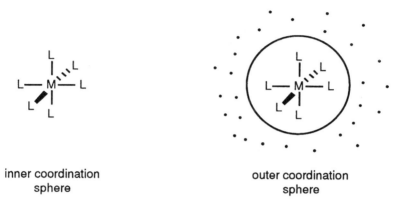

inner coordination
sphere

outer coordination
sphere

First we define two terms that are frequently encountered in discussions of inorganic mechanisms, *inner sphere* and *outer sphere*. The inner coordination sphere of a metal ion refers to the ordered array of ligands that bind directly to the metal center. The positive charge on the metal can also impose some order on solvent molecules, ligands, or counterions, in the region that surrounds this inner coordination shell. This quasi-ordered domain is referred to as the outer coordination sphere. When we speak of substitution reactions of metal complexes, it is understood that this results in an exchange, or change, in the ligands of the inner coordination sphere. However, the outer coordination sphere may also play an important role in the reaction pathway.

A general scheme (Eigen, Tamm, Wilkins mechanism) for ligand substitution of aquo complexes is summarized below, where K_{os} is the formation constant for the outer sphere complex,[15] and k_i is the rate of exchange of ligand L for an inner sphere water.

$$\left(M^{n+}-L\right)\ \left(L'\right)\ \underset{}{\overset{K_{os}}{\rightleftharpoons}}\ \left(M^{n+}-L\right)\left(L'\right)\ \overset{k_i}{\longrightarrow}\ \left(M^{n+}-L'\right)\left(L\right)\ \longrightarrow\ \left(M^{n+}-L'\right)\ +\ \left(L\right)$$

It may be shown that, in the limit of excess L', the observed rate constant k_{obs} is given by the following equation:[16]

$$k_{obs} = k_i K_{os}$$

Interchange mechanisms may be termed *dissociative* (I_d) or *associative* (I_a), according to whether the rate-determining step (rds), the interchange step (k_i), is dissociative or associative. In dissociative mechanisms, the coordination number of the metal center formally decreases, whereas it increases for an associative mechanism. Dissociative (D) or associative (A) pathways can be thought of in terms of the more familiar examples of S_N1 or S_N2 pathways in organic chemistry. The distinction between I_d and D, or I_a and A, rests with the preequilibrium step. Pure D or A mechanisms do not proceed through the preequilibrium outer sphere complex illustrated earlier.

$$ML_6^{n+} \underset{rds}{\rightarrow} ML_5^{n+} + L \underset{-L}{\overset{L'}{\rightarrow}} ML_5L' \qquad \text{dissociative (D)}$$

$$ML_6^{n+} + L' \underset{rds}{\rightarrow} ML_6L'^{n+} \rightarrow ML_5L' + L \qquad \text{associative (A)}$$

One of the simplest reactions to consider in aqueous solution is that of solvent exchange.

$$M(H_2O)_x^{n+} + H_2O^* \rightarrow M(H_2O)_{x-1}(H_2O^*)^{n+} + H_2O$$

Eigen has shown that, for most metal ions, dissociative loss of H_2O is rate limiting.

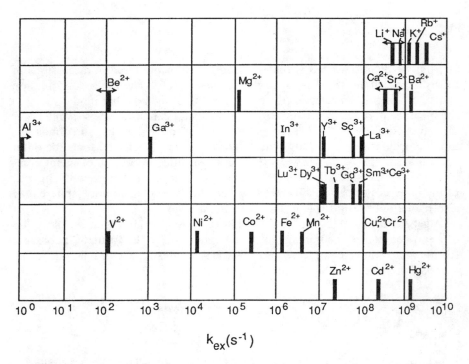

Figure 1.9 Solvent exchange rates for inner-sphere water molecules in M^{n+}(aq). The rows show groups of ions with similar chemistries. [Adapted from H. Diebler et al., *Pure Appl. Chem.*, **20**, 93–115 (1969).]

$$M(H_2O)_x^{n+} \rightarrow M(H_2O)_{x-1}^{n+} + H_2O$$

Substitution reactions of aquo metal complexes[17] of the alkali or alkaline earth metals tend to proceed by loss of H_2O via D or I_d mechanisms, and so k_{obs} is equivalent to the solvent exchange rate. Such reactions tend to proceed fairly rapidly (Figure 1.9) with a systematic variation of rates that can be readily understood on the basis of the size and charge of the cation. Loss of a ligand from a metal ion is favored by a large cation of low charge, which results in a weaker metal–H_2O bond. The water exchange rates for divalent transition-metal ions vary from 10^2 to 10^8 s^{-1} and are greatly influenced by ligand field stabilization. Loss of LFSE from formation of intermediates of distinct geometry results in large activation energies for exchange. The effect is most evident in ions, such as Cr^{3+} and Co^{3+}, with the largest LFSE (d^3 and low-spin d^6 configurations). The rate constants for Cr^{2+}(aq) (d^4) and Cu^{2+}(aq) (d^9) are much higher than expected as a result of Jahn–Teller distortion.[18] This weakens the coordination of the two axial water molecules (assuming octahedral coordination) and provides a pathway for rapid exchange. There is more associative character to the left of the transition series, since these ions are larger and there are more empty d-orbitals to accept electrons from incoming nucleophiles. Table 1.8 summarizes the general trends in mechanistic character for substitution reactions of first-row transition-metal complexes.

A metal complex is described as "inert" or "labile" according to whether it reacts slowly ($t_{1/2} \geq 1$ min) or rapidly ($t_{1/2} < 1$ min). Clearly, this is an arbitrary definition.

Summary of Sections 1.6–1.7

1. d-orbitals lose their degeneracy in a ligand field according to the geometry of the ligand set. As a result of this energy splitting, a transition metal ion may be stabilized by an amount that depends on the electron configuration, the identity of the ligands, and the ligand geometry. Only d^0, high-spin d^5, and d^{10} configurations have no ligand field stabilization energy. LFSE influences many physicochemical properties of transition metals: ionic radii, heats of hydration and complex formation, reduction potentials, ligand exchange kinetics, and so on.

2. Ligand exchange at a metal center may be described as *associative* (A) or *dissociative* (D), according to whether the coordination number at the metal center increases or decreases during the rate-limiting step. The general *interchange* mechanism is characterized by formation of an outer-sphere preequilibrium complex and ranges from I_a to I_d.

Review Questions

- Draw d-orbital energy level diagrams for the following complex ions. Show the electrons and their relative spins. Label each as high or low spin.

    $FeCl_4^{2-}$ $Fe(CN)_6^{3-}$ $Fe(H_2O)_6^{2+}$ $Mn(H_2O)_6^{2+}$ $Mg(H_2O)_6^{2+}$

Table 1.8 Mechanistic Preferences for Substitution Reactions of Metal Cations

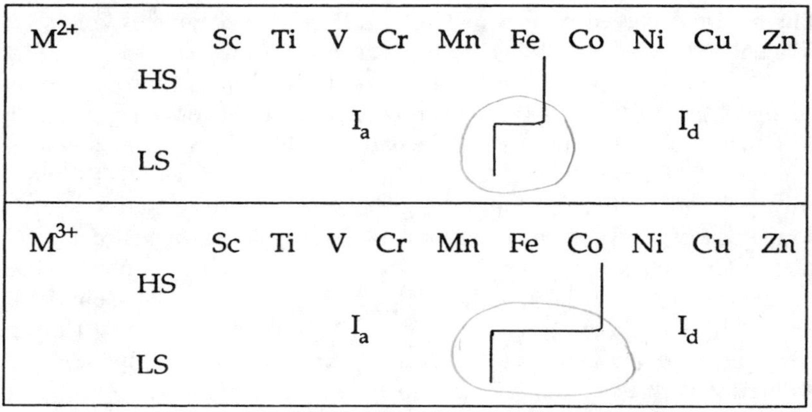

M^{2+}		Sc	Ti	V	Cr	Mn	Fe	Co	Ni	Cu	Zn
	HS										
				I_a						I_d	
	LS										

M^{3+}		Sc	Ti	V	Cr	Mn	Fe	Co	Ni	Cu	Zn
	HS										
				I_a						I_d	
	LS										

- Why can d–d energies be more readily estimated from $Co(III)(NH_3)_5X$ absorption spectra than from $Ni(II)(NH_3)_5X$ spectra?
- The nickel complex $Ni(CN)_4^{2-}$ is thermodynamically stable ($\log \beta_4 = 30.5$) but is kinetically labile (CN^- exchange rate $> 10^{-2}\,s^{-1}$). Discuss the apparent contradiction here.
- Rationalize the relative magnitudes of the solvent exchange rates for the metal ions in Figure 1.9.
- Explain why the self-exchange of $[Co(NH_3)_5Cl]^{2+}$ with $^*Cl^-$ is slow, whereas the self-exchange of $[Co(NH_3)_5Cl]^+$ with $^*Cl^-$ is fast. Be succinct!

1.8 Electron-Transfer Reactions

The development of the theoretical and experimental aspects of modern coordination chemistry has been greatly dependent on the study and/or use of electron-transfer (ET) reactions between inorganic complexes. These reactions may be classified in several ways, one of the simplest being consideration of the net change in oxidation state of each metal ion. Electron-transfer reactions are described as complementary when there are equal numbers of oxidants and reductants.

$$Fe(CN)_6^{3-} + Ru(NH_3)_6^{2+} \rightarrow Fe(CN)_6^{4-} + Ru(NH_3)_6^{3+}$$

Such processes typically involve the transfer of one electron. Noncomplementary reactions, such as that shown here, involve unequal numbers of oxidants and reductants because the change in redox state at each metal center is different.

$$Cr^{6+} + 3Fe^{2+} \rightarrow Cr^{3+} + 3Fe^{3+}$$

Biological redox reactions cover both categories. Noncomplementary reactions between inorganic ions proceed by a series of bimolecular steps, since the probability of a termolecular or higher-order collision is extremely small. A distinction may be made between inner-sphere and outer-sphere electron-transfer processes. An inner-sphere ET pathway can be defined as one where the inner coordination sphere of at least one reactant is changed in attaining the transition state for electron transfer. Inner-sphere reactions will occur only if one of the reactants is labile and the other contains a potential bridging ligand (i.e., a ligand in possession of a free electron pair). The classic example of the reaction between $[Co(NH_3)_5Cl]^{2+}$ and $Cr^{2+}(aq)$ can be summarized as follows. (1) The chloride ligand on the *inert* Co(III) ion possesses an available electron pair that can substitute for an aquo ligand on *labile* Cr(II). (2 and 3) After electron transfer from Cr(II) to Co(III), a labile Co(II) ion and inert Cr(III) ion are generated.

(1)

$Cr(II)(H_2O)_6$ + $Co(III)(NH_3)_5Cl$ → $[(H_2O)_5Cr(II)ClCo(III)(NH_3)_5]$
labile   inert   precursor complex

(3)   ↓ (2)

$Cr(III)(H_2O)_5Cl$ + $Co(II)(NH_3)_5(H_2O)$ ← $[(H_2O)_5Cr(III)ClCo(II)(NH_3)_5]$
inert   labile   successor complex

↓ H^+, H_2O

$Co(II)(H_2O)_6$ + 5 NH_4^+

When the complex dissociates, the chloride remains bound to the *inert* Cr(III) center while the labile $Co(II)(NH_3)_5H_2O$ species is readily hydrolyzed.[19] In a classic experiment, Taube demonstrated the transfer of Cl^- as a bridging ligand by use of radiolabeled $*Cl^-$. The exchange of $*Cl^-$ from $Co(NH_3)_5*Cl$ to $Cr(H_2O)_5*Cl$ is almost quantitative. Inner-sphere electron transfer is characterized by the penetration of one redox center into the inner coordination sphere of the other, and so the electron-transfer rate (k_{ET}) cannot be faster than the rate of substitution (k_{sub}). If $k_{ET} > k_{sub}$, the reaction must proceed by an outer-sphere pathway. Any one of steps 1, 2, or 3 may be rate limiting. If step 1 is limiting, $k_{ET} \sim k_{sub}$. If step 2 is limiting, it may be possible to observe the precursor complex. If step 3 is limiting, it may be possible to observe the precursor and/or successor complexes (Figure 1.10).

The illustrative examples shown in Figure 1.10 employed two metal centers. In biological redox chemistry, a more typical example might show the interaction of a metal site with a substrate molecule. For example, the reduction of nitrate to nitrite by the molybdoenzyme nitrate reductase might proceed by transfer of two electrons from Mo(IV) to nitrate after the latter binds directly to the molybdenum center. Note that, in this case, the electron-transfer event is coupled to a net chemical change in the inner coordination spheres of each reagent.

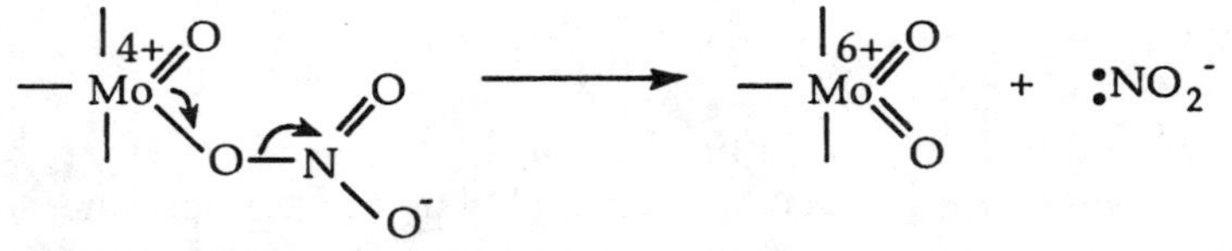

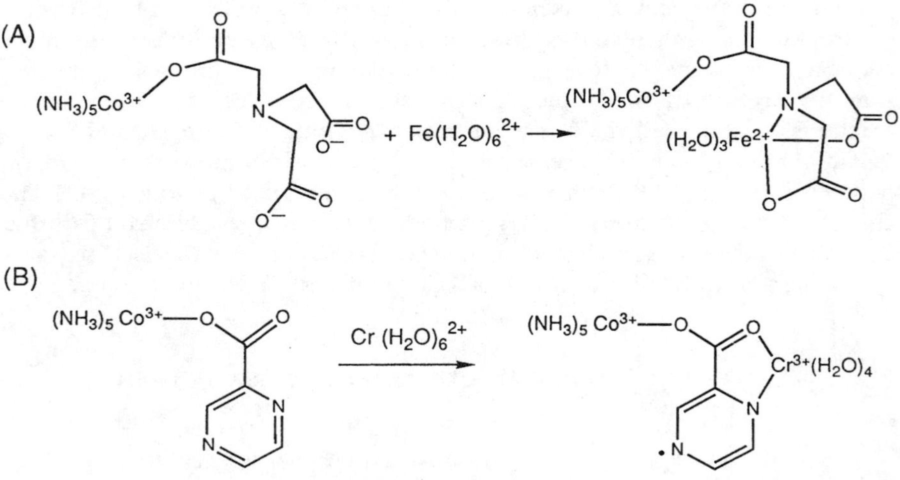

Figure 1.10   Intermediates identified in electron-exchange reactions. (A) Precursor complex formation. The electron remains on Fe^{2+} and can be detected spectroscopically. (B) An example of a stepwise reaction. The nitrogen-radical intermediate can be detected by EPR (electron paramagnetic resonance) spectroscopy.

In the following reaction, both reactants are substitutionally inert; neither has a potential bridging ligand (no free electron pairs).

$$Co(phen)_3^{3+} + Ru(NH_3)_6^{2+} \rightarrow Co(phen)_3^{2+} + Ru(NH_3)^{3+}$$

The reaction must proceed by an outer-sphere mechanism where each complex retains its own ligand set and electron transfer occurs via an outer-sphere complex. Outer-sphere reactions are relatively simple (no bond making or breaking); consequently, they have been the subject of extensive experimental and theoretical study. The rate of electron transfer between two redox centers may be estimated from knowledge of the individual electron-exchange rates for the reactants and the difference in their electrochemical reduction potentials. Consider the following equations where k_{12} is the cross-reaction rate, k_{11} is the self-exchange rate for $Fe(phen)_3^{3+/2+}$, k_{22} is the self-exchange rate for $Ru(NH_3)_6^{3+/2+}$, and K_{12} is the equilibrium constant for the cross-reaction:

$$Fe(phen)_3^{3+} + Ru(NH_3)_6^{2+} \rightleftharpoons Fe(phen)_3^{2+} + Ru(NH_3)_6^{3+} \qquad (k_{12}, K_{12})$$

$$Fe(phen)_3^{3+} + Fe(phen)_3^{2+} \rightleftharpoons Fe(phen)_3^{2+} + Fe(phen)_3^{3+} \qquad (k_{11})$$

$$Ru(NH_3)_6^{3+} + Ru(NH_3)_6^{2+} \rightleftharpoons Ru(NH_3)_6^{2+} + Ru(NH_3)_6^{3+} \qquad (k_{22})$$

The following simple relationship holds,[20] where $f_{12} \sim 1$ is a constant.

$$k_{12} = (k_{11} \, k_{22} \, K_{12} f_{12})^{1/2}$$

Provided the two species are similar with regard to their charge type and possess similar inner-sphere ligands, this equation has proved remarkably useful and generally predicts rate constants that are accurate to within an order of magnitude. The equation can also be used to calculate K_{12} (and therefore redox potentials; see Chapter 2) from rate data. The relationship typically breaks down if either of the preceding assumptions is not valid. For example, the third and fourth entries in Table 1.9 have ions of similar charge type; however, the ligands are very different (phen versus H_2O). In the self-exchange reactions the association of $Fe(phen)_3^{2+}$ and $Fe(phen)_3^{3+}$ is favored by van der Waals interactions between the planar aromatic ligands (see Appendix 1). In the case of $Fe(H_2O)_6^{2+/3+}$, self-exchange is favored by dipole–dipole interactions from hydrogen bonding. Both of these favorable interactions are excluded in the cross-reaction involving $Fe(phen)_3^{3+}$ and $Fe(H_2O)_6^{2+}$, and so the observed rate is smaller than the predicted rate. In later examples, where reactants of opposite charge are employed, the electrostatic repulsive interactions of the self-exchange reactions are replaced by favorable attractive forces for the cross-reaction, and so $k_{12}(obs) > k_{12}(calc)$. These types of interactions are refered to as *work terms,* since they influence the energetics of outer-sphere complex formation.

Summary of Section 1.8

1. Electron-transfer reactions are defined as *inner sphere* if one reactant penetrates the inner coordination sphere of the other and as *outer sphere* otherwise. If $k_{ET} > k_{sub}$, the reaction must be outer sphere.
2. For outer-sphere electron transfer, the Marcus relationship for cross-reactions is

$$k_{12} = (k_{11}k_{22}K_{12}f_{12})^{1/2}$$

Table 1.9 Comparison of Observed and Calculated Rate Constants Using the Marcus Equation

Species 1	Species 2	K_{12}	k_{11}	k_{22}	k_{12} (obs)	k_{12} (calc)
$Ru(H_2O)_6^{3+}$ +	$Ru(H_2O)_6^{2+}$	1	—	—	20	60
$Ru(NH_3)_6^{2+}$ +	$Ru(NH_3)_6^{3+}$	1	—	—	7×10^3	1×10^5
$Fe(phen)_3^{3+}$ +	$Fe(H_2O)_6^{2+}$	2.5×10^5	1×10^6	4	3.7×10^4	5.6×10^5
$Co(phen)_3^{3+}$ +	$V(H_2O)_6^{2+}$	4×10^{10}	5	1×10^{-2}	3.8×10^3	2.3×10^4
$Co(phen)_3^{3+}$ +	$Fe(CN)_6^{4-}$	5	44	1.9×10^4	6×10^6	2.1×10^3
$HMnO_4$ +	$Mo(CN)_8^{4-}$	1.2×10^{-4}	1.1×10^3	1.2×10^{-4}	1.9×10^7	2.4×10^7
MnO_4^- +	$Fe(CN)_6^{4-}$	2×10^2	4×10^3	1.9×10^4	2.5×10^4	5×10^4

All rates are in units of $(M^{-1}s^{-1})$. phen = 1,10-phenanthroline. In the first two examples k_{12} corresponds to the self-exchange rate since $k_{11} = k_{22} = k_{12}$.

Review Question

• From a study of the reduction of the copper protein azurin by $Cr(H_2O)_6^{2+}$, it was observed that the reduced protein contained a tightly bound trivalent chromium. Provide an explanation for this observation.

[*Proc. Natl. Acad. Sci. USA 79*, 4190 (1981)]

PART B.   FUNDAMENTALS OF REACTION KINETICS AND THERMODYNAMICS

1.9   Rate Laws and Rate Constants

A reaction *mechanism* is a detailed picture of the way reactant species are turned into products: that is, which bonds are broken and formed, and the sequential order in which these bonds are broken and formed. Because kinetics is the study of how quickly (or slowly) reactions proceed, and is related to the speed of bond cleavage and formation, kinetic studies can provide valuable insight into mechanistic problems. This contrasts with reaction thermodynamics, which provides no such insight, being concerned only with the initial and final energy states (for reactants and products, respectively) of a chemical reaction. It is important to keep in mind that a proposed mechanism must be consistent with kinetic data. However, kinetic results alone would not prove that a reaction actually followed a particular type of mechanism, but only that these results were consistent with such a mechanism. The reaction mechanisms shown in other chapters in the text have not necessarily been "proven," but, in most cases, they are consistent with available kinetic data, and so we believe that these mechanisms as written do, in fact, provide a reasonable picture of what is actually happening. It is possible, of course, that one could write a different mechanism that would be fully consistent with all the available kinetic data. This is especially true of reactions (enzymatic or otherwise) that have not been extensively studied.

To evaluate a reaction mechanism by use of reaction kinetics, we must derive a rate law. The observed rate of a chemical reaction will obviously depend on the identity of reacting chemicals but will also vary with other parameters such as temperature and concentration. So far we have not clearly distinguished the terms rate and rate constant and have not clearly explained how these rate constants (k) are determined. In the following sections we will clear up these two points.

1.9.1   First-Order Rate Law

Consider a reaction that we examined previously, where the first step is the slow rate-determining step (rds).

$$ML_6^{n+} \xrightarrow[\text{rds}]{} ML_5^{n+} + L \xrightarrow[-L]{L'} ML_5L'^{n+} \qquad \text{dissociative (D)}$$

We can monitor the rate of the reaction by measuring the change in concentration of either the reactant ML_6^{n+} or the final product $ML_5L'^{n+}$, over a specified period of time (Δt), and so, by definition,

$$\text{rate} = -\frac{\Delta[ML_6^{n+}]}{\Delta t} = \frac{\Delta[ML_5L'^{n+}]}{\Delta t}$$

Note that the symbol Δ indicates a change in the parameter that follows. So Δt represents a change of time, $\Delta[ML_6^{n+}]$ represents a change of reactant concentration, etc. In the limit of an infinitely small change Δ, we use the differential form $d[ML_6^{n+}]$ and dt. For a dissociative reaction, the rate-determining-step depends only on the concentration of ML_6^{n+} and is independent of L'. Doubling the $[ML_6^{n+}]$ would double the observed rate of disappearance of $[ML_6^{n+}]$, but changing $[L']$ would have no effect, and so

$$\text{rate} = -\frac{d[ML_6^{n+}]}{dt} \, \alpha \, [ML_6^{n+}]$$

Note the appearance of the negative sign before $-d[ML_6^{n+}]/dt$ because ML_6^{n+} is a reactant that is being consumed. Because rate must be a positive quantity, we need the negative sign whenever we are monitoring the disappearance of a reactant. Replacing the proportionality sign with a constant, gives

$$-\frac{d[ML_6^{n+}]}{dt} = k_1 [ML_6^{n+}] \qquad\qquad \textit{differential rate laws} \quad (1.1)$$

In this case there is only one concentration term in the rate law. The reaction follows a first-order rate law, and the constant k_1 is a first-order rate constant. Rate has units of M s^{-1}, and so the first-order rate constant has units of s^{-1}. It is more important for us to know the rate constant k_1 for a reaction than the absolute rate. The latter will vary with the concentrations of reactants; however, the rate constant reflects the inherent activity of molecules that follow a well-defined mechanism under particular conditions of temperature and pressure. How, then, do we determine the rate constant?

To evaluate the rate constant, we need an expression that relates the concentration of a reactant or product to the period of time over which the reaction has taken place. Integrating the differential form of the rate expression (1.1) between the limits of time $t = 0$ (the start of the reaction) and a later time t, yields an expression for the concentration of ML_6^{n+} at any time t. In brief,

$$-\frac{d[ML_6^{n+}]}{[ML_6^{n+}]} = k_1 \, dt$$

For simplification, let $Y = [ML_6^{n+}]$. Integrating gives,

$$-\int_0^t \frac{dY}{Y} = k_1 \int_0^t dt$$

and so

$$-\ln \frac{Y_t}{Y_0} = k_1 t$$

This is normally rearranged to the exponential form

$$Y_t = Y_0 \exp(-k_1 t)$$

or

$$[ML_6^{n+}]_t = [ML_6^{n+}]_0 \exp(-k_1 t) \qquad \textit{integrated rate law} \qquad (1.2)$$

A plot of $[ML_6^{n+}]$ versus time is shown in Figure 1.11. Note that the rate at any time t is really the gradient of the $[ML_6^{n+}]_t$ versus t plot at that particular time. The rate constant k_1 is usually obtained by fitting the raw data to equation (1.2) with a personal computer.

1.9.2 Second-Order Rate Law

Consider an associative reaction,

$$ML_6^{n+} + L' \xrightarrow[\text{rds}]{} ML_6 L'^{n+} \rightarrow ML_5 L'^{n+} + L \qquad \text{associative (A)}$$

Again, the first step is the rate-determining-step. If we were to double the concentration of L', we would expect the observed rate of reaction to double, because the probability of L' meeting and reacting with a molecule of ML_6^{n+}

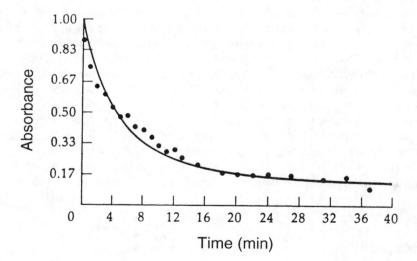

Figure 1.11 Reaction profile showing the variation of $[ML_6^{n+}]$ (proportional to absorbance) with time. The gradient at any specific time is given by equation (1.1). Clearly, as the reaction proceeds, the concentration of ML_6^{n+} decreases, and the rate decreases, although the rate constant is *constant*!

would double. By similar reasoning, if we double the concentration of ML_6^{n+}, then, we would again expect the rate to increase by a factor of two. If we double both the concentrations of L' and ML_6^{n+} at the same time, then, we would expect the reaction rate to increase by a factor of four; that is, the rate of reaction is proportional to the concentrations of both reactants, and so,

$$- \frac{d[ML_6^{n+}]}{dt} = k_2\,[L']\,[ML_6^{n+}] \tag{1.3}$$

Because the rate law contains two concentration terms, it is referred to as a second-order rate law, and the rate constant is a second-order rate constant (k_2). The units for the rate and the rate constant follow from the basic definitions, and it is clear that the second-order rate constant k_2 has units of $M^{-1}\,s^{-1}$.

Again, this differential form of the rate equation can be integrated to yield an analytical expression that defines the concentration of ML_6^{n+} and L' at a time t after the start of the reaction. However, this expression is mathematically much more complex and is not convenient to use, and so we normally carry out the reaction with an excess of one or the other reactant, using at least a tenfold excess. Under these conditions, the rate equation reduces to the form of a first-order equation, because the change of concentration of the species in excess is assumed to be negligibly small, and this concentration is taken to be constant. Such an experiment is said to be carried out under *pseudo-first-order conditions*. So, for example, if $[L']$ was in excess, then, the rate equation would simplify to

$$- \frac{d[ML_6^{n+}]}{dt} = k_{obs}\,[ML_6^{n+}]$$

where $k_{obs} = k_2\,[L']$. This is the same form as the first-order equation described earlier, and the value of k_{obs} could be obtained in a similar manner. By determining k_{obs} in a series of experiments with differing $[L']$, each carried out under pseudo-first-order conditions, the value of k_2 could be obtained from a linear fit of k_{obs} versus $[L']$.

Rate constants provide a lot of information about the influence of steric, stereochemical, and electronic (sometimes summarized as stereoelectronic) factors underlying the reactivity of a molecule. Reactivity trends can be empirically understood in terms of the hard/soft concepts, that we defined earlier in Section 1.3, and the electronic structures of metal ions as discussed in Sections 1.6 and 1.7. As a general rule, hard nucleophiles react most readily with hard metal centers, and soft nucleophiles with soft metal centers. Although these ideas are fairly qualitative, they are extremely useful. Realize that the relative ordering of nucleophiles and leaving groups may vary according to the hard/soft character of the center. These principles also explain why nucleophilicity does not necessarily follow the basicity of a molecule, because basicity normally refers to the affinity for H^+ (a hard center), whereas nucleophilicity usually refers to attack at a softer center.

Summary of Section 1.9

1. Kinetic data provide support for a reaction mechanism, but do not prove it.
2. Rate constants reflect the reactivities of molecules. By comparing a lot of data for a large number of reactions, good nucleophiles and leaving groups can be identified and general trends in reactivity can be established.
3. For a first-order rate law, the differential and integral forms are, respectively,

$$-\frac{d[ML_6^{n+}]}{dt} = k_1 [ML_6^{n+}]$$

and $[ML_6^{n+}]_t = [ML_6^{n+}]_0 \exp(-k_1 t)$

4. For a second-order rate law the differential and integral forms are, respectively,

$$-\frac{d[ML_6^{n+}]}{dt} = k_{obs} [ML_6^{n+}], \quad \text{where } k_{obs} = k_2 [L'].$$

and $[ML_6^{n+}]_t = [ML_6^{n+}]_0 \exp(-k_{obs} t)$

Review Question

• Try to derive the integrated form of the second-order rate law when the pseudo-first-order conditions do not hold; that is, the concentrations of both ML_6^{n+} and L' may vary during the course of the reaction.

[Moore, J. W. and R. G. Pearson, *Chemical Kinetics*, 3rd ed., pp. 22–25, Wiley–Interscience, 1981]

1.10   Thermodynamics and Equilibrium

1.10.1   Chemical Equilibrium

Thermodynamics is the area of chemistry that relates to the intrinsic energies of molecules. Let us assume that a molecule can exist in one of two states, for example, the potassium-bound and metal-free form of valinomycin or enniatin, illustrated schematically as

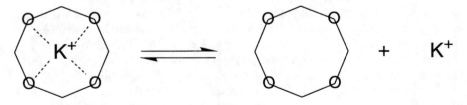

which we considered earlier in Section 1.4. The intrinsic energy of the bound form is lower, and so one might expect this to be the most abundant form in solution. In other words, the energy minimum for the potassium complex is lower than that of the free cation and ligand. In solution, the bound and free forms are interconverting. However, under a defined set of solution conditions the ratio of the two forms will be constant. The bound and free forms are said to be in reversible equilibrium. If we were somehow able to trap the bound form (perhaps as the crystalline solid) and dissolve it in water, we would then find that some of the potassium ions would dissociate to yield some metal-free ligand. Eventually, the rate of dissociation would be balanced by the rate of association of K^+ and the valinomycin ligand, and equilibrium would again be established. Earlier in Section 1.4, we defined an equilibrium or stability constant (K), which is a parameter that defines the relative concentrations of the solution species present under equilibrium conditions. The equilibrium constant K provides us with a quantitative measure of how much of the bound and free forms of the ligand are present under a well-defined set of solution conditions. The constant K is defined as

$$K = \frac{\text{product of the concentrations of species on the right-hand side of the equilibrium equation}}{\text{product of the concentrations of species on the left-hand side of the equilibrium equation}}$$

$$= \frac{[\text{free}][K^+]}{[\text{bound}]}$$

Note that, if $K = 1$, then, at equilibrium, $[\text{free}][K^+] = [\text{bound}]$. If $K \leq 1$, then, $[\text{free}][K^+] \leq [\text{bound}]$, and, if $K \geq 1$, then, $[\text{free}][K^+] \geq [\text{bound}]$. Also observe that, if we write the reaction in reverse, the equilibrium constant is inverted to give a new constant, $K'(= 1/K)$.

$$K' = \frac{[\text{bound}]}{[\text{free}][K^+]} = \frac{1}{K}$$

We have just defined an equilibrium as the state where the rates of the interchange of species on the left- and right-hand sides of the equilibrium equation are equal. Both the forward and reverse steps have an associated rate constant, k_1 and k_{-1}, and so, the equilibrium equation can be written in a form that shows the rate constant associated with the forward and reverse directions.

$$[\text{bound}] \overset{k_1}{\underset{k_{-1}}{\rightleftharpoons}} [\text{free}][K^+]$$

At equilibrium,

forward rate = reverse rate

and so, by definition,

$$k_1 [\text{bound}] = k_{-1} [\text{free}][K^+]$$

Rearranging yields

$$k_1/k_{-1} = [\text{free}][K^+]/[\text{bound}] = K$$

that is, we can define the equilibrium constant as a ratio of rate constants. This can sometimes prove to be of value. If two of the constants K, k_1, or k_{-1} are known, then the third can be calculated.

One might expect that the relative concentrations of the equilibrium species might depend on their relative stabilities; that is, on their intrinsic energies, there should exist a relationship between the equilibrium constant K and the energy state of molecules. This is indeed the case, and is discussed further in the next section.

1.10.2 Free Energy and Chemical Equilibrium

When we talk about the energy of a molecule, we are really referring to the free energy (ΔG) of that molecule. The free energy of a collection of molecules will vary with temperature, volume, pressure, and the number of molecules under consideration, and so it is often convenient to refer to a set of standard conditions for comparison with other molecules. These standard conditions are defined as *one mole of the molecule in its normal physical state at 25 °C, 1 atm pressure*. The standard molar free energy is designated $\Delta G°$. Because we will seldom be considering reactions under these conditions, we will typically use a general molar free energy ΔG, but the reaction conditions must be carefully defined.

When considering a reaction or any equilibrium state, we must consider the total free energies of the reactant and product species (Figure 1.12). The preferred state is that with the lowest free energy. If the free energy change for the reaction (reactants → products) is negative, that is, $\Delta G < 0$, then the reaction is said to be favorable. If the free energy change is positive ($\Delta G > 0$), then the reaction is said to be unfavorable. The equilibrium lies to the right for a favorable reaction, and the equilibrium lies to the left for an unfavorable reaction. The more favorable the reaction, the larger the magnitude of the negative free energy change. As expected, a relationship exists between the reaction free energy and the equilibrium constant, and this is written as,

$$\Delta G = -RT \ln K \qquad\qquad R = 1.987 \text{ cal}/K \text{ mol}$$

where ΔG is the free energy change for the reaction, R is the gas constant (8.314 J K^{-1}), T is the temperature in degrees Kelvin, and K is the equilibrium constant for the reaction at that temperature. This equation is one of the most powerful in chemistry because it allows us to evaluate the equilibrium concentrations of reactant and product molecules from a knowledge of the free energy change (ΔG). Note that it is usually best to convert all temperatures to units of Kelvin when attempting problems (K = °C + 273.2).

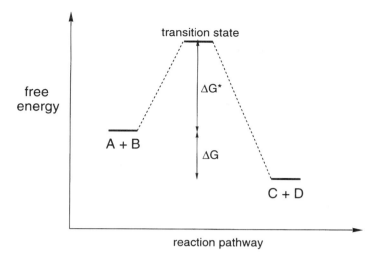

Figure 1.12   Free energy profile of a reaction showing the transition state, the activation energy ΔG^*, and the reaction free energy ΔG. The label on the x axis is a rather loose description that signifies the making and breaking of bonds as we move from reactant to product molecules.

1.10.3   Enthalpy and Entropy

The free energy of a molecule (G) is comprised of two distinct terms, the enthalpy (H) and entropy (S). Enthalpy is dominated by factors that contribute to various kinds of bonding interactions (covalent, ionic, hydrophobic, and hydrogen bonding) within a molecule and between molecules. The more extensive these bonding interactions are, the larger the intrinsic enthalpy of a molecule. For that reason, larger molecules tend to possess larger intrinsic enthalpies than smaller molecules. For example, the heat of formation (ΔH_f) for crystalline $CoBr_2$ at 298 K is -52.8 kcal mole^{-1}, but is -216.4 kcal mole^{-1} for the complex $[Co(NH_3)_6]Br_2$. Also, the enthalpy is found to increase for $[Co(NH_3)_6]Br_3$, with $\Delta H_f \sim -239.7$ kcal mole^{-1}, and for $[Co(NH_3)_6](NO_3)_3$, with $\Delta H_f \sim -306.4$ kcal mole^{-1}. In the case of $Cu(ClO_4)_2 \cdot 6H_2O$, $\Delta H_f \sim -460.9$ kcal mole^{-1}.

In contrast to the enthalpy term, entropy is not an energy parameter. Rather, it is a proportionality constant that reflects the degree of order within a molecule and its extrinsic environment and defines how widely the energy in a molecule is distributed (as translational, vibrational, or rotational motions) at a specific temperature: that is,

$$\text{energy} = S \times T$$

Factors contributing to the entropic energy of a molecule include its physical state (solid, liquid, gas, solution), its influence on solvent structure (for dissolved substances), and the intrinsic flexibility and mobility of the molecular framework. Note that, when we compare the same substance under different conditions, the entropy varies according to the physical state. For example, the entropy of phosphorus varies as $S_{gas} > S_{liq} > S_{solid}$, with values of 39.0, 10.2, and 5.4 cal K^{-1} mole^{-1}, respectively. Also observe that more complex molecules with a larger number of atoms and bonds generally show larger entropy values because these molecules possess more *degrees of freedom* as a result of their ability to rotate around bonds. For example, the entropy of hydrogen gas is 31.0 cal K^{-1} mole^{-1}, whereas the entropy for gaseous hydrogen peroxide is 55.6 cal K^{-1} mole^{-1} and is 77.7 cal K^{-1} mole^{-1} for $[Co(NH_3)_6]Br_3$ in the crystalline state.

A simple equation relates G, H, and S, namely,

$$G = H - TS$$

When changing from an initial to a final state, the relationship is written

$$\Delta G = \Delta H - T\Delta S$$

Reactions with a negative enthalpy change ($\Delta H < 0$ kcal mole^{-1}) are described as *exothermic* (from the Latin for heat given out) and are *endothermic* (from the Latin for heat taken in) if the reaction is accompanied by a positive enthalpy change ($\Delta H > 0$ kcal mole^{-1}).

1.11 Catalysis

1.11.1 Activation Energies.

We have introduced some basic concepts in kinetics and thermodynamics, and now we must look at the relationship between them. First, let us examine what kinetics does not tell us about thermodynamics, and vice versa. Surprisingly, even if a reaction is very favorable ($\Delta G \ll 0$), it may proceed very slowly, if at all. This turns out to be important because almost all carbon-based molecules could form more stable products by combustion with oxygen. Obviously spontaneous ignition is rare! For the resulting reaction to be fast, it is not enough by itself for the products of a reaction to be much lower in energy than the reactants, that is, kinetics and thermodynamics relate to different factors. Consider the free energy diagram in Figure 1.12. When reactant molecules are converted to product molecules the reaction must proceed through a transient state termed the transition state. The free energy of the transition state is higher than that of either the reactant or product states. To reach the transition state, energy must be supplied to the reactants from the immediate environment, and energy is subsequently released during product formation. It is the activation free energy ΔG^* that defines the rate of a reaction, whereas the free energy change ΔG defines the equilibrium concentration of reactants and products. Slow reactions possess large activation free energies and fast reactions have a small ΔG^*, irrespective of the magnitude of the free energy change ΔG for the reaction. A relationship (the Eyring equation) exists between the kinetic rate constant k, and the activation free energy ΔG^*:

$$k = (\mathbf{k}T/h) \exp(-\Delta G^*/RT) \tag{1.4}$$

where $\mathbf{k}$ is the Boltzmann constant, h is Planck's constant, and other parameters were defined earlier. This equation implies that an increase in temperature will increase the reaction rate. To understand the reason for this, consider the plot in Figure 1.13. This shows the probability that a molecule will have a free energy (G) in a large population of molecules. Only those molecules with combined energies greater than or equivalent to ΔG^* will be able to overcome the barrier (ΔG^*) shown in Figure 1.12. By increasing the temperature, we shift the distribution of energies to the right, that is, there is a greater probability that a molecule will have sufficient energy to overcome ΔG^*.

The activation free energy can be written in terms of enthalpic and entropic components,

$$\Delta G^* = \Delta H^* - T\Delta S^*$$

reflecting bond making/breaking (ΔH^*) and changes in solvation and/or disorder (ΔS^*) during the approach to the transition state. For the reverse reaction, the barrier ($\Delta G^* - \Delta G$) defines the activation energy. We can rewrite the equation for the rate constant in a form that will help us to determine both the ΔH^* and ΔS^* terms. By taking natural logarithms of each side we obtain

$$\log k = \log(\mathbf{k}T/h) - \Delta G^*/RT$$

Substituting for ΔG^* and rearranging gives

$$\log(kh/\mathbf{k}T) = \Delta S^*/R - \Delta H^*/RT$$

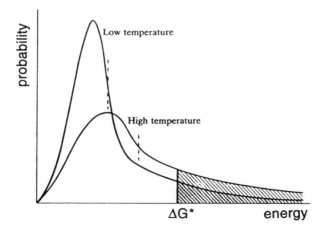

Figure 1.13 Free energy distribution of a collection of molecules. The average free energy is indicated by the vertical dashed line. At higher temperatures the distribution moves to higher free energies, as does the average value (indicated by the vertical dotted line).

and so a plot of log $(kh/\mathbf{k}T)$ versus $1/T$ should give a straight line with slope equal to $-\Delta H^*/R$ and an intercept on the y axis with a value equal to $\Delta S^*/R$.

1.11.2 Catalysis

We have seen that the rate of approach to chemical equilibrium is not defined by the free energy difference ΔG but by the activation free energy ΔG^*. The rate constant is a kinetic parameter. To increase the rate, we must lower the activation energy ΔG^* for the reaction. This is a function of a *catalyst*. A catalyst is a chemical species that becomes involved in the reaction but at the end of the reaction is chemically unchanged. The role of the catalyst is to provide an alternative reaction pathway with a lower activation barrier. Figure 1.14 shows that this pathway may involve more steps and additional reaction intermediates, but ΔG^* is smaller and the reaction proceeds more rapidly. Note also that the thermodynamic free energy difference ΔG between the reactants and products is the same for both the catalyzed and uncatalyzed reaction. The equilibrium constant remains the same. The catalyst does not result in the formation of more product but does allow the equilibrium concentration to be attained more rapidly and, usually, under gentler conditions of temperature and/or pressure.

The phenomenon of catalysis is very important in chemistry because many chemical reactions would take a very long time without assistance. This is also true of living systems, where reactions must take place rapidly, but at the relatively low temperature of 37 °C. Also, without catalysis, many chemical manu-

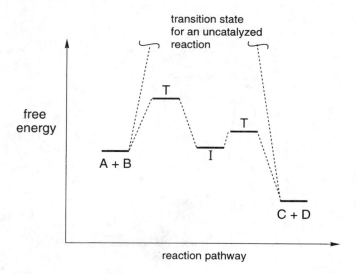

Figure 1.14 The role of a catalyst is to provide an alternative pathway with lower activation barriers. Reactions often proceed through one or more intermediate states I, each with its own transition state T. The energies of the reactant and product states are unchanged.

facturing processes would be unworkable, requiring extremes of temperature to attain acceptable rates that were physically and economically untenable. It is important to realize that all catalysts, biological or not, work in precisely the same manner, by providing an alternative reaction pathway with a lower ΔG^* for transforming reactant molecules into product molecules.

Solvent conditions, or the polarity of the reaction medium, can greatly influence activation barriers. Polar solvents significantly stabilize transient charged intermediates and lower the activation barrier (ΔG^*). Similarly, the stabilization of polar reactant or product molecules tends to lower the energy of these states quite dramatically, with the result that the activation barrier is greater and the reaction proceeds more slowly. Clearly, a change of solvent environment may dramatically alter the kinetics of a reaction, and the equilibrium concentration of reactants versus products may also be significantly perturbed if one or the other becomes stabilized or destabilized relative to the other.

Nature uses a variety of large molecules called *enzymes* to catalyze biochemical reactions and similar ideas contribute to the design and function of these biological catalysts. For example, the interior active site of an enzyme is often a pocket of relatively low polarity. As such, the reactant molecules are more reactive by pathways that minimize charge buildup. Alternatively, the active site pocket may contain charged centers or metal ions that serve to stabilize charged reaction intermediates. In either case, the enzyme binds the reactant molecule or molecules and provides an alternative reaction pathway with a lower activation energy. Because most biochemical reactions must take place at a relatively mild temperature (~ 37 °C) and pH ~ 7.3, catalysts are essential if the biochemical reactions are to take place at the rates required to sustain life. The picture shows a schematic illustration of the active site of an enzyme called methane monooxygenase. We will have a closer look at this enzyme in Chapter 5.

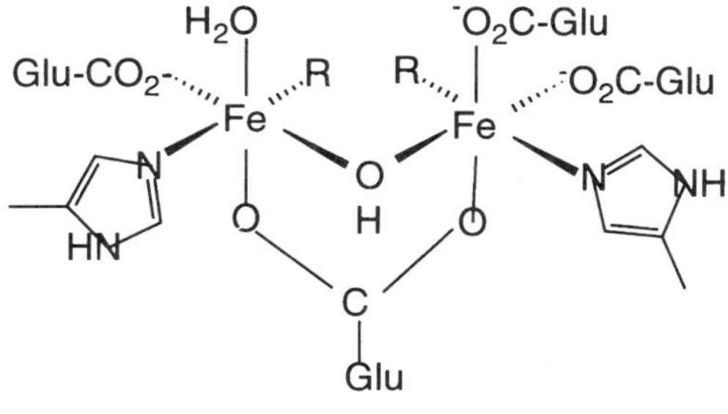

Catalytic site of methane monooxygenase

Methane monooxygenase serves to catalyze the oxidation of methane to methanol by cleavage of a carbon–hydrogen bond, and insertion of hydroxyl.

$$CH_4 + O_2 + 2e^- + 2H^+ \rightarrow CH_3OH + H_2O$$

Normally, this reaction is extremely difficult, with a high activation barrier; however, the enzyme effects rapid turnover of CH_4 to CH_3OH and is of great interest to petrochemical companies because the reaction turns a rather inexpensive material into a product of far greater value.

Summary of Sections 1.10 and 1.11

1. The free energy change for a reaction (ΔG) and the equilibrium constant defining a reaction are related through the equation,

$$\Delta G = -RT \ln K = \Delta H - T\Delta S$$

2. The rate of a chemical reaction is dependent on the magnitude of the activation energy ΔG^*. A catalyst provides an alternative reaction pathway with a lower ΔG^*.
3. Only a fraction of the molecules in each reaction mixture will have sufficient energy to overcome the activation barrier ΔG^*. Increasing the temperature typically increases the probability that a molecule will have energy $\geq \Delta G^*$.
4. The relationship between the rate constant and activation free energy is given by

$$\log k = \log (\mathbf{k}T/h) - \Delta G^*/RT$$

5. Catalysts do not change the thermodynamic driving force (i.e., ΔG), and so they do not alter the equilibrium state of a reaction. However, they do increase the rate at which equilibrium is achieved.
6. Catalysts participate in the reaction, but are regenerated, and are chemically unchanged at the end of the reaction cycle.

PART C. FUNDAMENTALS OF BIOCHEMISTRY

1.12 Biological Ligands

1.12.1 Peptides and Proteins

Proteins constitute one of the basic functional units in biology. Their diverse roles include the transport of neutral molecules, ions and electrons (e.g., the O_2 carrier hemoglobin or the redox protein cytochrome c); the formation of biological materials (e.g., muscle, hair, bone); or the regulation of activity by binding to other proteins (e.g., calmodulin). Many other functions will emerge later in the text. Proteins that catalyze chemical reactions are called *enzymes*. As in any chemical reaction the enzyme catalyst provides an alternative reaction pathway with a lower activation barrier. Proteins and enzymes that bind metal ions are called metalloproteins and metalloenzymes, respectively. Metalloproteins fall into two general classes: (1) The metal ion is an integral part of the protein and is retained during typical isolation and purification procedures; (2) the metal serves as a cofactor that is required for activity but is not normally associated with the protein. In each case the metal ion may bind directly to the protein as an isolated center or as a more elaborate metal complex in association with an organic ligand (e.g., heme, vitamin B_{12}). Examples of each are illustrated in Figure 1.15. Whatever the example under consideration, it is likely that at least one of the ligands binding the metal ion will be an amino acid residue. Amino acids form the building blocks for peptides and proteins. There are 20 common amino acids and some additional ones (γ-carboxyglutamate and hydroxyaspartate) that are used in Ca^{2+}-binding proteins. All possess an L rather than a D configuration at the chiral center, where R is one of the side chains shown in Figure 1.16. Typically, the amino acid side chains provide the ligating atoms;

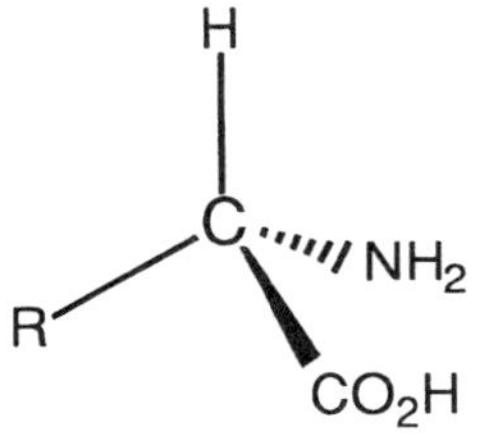

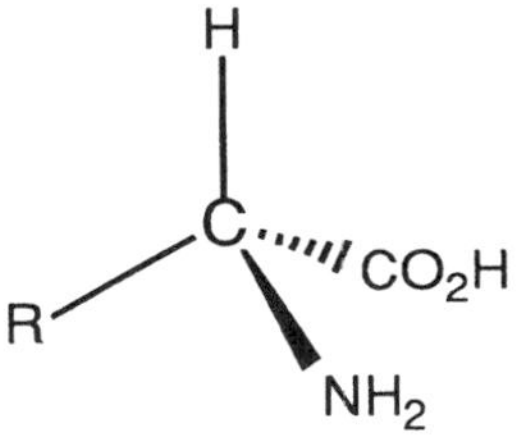

however, deprotonated amide nitrogen or carbonyl oxygens are also used. The structural formulas, abbreviations, and ionization constants (pK_a's) for side-

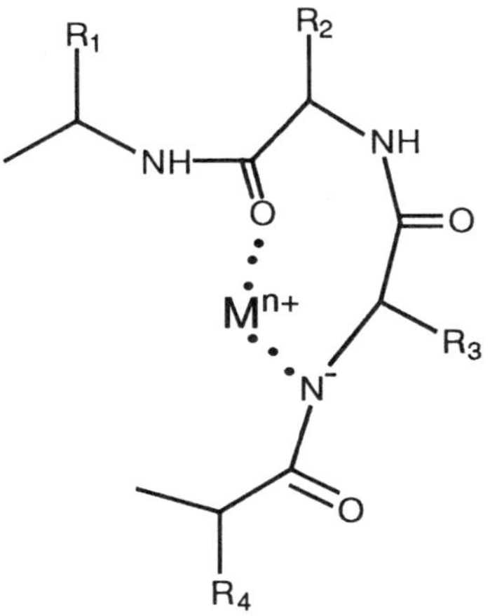

chain functionality of the 20 common amino acids are shown in Figure 1.16. Figure 1.17 shows how the protein backbone is formed from this basic set of amino acids by formation of amide links between the amino and carboxylic acid functionality. An amino acid unit within a peptide chain is referred to as an *amino acid residue*.

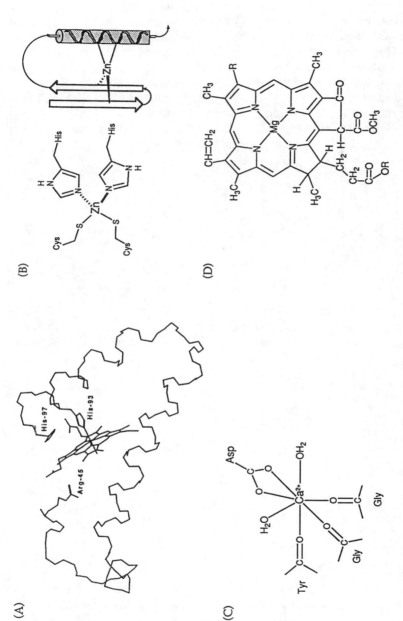

Figure 1.15 Coordination modes for metal binding to metalloproteins and peptides. (A) The heme prosthetic center and a portion of the backbone in myoglobin. (B) Bound Zn^{2+} in a zinc finger. On the right the portion of the protein backbone that forms the "finger" is traced. Figure 1.19 gives more details on such schematic diagrams. (C) The metal-binding domain of a Ca^{2+}-activated enzyme (phospholipase A_2) showing coordination of a chelating carboxylate, two water molecules, and three backbone carbonyls. (D) Chlorophyll from the light-harvesting complex of the photosynthetic reaction center.

Amino Acid (three- and one-letter codes, mol. wt.)	Structure	pK$_a$ of sidechain	Amino Acid (three- and one-letter codes, mol. wt.)	Structure	pK$_a$ of sidechain
Nonpolar			**Neutral - polar**		
Alanine (Ala, A, 89)	CH$_3$—CH COO$^-$ NH$_3^+$	—	Asparagine (Asn, N, 132)	H$_2$N—CO—CH$_2$—CH COO$^-$ NH$_3^+$	—
Glycine (Gly, G, 75)	H—CH COO$^-$ NH$_3^+$	—	Cysteine (Cys, C, 121)	HS—CH$_2$—CH COO$^-$ NH$_3^+$	8.35
Isoleucine (Ile, I, 131)	CH$_3$—CH$_2$—CH—CH COO$^-$ CH$_3$ NH$_3^+$	—	Glutamine (Gln, Q, 146)	H$_2$N—CO—CH$_2$—CH$_2$—CH COO$^-$ NH$_3^+$	—
Leucine (Leu, L, 131)	CH$_3$ CH—CH$_2$—CH COO$^-$ CH$_3$ NH$_3^+$	—	Serine (Ser, S,105)	HO—CH$_2$—CH COO$^-$ NH$_3^+$	—
Methionine (Met, M, 149)	CH$_3$—S—CH$_2$—CH$_2$—CH COO$^-$ NH$_3^+$	—	Threonine (Thr, T, 119)	CH$_3$—CH—CH COO$^-$ OH NH$_3^+$	—
Phenylalanine (Phe, F, 165)	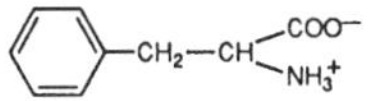 CH$_2$—CH COO$^-$ NH$_3^+$	—	Tryptophan (Trp, W, 204)	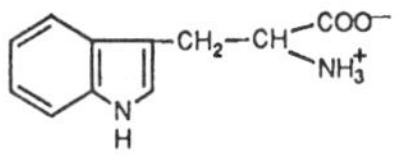 CH$_2$—CH COO$^-$ NH$_3^+$	—
Proline (Pro, P,115)	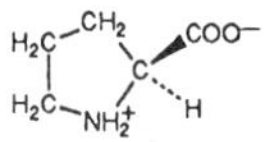 H$_2$C H$_2$C C—COO$^-$ NH$_2^+$ H	10.64			
Valine (Val, V, 117)	CH$_3$ CH—CH COO$^-$ CH$_3$ NH$_3^+$	—			
Acidic			**Basic**		
Aspartic acid (Asp, D,132)	HOOC—CH$_2$—CH COO$^-$ NH$_3^+$	3.90	Histidine (His, H,155)	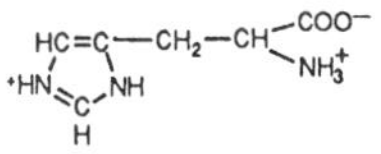 HC=C—CH$_2$—CH COO$^-$ $^+$HN C NH H NH$_3^+$	7.04
Glutamic acid (Glu, E,147)	HOOC—CH$_2$—CH$_2$—CH COO$^-$ NH$_3^+$	4.07	Arginine (Arg, R,174)	H$_2$N—C—NH—CH$_2$—CH$_2$—CH$_2$—CH COO$^-$ $^+$NH$_2$ NH$_3^+$	12.48
Tyrosine (Tyr, Y,181)	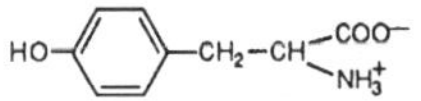 HO CH$_2$—CH COO$^-$ NH$_3^+$	10.13	Lysine (Lys, K,146)	H$_3$N$^+$—CH$_2$—CH$_2$—CH$_2$—CH$_2$—CH COO$^-$ NH$_3^+$	10.79

Figure 1.16 Twenty common amino acids.

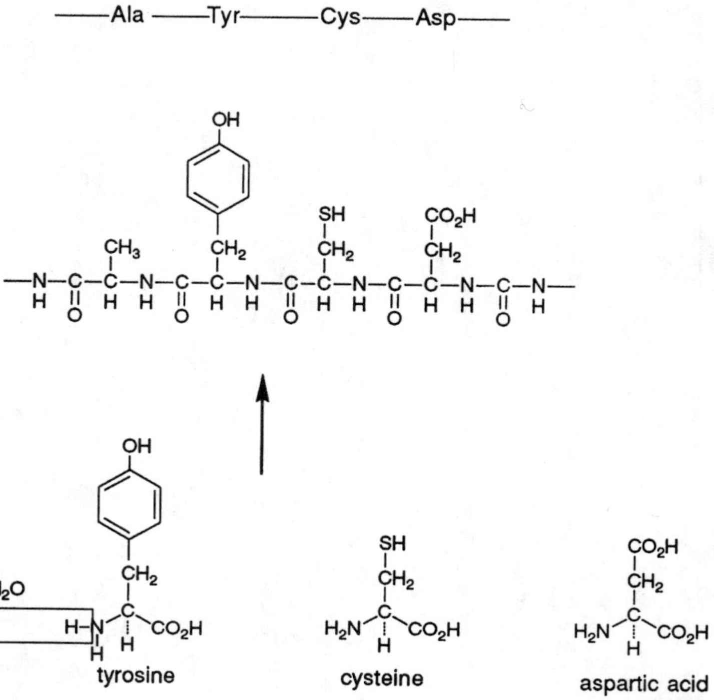

——Ala ——Tyr———Cys——Asp——

Figure 1.17 Synthesis of a polypeptide chain by amide (peptide) bond formation through condensation of amino and carboxylate functional groups from distinct amino acids. Each amino acid within the peptide chain is called a *residue*. The unique side chains for each of the 20 amino acids convey specific structural, catalytic, and functional properties to the polypeptide (or protein) molecule.

Amino acid residues can also be classified according to the HSAB theory described earlier, which, in part, provides a criterion for selectivity among metal ions.

Hard: Glu, Asp, Tyr, Ser, Thr
Intermediate: His
Soft: Cys, Met

Coordination numbers and geometries at metal-binding sites reflect the trends noted in Table 1.3, although distortions are common as a result of the structural flexibility of the protein backbone and the possibility of positioning ligands in defined orientations.

1.12.2 Protein Structure

Proteins and enzymes display varying degrees of structural order (Figure 1.18). The *primary* structure of a protein defines the sequence of amino acids and the

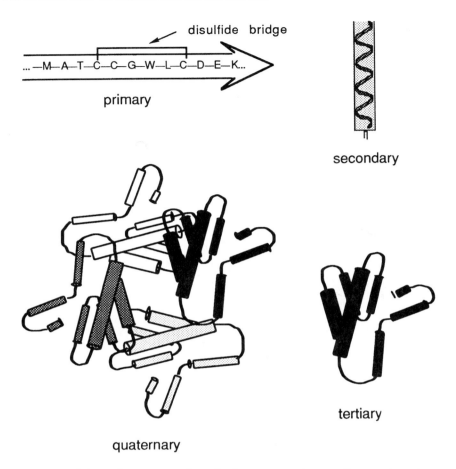

Figure 1.18 Schematic representation of primary, secondary, tertiary, and quaternary protein structures. Several types of secondary structure are illustrated further in Figure 1.19. The tertiary structure consists of a variety of secondary structural elements. Occasionally, several protein units combine through noncovalent hydrophobic or electrostatic interactions to form a higher order quaternary structure.

positions of disulfide links between cysteines. It defines all the covalent interactions in the protein. The polypeptide will fold to form elements of *secondary* structure that are regularly repeating conformations of the peptide backbone (e.g., α-helices, β-pleated sheets, reverse turns) formed by specific H-bond interactions between backbone amide N–H protons and carbonyls (Figure 1.19). Figure 1.18 also shows how these structural elements form a higher order *tertiary* structure that imposes a longer range ordering of residues and secondary structural motifs. This defines the three-dimensional conformation of the active or native form of the protein and is characterized by folding patterns that bring together amino acid residues that are far apart in the primary structure. The

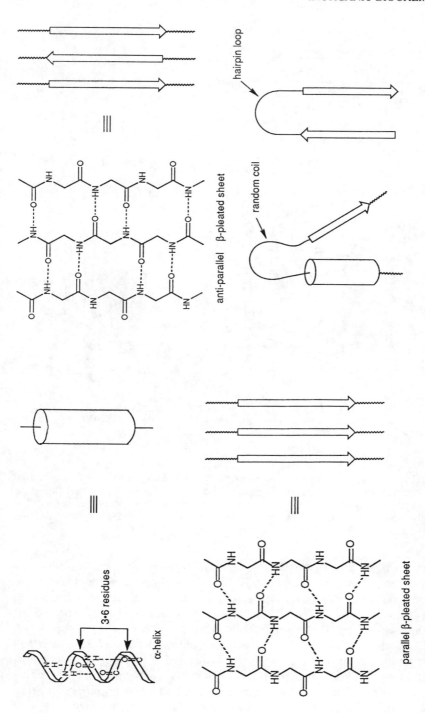

Figure 1.19 Examples of secondary structure and the schematic notation used to represent each structural form. Regions of the peptide backbone that connect α-helices or β-sheets are often called *random coil*. The short sequence linking antiparallel β-sheets is termed a *hairpin loop* or *reverse turn*.

tertiary structure is stabilized by hydrogen bonds, salt bridges, disulfide linkages, and hydrophobic interactions. Again, the information that defines this structure is encoded in the sequence of amino acid side chains. Understanding the relationship between primary, secondary, and tertiary structure is the essence of the protein folding problem. Several polypeptide chains may aggregate to form a complex of several protein subunits (*quaternary* structure) held together by noncovalent interactions. Such a multimeric protein may contain identical or distinct component proteins and is named accordingly. For example, hemoglobin is an $\alpha_2\beta_2$ tetramer. It contains two α-subunits and two β-subunits. Proteins may be further modified (see Figure 1.25) by attachment of sugar chains (glycoproteins), lipids (lipoproteins), fatty acids (myristylation), or phosphorylation of amino acid residues.

A variety of distinct molecular species may be required for the proper function of a protein or enzyme. Binding sites for metal ions, substrate molecules, prosthetic centers, and coenzymes are defined by the conformation of the protein backbone and the side chains of the amino acid residues. A prosthetic center can be defined as a metal–ligand complex that is firmly bound to the protein and is normally isolated in association with it. In contrast, coenzymes are often less firmly associated and may not be isolated with the protein or enzyme. In all cases, one can imagine the protein backbone providing a pocket for these prosthetic groups or coenzymes that protects them from the external environment, with the side chains helping to bind the centers and regulate their physicochemical properties by direct coordination or electrostatic interactions.

1.12.3   Nucleotides and Nucleic Acids

Genetic information is stored in polymers of nucleic acids. Nucleic acids are constructed from one of five bases that are derived from purine (adenine, guanine) or pyrimidine (cytosine, thymine, or uracil) and are attached to a ribose sugar ring by an *N*-glycosidic linkage (Figure 1.20). Thymine is found in DNA, and uracil is in RNA. The combination of base and ribose is termed a *ribonucleoside*. If one of the 2′, 3′, or 5′ hydroxyls is phosphorylated, the suffix -*tide* is used (e.g., *ribonucleotide*). If the sugar ring lacks the 2′-OH, the appropriate names are *deoxyribonucleoside* or *deoxyribonucleotide*. RNA (ribonucleic acid) and DNA (deoxyribonucleic acid) are constructed from these building blocks to form a polymeric chain formed by phosphodiester linkages between phosphate and the 3′ and 5′ hydroxyls of ribose.

1.12.4   Polynucleotide Structure

DNA typically exists in the form of a double-stranded (ds) structure formed by hydrogen bond formation between specific base pairs (Figure 1.21). Watson–Crick base pairing is most common, but Hoogsteen interactions are also found. The two sugar–phosphate backbones twist in a helical conformation around the central stack of base pairs, generating a major and minor groove. The back-

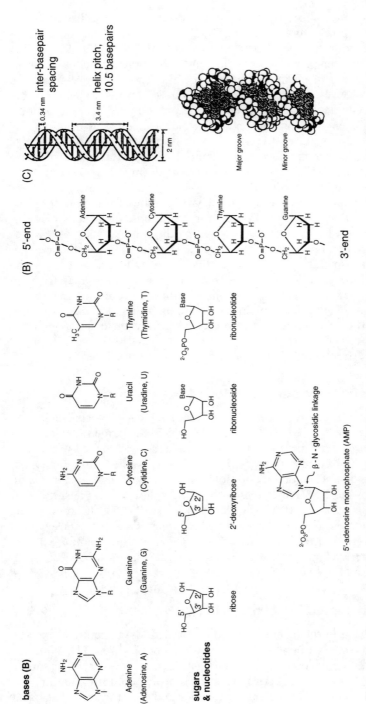

Figure 1.20 Structural units of the nucleic acids. (A) Purine and pyrimidine bases, ribose sugar, and nucleotides. The names in parentheses below the bases are the names of the corresponding nucleosides or nucleotides (i.e., base + ribose-phosphate). The 2′, 3′, and 5′ carbons on ribose are indicated. (B) Base attached to the deoxyribose–phosphate backbone. A single strand of DNA. Double-stranded DNA is formed by specific hydrogen bond patterns formed between complementary base pairs (see Figure 1.21). (C) Structural features of B-DNA. The vertical rise of 3.4 nm for a complete 360° turn is accommodated within 10.5 base units. (Adapted from R. L. P. Adams, J. T. Knowler, and D. P. Leader, *The Biochemistry of the Nucleic Acids*, Chapman and Hall, 1986).

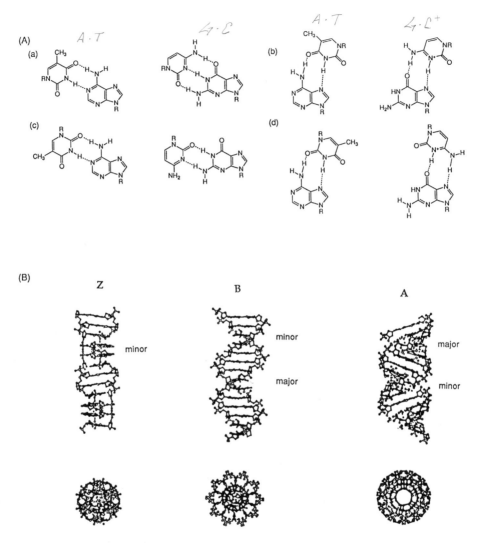

Figure 1.21 (A) Hydrogen bonding leads to pairing between specific bases (A·T and G·C). This connects two single-stranded nucleotide chains (Figure 1.20) to form a double-stranded polynucleotide that adopts a helical structure, the "double helix." R represents the point of attachment to the ribose–phosphate backbone. Two common base-pair patterns include Watson–Crick (a) and Hoogsteen (b). Reversed Watson–Crick (c) and reversed Hoogsteen (d) can also be found. (B) Common backbone conformations adopted by double-stranded DNA. Major and minor grooves are indicated. The major groove of A-DNA is wider and deeper relative to the B-conformation. In Z-DNA the major groove is converted to a shallow convex surface. Only a deep minor groove remains.

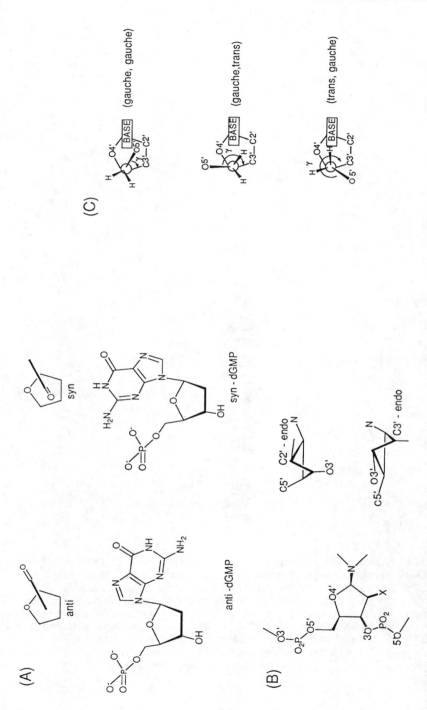

Figure 1.22 Graphic representation of key structural parameters detailed in Table 1.10. See also Figure 1.20. (A) *Syn/Anti* orientation of the base relative to the ribose ring. (B) The *C2'-endo* and C3'-*endo* sugar pucker angles. (C) The terms *gauche* and *trans* define the C5'-O5' conformation relative to the ribose ring.

bone may adopt several conformations, several of which are shown in Figure 1.21. Conformations A and B adopt a right-handed helix, whereas Z is an example of a left-handed helix. Typically, DNA adopts a B conformation, which may convert to Z-DNA under conditions of high salt or in the presence of organic solvents. The Z-configuration is particularly favorable for sequences containing alternating GC nucleotides. An A-conformation is normally associated with double-stranded RNA or RNA–DNA hybrids. Structural parameters associated with each form are noted in Table 1.10 and are illustrated in Figure 1.22.

RNA normally exists in the form of a single strand (ss), but this may fold to form a secondary structure through specific base pairing,[21] which, in turn, can fold into a higher order tertiary structure. The negatively charged ribose–phosphate backbone of the major and minor grooves of ds DNA and RNA play host to a variety of positively charged species (metals, ligands, protein side chains). Alkali and alkaline earth metals tend to coordinate to the oxyligands (phosphate, sugar hydroxyls, and carbonyl functionality), whereas softer transition metals preferentially coordinate to base nitrogen atoms. The principal metal-binding domains on nucleotides are illustrated in the following.

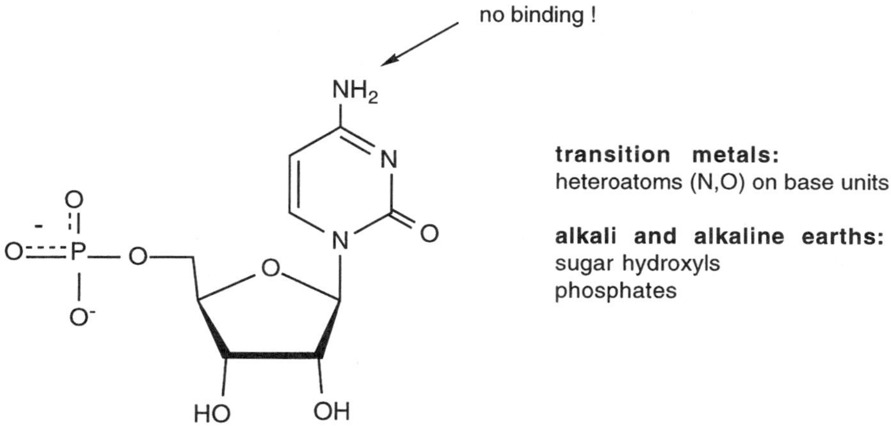

no binding !

transition metals:
heteroatoms (N,O) on base units

alkali and alkaline earths:
sugar hydroxyls
phosphates

Summary of Section 1.12

1. Proteins and enzymes are constructed from a collection of 20 amino acids by formation of amide links between the amino and carboxylic acid functionality. Metalloproteins and metalloenzymes bind metal ions, either free or in association with an organic ligand. These are termed *cofactors* (weakly bound) or *prosthetic centers* (tightly bound).

2. Five nucleic acids (abbreviated A, T, U, G, C) form the building blocks for RNA and DNA,[25] each consisting of a ribose–phosphate ring connected to

Table 1.10   Structural Parameters Associated with A-, B-, and Z-DNA

Property	A-DNA	B-DNA	Z-DNA
Helix handedness	Right	Right	Left
Rotation per base pair	32.7°	34.6°	30°
Base pairs per turn	≈11	10.4	12
Inclination of base pair to helix axis	19°	1.2°	9°
Rise per base pair along helix axis (Å)	2.3	3.3	3.8
Pitch (Å)	24.6	34.0	45.6
Diameter (Å)	25.5	23.7	18.4
Conformation of glycosidic bond	*anti*	*anti*	*anti* at C, *syn* at G

Adapted from R. Dickerson et al., *Science, 216*, 475–485 (1982).

a purine or pyrimidine base by an *N*-glycosidic linkage. Metal ions may bind to the phosphate, sugar hydroxyls, or base heteroatoms.

3. Proteins can be described in terms of their primary, secondary, tertiary, and quaternary structures. This information is defined by the sequence of the amino acid side chains. Common secondary structural features include α-helices and β-sheet.

4. Polynucleotides are cross-linked through specific base-pair patterns [A–T, (A–U for RNA), G–C]. DNA may adopt a variety of backbone conformations, including A-, B-, and Z-conformations. Double-stranded RNA adopts an A-conformation.

Review Questions

- How is a β-turn stabilized?
[Stryer, L. *Biochemistry*, 4th ed., pp. 30–32, Freeman, New York, 1995]
- The propensity for metal binding by basic amino acid side chains in aqueous solutions of metalloproteins varies as His > Lys > Arg. Similarly, for acidic groups, the order is Glu ~ Asp > Tyr. Provide a simple explanation for these trends.
- One normally thinks of the rate of diffusion as a constant for a specific solution medium at a well-defined temperature. The diffusion rate constant for an ion moving through an aqueous solution is ~ 10^9 s^{-1}. However, it is ~ 10^7 s^{-1} for a protein. Provide an explanation for this fact.

1.13   The Relationship Between Nucleotide and Protein Sequence

The information encoding a protein sequence is held in the ordering of the bases in DNA. Each amino acid is uniquely determined by a series of three bases called a *codon*. (For reference, a complete tabulation of codon usage is listed in

Appendix 9.) More than one codon may exist for a particular amino acid (e.g., TCC = serine, AGC = serine, GCA = alanine, and so on). There are also start and stop codons that define where the reading of the DNA sequence should begin and end for a particular polypeptide chain.

A *gene* is a piece of DNA that encodes a complete protein. Figure 1.23 summarizes the important steps in protein synthesis. The DNA sequence is *transcribed* (or read) by an enzyme called *RNA polymerase* (RNAP). RNAP binds close to the start codon and reads through the DNA sequence until it reaches a termination point.[22] The polymerase synthesizes a strand of messenger RNA (mRNA) that is complementary to the DNA.[23] This process of reading the DNA sequence to give mRNA is termed *transcription*. The mRNA carries the information (or blueprint) required for protein synthesis to another region of the cell where a molecular machine called the *ribosome* constructs the protein from amino acid building blocks. The ribosome brings together the mRNA and tRNAs (transfer RNA). The latter have specific amino acids covalently bonded to their 3'-ends. Each tRNA (carrying an amino acid) possesses a short sequence of 3-bases that complements the bases in the codon on the mRNA. In this way, the amino acids are delivered to the site of protein synthesis in the correct order and are coupled together by formation of amide (or peptide) linkages to form the protein. This process of using the information encoded in the mRNA to produce a polypeptide is called *translation*. The stop codon does not bind a tRNA, and so the growing polypeptide chain is released.

1.14 Cell Biology

To appreciate fully the nature of the problems in the field of inorganic biochemistry, we must learn a few fundamental principles of molecular and cell biology. This allows us to identify and understand better the most important problems to be tackled. In this section we shall address a few key points that provide an important framework for future chapters.

From a cursory glance at the literature, a reader might form the impression that enzymes, proteins, and a variety of other biological macromolecules are ubiquitous, in the sense that they not only are found in every cell but also are located everywhere within the cell. This, of course, is not the case. Apart from single-cell life forms, higher organisms are multicellular. In mammals, the cells that form skin are different from liver cells, brain cells, muscle, and so on. Although there are many common aspects to the biochemistry within these cells, each has its own individuality. Interestingly, every cell in a multicellular organism (e.g., liver, muscle, spleen, etc., in mammals) possesses the same copy of genomic DNA, and so specific genes must be preferentially activated within each cell. A more fundamental division of cell type lies in the distinction between prokaryotic and eukaryotic cells. The former consist of bacteria, archaebacteria, and blue-green algae (cyanobacteria) and are all single-cell species. The archaebacteria comprise all known methanogenic bacteria and halophilic and ther-

(1) Transcription

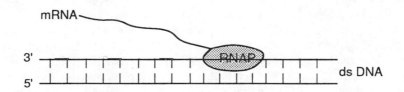

(2) Translation

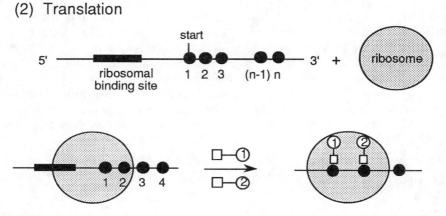

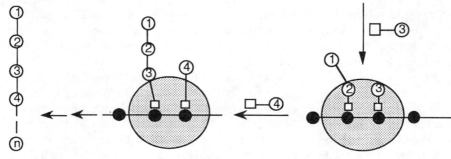

Figure 1.23 How a protein is made. (1) RNA polymerase (RNAP) synthesizes a complementary single strand of mRNA from the information encoded by dsDNA. Ribo-
• nucleotide triphosphates (NTPs) are joined in the correct order to produce the mRNA.
(2) The 5'-end contains a short sequence before the start codon that serves to anchor the
RNA to the ribosome. Codons are represented by (●). Amino acids (○) are carried to
the appropriate codon by a transfer RNA (□). The peptide backbone is formed by
peptide bond formation between amino acids held in place on the ribosome by their
respective tRNA molecules. The result of successive peptide bond formations is the synthesis of a growing protein chain.

mophilic organisms. Eukaryotes, including plants, animals, fungi, and protozoa, possess a discrete membrane-bound nucleus within the cell that contains the genomic DNA.

A breakdown of cell morphology for prokaryotes and eukaryotes is given in Figures 1.24 and 1.25. Prokaryotic cells are surrounded by outer and inner (plasma or cytoplasmic) membranes. The region between these two membranes is called the *periplasmic space.* The part of the cell enclosed by the inner membrane is termed the cytoplasm; the cytosol is the aqueous medium in which the cellular DNA, proteins, and metabolites are suspended or dissolved. On the exterior of the cell are three appendages that are formed from protein aggregates: flagella (responsible for cell motion), pili (sexual conjugation), fimbriae (adhesion to surfaces).

Eukaryotic cells contain many specialized organelles that are separated from the cytosol by membranes. These membrane-bound organelles perform a variety of roles: mitochondrion (aerobic respiration), golgi complex (processing of se-

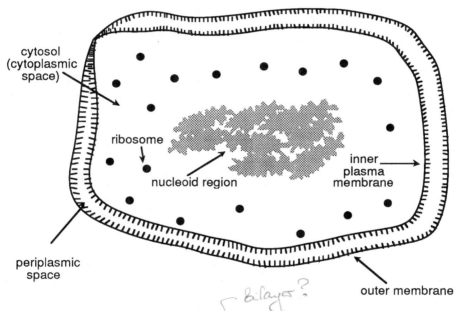

Figure 1.24   Morphology of prokaryotes. There are no intracellular compartments. The outer lipid membrane is also covered by a layer of teichoic acids and peptidoglycans (see Sections 3.1, 3.2, and 6.6). The inner plasma membrane is composed of lipids and membrane-spanning proteins that mediate the transfer of minerals and nutrients to and from the cell. The portion of the cell surrounded by the inner membrane is termed the *cytoplasm* and the aqueous medium is called the *cytosol.* The DNA within the cell is localized in the nucleoid region but is not surrounded by a membrane. Other than the genomic DNA, bacteria often possess smaller circular sequences of DNA called *plasmids* that are located in the cytosol. Each bacterial cell contains several thousand ribosomes for protein synthesis.

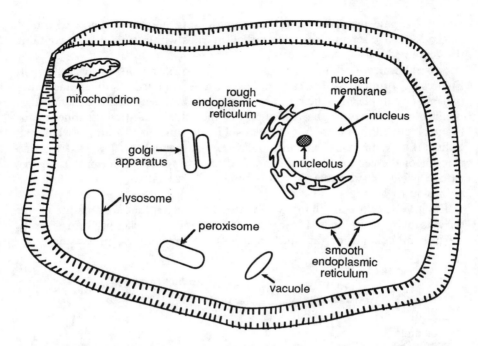

Figure 1.25   Morphology of eukaryotes. A typical mammalian cell is shown. In contrast to prokaryotes, eukaryotic cells contain internal compartments in the cytosol. The nucleus encloses the genomic DNA and is contained by a membrane that contains pores to facilitate the movement of mRNA to the ribosomes. Other compartments handle specific cellular functions: mitochondrion (metabolism, energy production, ATP synthesis); lysosome (digestion of peptides and nucleic acids); golgi apparatus (protein modification to form glycoproteins and lipoproteins); rough endoplasmic reticulum (coated with ribosomes for protein synthesis; it also makes membrane lipids and secretory proteins); peroxisomes (compartments where oxidation reactions are carried out and toxic by-products such as H_2O_2 are eliminated); smooth endoplasmic reticulum (lipid synthesis, protein modification). Other cell types may contain specialized compartments (e.g., chloroplasts in plant cells that contain the apparatus for photosynthesis).

cretory proteins), chloroplast (photosynthetic unit in plants and algae), ribosome (protein synthesis), endoplasmic reticulum (protein processing and transport), and sarcoplasmic reticulum (Ca^{2+} storage in muscle cells). The nucleus contains the chromosomal DNA and is surrounded by a membrane termed the *nuclear envelope*. This contains membrane-bound proteins that control the movement of proteins and mRNA to and from the nucleus. (See the discussion of replication, transcription, and translation in Section 1.13.) Many of the cellular ribosomes are embedded in a membrane (rough endoplasmic reticulum, ER) that meshes with the nuclear envelope. The smooth endoplasmic reticulum and the golgi apparatus modify proteins after synthesis at the ribosome. For example, carbohydrate fragments and lipids may be added to give glycoproteins and lipoproteins. Lysosomes contain a variety of digestive enzymes that break down proteins, nucleic acids, and sugars into simpler molecular fragments that can be reused in cellular metabolism. In bacteria, these digestive enzymes are commonly located in the periplasm between the outer and inner membranes. Clearly, mechanisms must exist to shuttle substrates and products to and from the cytosol. Peroxisomes contain oxido-reductase enzymes that carry out oxidation reactions on organic substrates in which H_2O_2 is a reactive by-product. Peroxisomes also contain enzymes that remove H_2O_2, and so protect the rest of the cell from oxidative damage. The mitochondrion is surrounded by a double membrane and is responsible for most of the energy production in the cell. It is here that amino acids, fatty acids, and carbohydrates are oxidized. The energy produced is stored in the form of chemical energy as adenosine triphosphate (ATP). The details of these reactions will be discussed in Chapter 5. Many proteins or enzymes within the cell are specifically located in one or other of these compartments. In the case of multicellular organisms, it is also possible to speak of the extracellular environment. This introduces the question of communication between cells, especially those of a different type (e.g., B-cells and T-cells associated with the immune response in blood plasma). The interaction between intra- and extracellular domains must be strictly regulated, and so certain proteins are located specifically in this domain.

Summary of Sections 1.13–1.14

1. The amino acid sequence for a protein is encoded in a DNA sequence called a *gene*. RNA polymerase reads this information (*transcription*) by synthesizing a complementary strand of messenger RNA. The mRNA travels to the ribosome, a molecular machine for protein synthesis (*translation*). Amino acids are brought to the ribosome, covalently bonded to transfer RNA. Each tRNA contains a group of three nucleic acids that match the codons on the blueprint mRNA, and so the polypeptide is synthesized in the correct order.
2. There are two types of cells, *prokaryotic* and *eukaryotic*. The latter are the more complex, because they contain discrete intracellular compartments that store genetic information, synthesize proteins, modify proteins, generate chemical energy, carry out oxidative or digestive chemistry, and so on. Membranes are an important aspect of cell structure.
3. Higher organisms (e.g., plants and mammals) are multicellular. Each contains many cells with different functions (e.g., skin, muscle, blood cells, etc.), but all contain the same copy of genomic DNA.

1.15 Molecular Biology define

1.15.1 Cloning

When a particular sequence of DNA is known to code for a specific enzyme or protein, it is possible to cut that piece of DNA from the larger genomic DNA by use of restriction enzymes and subsequently paste it into a small circular DNA called a plasmid. Plasmids may be replicated in bacterial cells, which are much easier to handle than higher organisms. The steps involved in cloning a gene (X) from genomic DNA into a plasmid (P) are illustrated in a very simplified diagram (Figure 1.26).

1.15.2   Site-Directed Mutagenesis

Site-directed mutagenesis is a powerful technique that enables the biochemist to change any residue in a protein to any other. In this way, the effect of that change on the physicochemical properties of the system may be examined. This can provide great insight into the role of specific residues in defining the structural, catalytic, binding, or regulatory properties of the protein. Figure 1.27 summarizes the procedures, which are made possible by the availability of a large repertoire of enzymes that carry out reactions on polynucleotides with great efficiency and specificity. It will be seen later that many of these enzymes require metal cofactors if they are to function properly. With reference to Appendix 9, it should be obvious that changing specific codons allows the substitution of one amino acid for another. This is the conceptual basis for the site-directed mutagenesis experiment. To effect these changes, the plasmid containing the gene of interest is removed from the bacterial cells—*Escherichia coli* is frequently used—and a series of enzymatic reactions is carried out to switch codons. The plasmid may then be transferred back into *E. coli* for expression.

1.15.3   Expression

Gene expression encompasses the two processes of transcription and translation whereby the information contained in a sequence of nucleic acids is turned into a protein molecule. The regulation of gene expression is an important problem that will be considered in more detail in Chapter 7. For the moment, we simply note that the initial transcription of DNA → mRNA by RNA polymerase requires a DNA binding site positioned close to the start codon (see Figure 1.23). This DNA sequence is termed a *promoter region*. RNA polymerase binds to the promoter before reading off the gene sequence. A strong promoter (i.e., one that results in high levels of expression and the production of large quantities of protein) has a high affinity for RNAP. We may clone our gene behind a variety of promoters. Each has distinct features that allow gene expression to be regulated in a number of ways (see Section 7.2).

Prokaryotic cells are valuable tools for studying biochemical mechanisms because they rapidly reproduce, may be readily grown in a defined medium, and have a relatively simple metabolism and organization of biochemical units within the cell. *Escherichia coli* is a gram-negative bacterium that has been studied extensively and is a good model for a variety of prokaryotic cells. Before closing this chapter we should note that many proteins from higher organisms

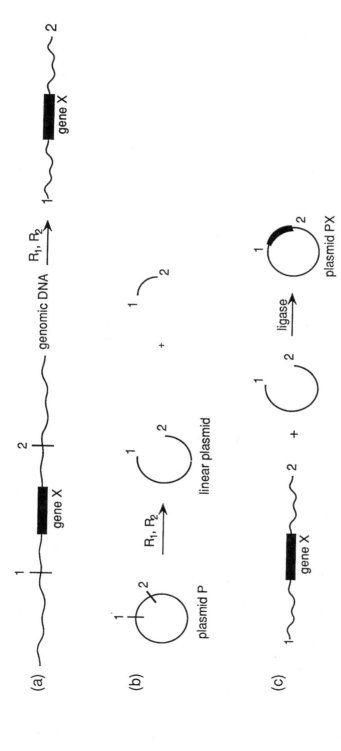

Figure 1.26 (a) A gene X is excised from genomic DNA by use of restriction enzymes (R_1, R_2) that cut double-stranded DNA at specific sites (1, 2). (b) The plasmid into which the gene is to be cloned is cut with the same enzymes to generate complementary ends. (c) Gene X can then be joined to the linearized plasmid by use of an enzyme (ligase) that joins the two ends in the correct orientation by formation of phosphodiester bonds.

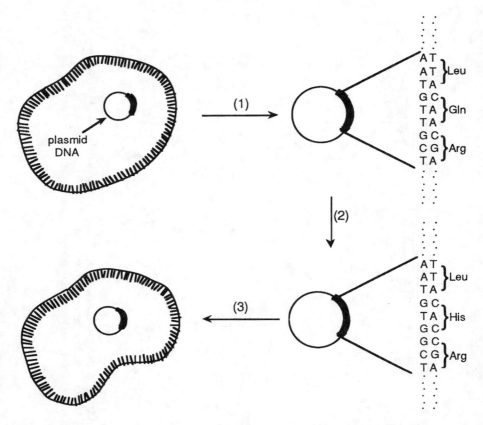

Figure 1.27 (1) The plasmid is isolated from the bacterium. A batch of cells is grown and harvested and the cellular contents are released after the inner and outer membranes are broken by chemical or physical methods (e.g., osmotic shock, sonication, French press). Plasmid DNA can be readily isolated. (2) A specific codon is changed by molecular biology techniques. (3) The new plasmid is incorporated back into the host cell. This procedure is called a *transformation*.

are not readily expressed in prokaryotes, which do not possess the molecular apparatus required to make chemical changes to a protein after expression (*posttranslational modification*). Eukaryotic cells are more complex than the simpler prokaryotic systems, which may lack important genetic apparatus. In these cases, yeast, baculovirus, or tissue cultures may be useful expression vehicles.

Summary of Section 1.15

1. *Cloning.* A specific length of DNA (perhaps a gene for a particular enzyme) may be cut from a larger piece of DNA by use of restriction enzymes. This is ligated into a circular plasmid for ease of manipulation.

2. *Mutagenesis.* A codon for a specific amino acid can be changed to another codon that translates to a different amino acid. In this way, the function of a specific amino acid in a protein can be tested.

3. *Expression.* A gene encoding a protein is cloned into an expression vector, which, in turn, is used to *transform* a host cell (typically *E. coli*) in which the protein will be synthesized (*expressed*).

Notes

1. For example, Mn(VII) is stabilized by four electronegative oxide ligands in MnO_4^-, whereas Ni(0) is stabilized by carbon monoxide in $Ni(CO)_4$. An important exception is Co(I) in vitamin B_{12}, which will be discussed in Chapter 5.

2. The exception of Mo (and to a lesser extent W) will be discussed in Chapter 5.

3. For some transition-metal ions [e.g., square planar Ni(II) complexes] the nature of the bonding (σ or π) is an important factor (see Section 1.6). Geometric restrictions on the ligand itself (e.g., planar protoporphyrin IX, Chapter 5) must also be considered.

4. Refer to R. G. Pearson, *J. Chem. Ed.* **45**, 581–587 (1968).

5. The word *stable* is unfortunately used in discussions of both kinetics and thermodynamics. Thermodynamic stability should never be confused with kinetic stability (or inertness). For example, $Ni(CN)_4^{2-}$ is thermodynamically stable (log β_4 = 30.5) but is kinetically labile (CN^- exchange rate $> 10^{-2}$ s^{-1}).

6. The concentration of species x will be denoted by the bracket rotation [x] throughout the text.

7. For example, see J. E. Prue, *J. Chem. Educ.* **46**, 12 (1969), and A. W. Adamson, *J. Am. Chem. Soc.*, **76**, 1578 (1969).

8. The concentrations of species should be represented by activities ($a_x = [x] \cdot \gamma_x$), where γ_x is an activity coefficient. $\gamma_x \sim 1$ only in very dilute solutions. This correction accounts for interactions among solute molecules that tend to decrease the availability, or effective concentration, of that species. For most practical purposes $[x] \neq a_x$, and can result in significant errors if ignored. *Thermodynamic stability constants* are usually determined by measuring stability constants in a series of solutions that contain a noncomplexing electrolyte, such as sodium perchlorate, and are extrapolated to zero ionic strength.

9. In a biological milieu many ligands compete for each metal ion. Moreover, the concentrations of available soluble metal ion will vary according to the solubility products (K_{sp}) with simple inorganic anions, $M^{n+}(aq)$, $+ L(aq) \rightleftharpoons ML(s)$, and $K_{sp} = [M^{n+}][L]$. The distribution of metal ions between a variety of ligands and solid precipitates will, therefore, depend on a large number of interrelated equilibria. This is referred to as the speciation of metal ions.

10. Defined for the reaction, $ML_x^{n+} + e^- \rightarrow ML_x^{(n-1)+}$. The more positive the potential, the more likely the oxidized center will accept the electron.

11. Common ligands are illustrated in Appendix 1.

12. Degenerate orbitals have equal energy.

13. According to Hund's rules, electrons will occupy a set of degenerate orbitals such that the number of unpaired spins is maximized and the electrons have their spins aligned.

14. Note, en = ethylenediamine; bipy = bipyridine; phen = o-phenanthroline. See Appendix 1 for structural formulas.

15. The comma in $[ML_6^{n+}, L']$ indicates formation of an outer sphere complex.

16. For a chelating bidentate ligand $L_1 - L_2$, it may be shown that $k = k_i K_{os}(1 + k_{2i}/k_{1i})$.

17. Complexes where at least one ligand is H_2O [e.g., $ML_5(H_2O)^{n+}$].

18. The electronic configuration of Cu^{2+} (d^9) may be written as $(d_{xy})^2(d_{yz})^2(d_{xz})^2(d_{z^2})^2 (d_{x^2-y^2})^1$. In an octahedral ligand field, the d_{z^2} and $d_{x^2-y^2}$ orbitals are nominally degenerate. However, as written, there is clearly more electron density located along the z axis (d_{z^2}) than exists along the x or y axis ($d_{x^2-y^2}$). This tends to repel the ligands along the z axis, weakens the coordination, and lowers the energy of d_{z^2}, leading to a further loss of oribtal degeneracy.

19. H. Taube, *Adv. Inorg. Chem. Radiochem., 1,* 1 (1959).

20. The Marcus cross-exchange equation. See Marcus and Sutin *Biochem. Biophys. Acta, 811,* 265–322 (1985) for an extensive review of electron-transfer processes in chemistry and biology.

21. Base pair specificity is less rigorous for RNA and exceptions from the normal A·U and G·C pairs are found.

22. A self-complementary sequence of RNA is formed (see note 23), which folds back on itself to form a termination loop that causes RNAP to dissociate from the DNA template.

23. A complementary sequence exactly matches a given sequence but uses the Watson–Crick base partner. For example, 3′-GCUAACUG-5′ is the complementary RNA strand to the DNA strand, 5′-CGATTGTC-3′.

Further Reading
Inorganic Solution Chemistry and Inorganic Biochemistry

Basolo, F., and R. G. Pearson. *Mechanisms of Inorganic Reactions,* Academic Press, 1968.

Burger, K. *Biocoordination Chemistry,* Ellis Horwood, 1990.

Burgess, J. *Ions in Solution,* Ellis Horwood, 1988.

Frausto da Silva, J. J. R., and R. J. P. Williams. *The Biological Chemistry of the Elements,* Oxford, 1991.

Harrison, P. M., and R. J. Hoare. *Metals in Biochemistry,* Chapman and Hall, 1980.

Hughes, M. N. Coordination compounds in biology, in *Comprehensive Coordination Chemistry* (Eds. G. Wilkinson, R. D. Gaillard, and J. A. McCleverty), Chap. 62.1, Vol. 6, Pergamon, 1987.

Hughes, M. N. *The Inorganic Chemistry of Biological Processes,* 2nd ed., Wiley, 1981.

Hughes, M. N., and R. K. Poole. *Metals and Microorganisms,* Chapman and Hall, 1989.

Kaim, W., and B. Schwederski. *Bioinorganic Chemistry: Inorganic Elements in the Chemistry of Life. An Introduction and Guide,* John Wiley & Sons, New York, 1994.

Lippard, S. J., and J. Berg. *Principles of Bioinorganic Chemistry,* University Science Books, Mill Valley, Calif., 1994.

Ochiai, E. I. *General Principles of Biochemistry of the Elements,* Plenum, 1987.

Prue, J. E. Ion pairs and complexes, free energies, enthalpies, and entropies, *J. Chem. Educ. 46,* 12, 1969.

Stadtman, T. C. Selenium Biochemistry, *Science, 183,* 915–922 (1974).

Werz, W. E. The Essential Trace Elements, *Science, 213,* 1332–1338 (1981).

Wilkins, R. G. *Kinetics and Mechanisms of Reactions of Transition-Metal Complexes.* 2nd ed. VCH, 1991.

Williams, D. R. Ed. *An Introduction to Bioinorganic Chemistry,* C. C. Thomas, 1976.

Williams, R. J. P. The evolution of bioinorganic chemistry. *Coordn. Chem. Rev. 100,* 573–610 (1991).

Biochemistry

Adams, R. L. P., J. T. Knowler, and D. P. Leader. *The Biochemistry of the Nucleic Acids,* Chapman and Hall, 1986.

Blackburn, G. M., and M. J. Gait, eds. *Nucleic Acids in Chemistry and Biology,* IRL Press, 1990.

Branden, C., and J. Tooze. *Introduction to Protein Structure,* Garland, 1991.

Cantor, C. R., and P. R. Schimmel. *Biophysical Methods* (Parts I, II, and III), Freeman, 1980.

Creighton, T. E. *Protein Structure and Molecular Properties,* Freeman, 1984.

Drlica, K. *Understanding DNA and Gene Cloning,* John Wiley, 1984.

Martin, D. W., P. A. Mayes, and V. W. Rodwell, eds. *Harper's Review of Biochemistry,* Lange, 1981.

Stryer, L. *Biochemistry,* 4th ed., Freeman, New York, 1995.

General Books and Review Series

Eichorn, G. L., and L. G. Marzilli, eds. *Advances in Inorganic Biochemistry,* Elsevier.
Frieden, E., ed. *Biochemistry of the Elements*, Plenum.
Gray, H. B., and A. B. P. Lever, Eds. *Physical Bioinorganic Chemistry,* Addison-Wesley,
Lippard, S. J., ed. *Progress in Inorganic Chemistry,* vols. 18 & 38, Wiley-Interscience.
Methods in Enzymology, Academic Press.
Siegel, H., ed. *Metal Ions in Biological Systems,* Dekker.
Spiro, T. G., ed. *Metal Ions in Biology*, Wiley-Interscience.
Structure and Bonding, Springer-Verlag.
Sykes, A. G., ed. *Advances in Inorganic and Bioinorganic Mechanisms*, Academic Press.
Trends in Biochemical Sciences, Elsevier.

Worked Problems

Question 1: Provide an explanation for the relative trends shown by the stability constants for a macrocyclic ligand. Why does the preferred metal depend on the solvent?

Metal Ion	$\log_{10}K_I$ in water	$\log_{10}K_I$ in methanol
Mg^{2+}	2.0	3.8
Ca^{2+}	4.1	4.3
Sr^{2+}	13.0	**6.1**
Ba^{2+}	**15**	5.9

Solution 1: In water, the smaller ions with the larger charge density are extensively solvated. The dehydration energy is not offset by binding to the macrocycle. As a result, the cation with lowest charge density binds most tightly. In this case solvation energy is the dominant factor. In methanol, however, the solvation energies are significantly lower. As a result, the range of values for $\log_{10}K_I$ is narrower, and other factors, such as optimal ionic radius, become important. In methanol, therefore, both solvation energy and size factors must by accounted for.

Question 2: An interesting experiment that can be demonstrated in the classroom is the decomposition of hydrogen peroxide, H_2O_2. This is a complex reaction, but can be summarized as

$$2H_2O_2 \rightarrow 2H_2O + O_2$$

This reaction occurs very slowly in aqueous solution, let's say with an observed first-order rate constant of 3.17×10^{-8} s^{-1} (i.e., only half of the sample has decayed in one year!). Hydrogen peroxide is formed as a reaction byproduct in living cells; however, H_2O_2 is a very reactive substance and must be removed before cellular damage results. For this reason, cells contain biological protein catalysts, called peroxidases, that decompose H_2O_2 by the reaction stated earlier. One can observe this by adding some bovine or chicken liver (chicken works best!) to a dilute aqueous solution of H_2O_2. Within a few seconds, a large frothy column will emerge as $O_2(g)$ is evolved. Let's say that the observed first order rate constant in this case is $1.6 \times 10^{-2} s^{-1}$ ($\sim$ 1 min). Calculate the apparent difference in activation energies for these two reactions and comment on the reaction mechanisms.

Solution 2: We use the equation that relates rate constant to activation energy; namely,

$$\log k = \log (kT/h) - \Delta G^*/RT$$

Rearranging this equation gives,

$$\Delta G^* = RT[\log (kT/h) - \log k] = RT[\log (kT/kh)]$$

and so ΔG^* can be calculated for the catalyzed and noncatalyzed reactions by substituting values for the constants and rate constants. At ambient temperature ($\sim$ 298 K)

$$\Delta G_{uncat}^* = 8.314 \text{ J K}^{-1} \text{ mole}^{-1} \times 298 \text{ K} \times [\log [(1.38 \times 10^{-23} \text{ J K}^{-1}$$
$$\times 298 \text{ K})/(3.17 \times 10^{-8} s^{-1} \times 6.626 \times 10^{-34} \text{ J s})]$$
$$= 115.8 \text{ kJ mole}^{-1} = 27.7 \text{ kcal mole}^{-1}$$
$$[\text{using the conversion 1 cal} = 4.184 \text{ J}].$$

$$\Delta G_{cat}^* = 8.314 \text{ J K}^{-1} \text{ mole}^{-1} \times 298 \text{ K} \times [\log [(1.38 \times 10^{-23} \text{ J K}^{-1}$$
$$\times 298 \text{ K})/(1.6 \times 10^{-2} s^{-1} \times 6.626 \times 10^{-34} \text{ J s})]$$
$$= 83.2 \text{ kJ mole}^{-1} = 19.9 \text{ kcal mole}^{-1}.$$

We have determined $\Delta\Delta G^* = \Delta G_{uncat}^* - \Delta G_{cat}^* = (27.7 - 19.9) \text{ kcal mole}^{-1}$ $= 7.8 \text{ kcal mole}^{-1}$. Clearly, the mechanisms for these two reaction pathways must be distinct, with different transition state energies. This reflects the role of a catalyst, whether biological or not: to provide an alternative pathway with a lower transition-state energy.

Problems

1. Sketch the two isometric structures of the pentaamminecobalt(III) complex of methyl sulfinate ($CH_3SO_2^-$). One of these forms may exist in an enantiomeric form. Explain.
2. Examine the plot of ionic radius versus atomic number for the given divalent metal ions. Provide an explanation for the trends shown by the experimental data, and state the significance of the dashed line.

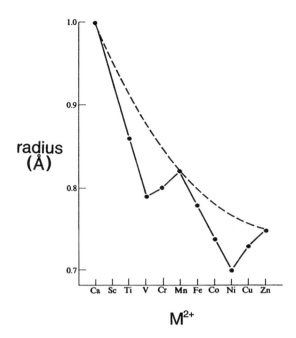

radius
(Å)

M^{2+}

[Jolly, W. L. *Modern Inorganic Chemistry*, McGraw-Hill, 1984, pp. 433–437]

3. The stepwise formation constants for reaction of Br^- ion with $Cd(H_2O)_6^{2+}$ to form a series of bromide substituted complexes are $K_1 = 2 \times 10^2$, $K_2 = 6$, $K_3 = 0.6$, $K_4 = 1.2$ for the consecutive addition of four bromide ions. Provide an explanation for the decrease in K_n as the number of bromide ions increases, and for the apparent break at K_4.

4. The stability constants and rates of formation of complexes between alkali metal ions and 18-crown-6 (a macrocycle with six ether oxygen ligand atoms) are noted below. Identify, discuss, and explain the apparent conflicting trends shown by these two parameters.

	Li^+	Na^+	K^+	Rb^+	Cs^+
$\log_{10} \beta_1$	1.5	4.6	6.0	5.2	4.6
$k_f \times 10^{-8}$	0.8	2.2	4.3	4.4	5.1

[Burgess, J., *Ions in Solution*, Halsted Press, John Wiley & Sons, 1988, pp. 139–145.]

5. What form will the rate law for substitution in square-planar complexes take if the solvolysis of the complex is rapid compared with ligand substitution?

[Burgess, J. *Ions in Solution*, Halsted Press, John Wiley & Sons, 1988, pp. 146–152.]

6. (a) The electron-transfer (ET) reaction noted below proceeds via an outer sphere pathway. Determine the electron exchange rate (k_{12}) at 298 K.

$$Fe(aq)^{2+} + Cr(aq)^{3+} \rightleftharpoons Fe(aq)^{3+} + Cr(aq)^{2+}$$
$$k_{11}(Fe) = 4 \ M^{-1} \ s^{-1}$$
$$k_{22}(Cr) = 2 \times 10^5 \ M^{-1} \ s^{-1}$$
$$E^\circ{}_{1/2} \ (Fe^{3+/2+}) = -0.77 \ V$$
$$E^\circ{}_{1/2} \ (Cr^{3+/2+}) = 0.41 \ V$$

(b) The electron-transfer reaction between $Cr(NH_3)_6^{3+}$ with $Cr(H_2O)_6^{2+}$ occurs very slowly ($k_{12} \sim 10^{-3} \ M^{-1} \ s^{-1}$). The solvent exchange rates for these two complexes are $\sim 10^{-10}$ and $10^9 \ M^{-1} \ s^{-1}$, respectively. Is the ET reaction likely to proceed via an inner or outer sphere pathway? Explain your reasoning carefully.

(c) In contrast, the reaction of $Cr(NH_3)_6^{2+}$ with $Cr(H_2O)_6^{3+}$ occurs much faster ($k_{12} \sim 6 \times 10^5 \ M^{-1} \ s^{-1}$). Provide an explanation for the difference in these two rates.

7. Examine the data for the two cross reactions represented by

$$A^+ + B \rightleftharpoons A + B^+$$

A^+	B	$k_{11}(A) \ M^{-1} \ s^{-1}$	$k_{22}(B) \ M^{-1} \ s^{-1}$	K_{12}	$k_{12}(calc) \ M^{-1} \ s^{-1}$	$k_{12}(obs) \ M^{-1} \ s^{-1}$
$IrCl_6^{2-}$	$Fe(DMP)^{2+}$	2.3×10^5	3×10^8	0.30	4.6×10^6	1.1×10^9
AmO_2^{2+}	NpO_2^+	2.4	96	6.4×10^7	5×10^4	2.5×10^4

DMP = 1,10-dimethylphenanthroline.

(a) Comment on the values of k_{11} and k_{22}.

(b) Comment on and explain the difference or similarity between $k_{12}(calc)$ and $k_{12}(obs)$ for each reaction.

(c) Determine the driving force in volts for each reaction.

8. A zinc finger is a DNA binding motif that is formed from a peptide sequence that binds a zinc ion in a tetrahedral site formed by cysteine and histidine residues. Cobaltous ion can compete for the zinc binding site as shown.

$$peptide\text{-}Zn^{2+} + Co(H_2O)_6^{2+} \rightleftharpoons Peptide\text{-}Co^{2+} + Zn(H_2O)_6^{2+}$$

The dissociation constants for zinc and cobalt binding to the peptide are 2.8×10^{-9} M and 3.8×10^{-6} M, respectively.

(a) Write an expression for the free energy difference for zinc and cobalt binding to the peptide in terms of the dissociation constants for each metal ion. Calculate the magnitude of this free energy difference.

(b) Assuming the Co^{2+} ion remains high spin in both octahedral and tetrahedral geometries and using $\Delta_o \sim 9300 \ cm^{-1}$ for $Co(H_2O)_6^{2+}$ and $\Delta_t \sim 4900 \ cm^{-1}$ for a tetrahedral N_2S_2 ligand environment, calculate the change in LFSE for the exchange equilibrium.

   (c) Comment on the implications of your answers to parts (a) and (b).

   (d) Would you expect Mn^{2+} to bind to the peptide more or less tightly than Co^{2+}.

   (e) Comment on the choice of Cys and His ligands for zinc finger metal-binding domains.

[Berg, J. M. and D. L. Merkle, *J. Am. Chem. Soc. 111,* 3759–3761 (1989)]

 9. You know the primary amino acid sequence for a low molecular weight electron-transfer protein. You would like to develop an overexpression system for this protein from an *E. coli* host. Propose how you would go about achieving this goal. Describe the individual steps in as much detail as you can.

[Eren, M. and R. P. Swenson, *J. Biol. Chem. 264,* 14874–79 (1989)].

10. The DNA sequences transcribed by RNA polymerase often end in a palindromic sequence [e.g., 5'-GCCGCCAGTTCGGCTGGCGGC-3'] followed by a sequence of A bases. Explain the significance of these two features.

[Stryer, L., *Biochemistry,* 4th ed., Freeman, New York, 1995, pp. 847–849.]

11. Draw the chemical structure of the short peptides that correspond to the following nucleotide sequences:

   (a) 5'-GCAATCGATCAA-3'

   (b) 3'-CGTAAATTTCTA-5'

CHAPTER

2

Experimental Methods

In this chapter we shall learn how a close analysis of the physical and chemical properties of a biological molecule may provide both structural information and mechanistic insight into its function. In particular, we shall review common spectroscopic, thermodynamic, and kinetic tools of the trade. Only the bare essentials of the most commonly used experimental techniques are presented here. However, the application of these methods is illustrated by specific examples throughout the text. Some readers may find it more useful to proceed directly on to chapter 3, returning to this chapter to clarify details of a particular technique or application.

PART A. PHYSICOCHEMICAL METHODS

2.1 Introduction to Spectroscopy

2.1.1 Electromagnetic Radiation and Energy States

Inorganic biochemists have found spectroscopic techniques to be of special utility in monitoring metal prosthetic centers. All forms of spectroscopy derive from the absorption or emission of electromagnetic radiation, so called because it consists of mutually perpendicular oscillating electric and magnetic fields (Figure 2.1). The energy of the radiation is related to the wavelength (λ) and frequency (ν) by a simple relationship [equation (2.1)], where h is Planck's constant

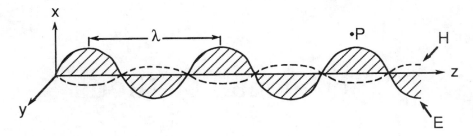

Figure 2.1   Electromagnetic radiation can be viewed as two mutually perpendicular oscillating electric (E) and magnetic (H) fields. The electric dipole moment (μ) is larger than the magnetic dipole moment (m) [$\mu \sim 10^4 \, m$]. The frequency ν is defined as the number of wave peaks that pass a defined point P per second.

(6.636×10^{-34} Js) and c is the speed of light (3×10^8 ms^{-1}). Since energy is often expressed in alternative units[1] it is recommended that some time be spent in becoming adept in handling conversions of energy units. Common conversion factors and other physicochemical constants and parameters are tabulated in Appendixes 10–12.

$$E = hc/\lambda = h\nu \tag{2.1}$$

Molecular energy levels represent specific *states* that reflect the overall structural, electronic, and motional properties of the molecule. Energy levels are quantized [i.e., they have discrete energies and require radiation of defined frequency (ν) to promote a molecule from a low-lying state to a higher energy state]. The ground state is of lowest energy, whereas excited states are of higher energy. If two or more states have the same energy, they are *degenerate*.

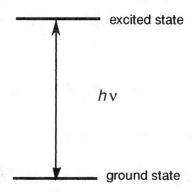

The total energy of a molecule or ion is comprised of contributions from many sources [equation (2.2)].

$$E_{total} = E_{translation} + E_{rotation} + E_{vibration} + E_{electronic} \\ + E_{electron\ spin} + E_{nuclear\ spin} + E_{nuclear\ levels} + \cdots \tag{2.2}$$

For example, absorbed energy may move an electron from a low-energy to a high-energy state (electronic absorption), reorient a nuclear moment in a magnetic field (nuclear magnetic resonance), or influence the vibrational motion of two or more bonded atoms (infrared). These and other examples will be discussed in more detail in later sections. Electromagnetic radiation forms a continuous spectrum, ranging from low-energy radio waves to high-energy γ-radiation. The wavelength and frequency of the radiation are related through equation (2.1). The energy separations between the various states represented by the terms in equation (2.2) vary greatly and cover the electromagnetic spectrum (Figure 2.2). This continuous spectrum can be divided into discrete regions, each of which is appropriate for a particular spectroscopy. For example, the energy required to excite an electron to a higher electronic state (ultraviolet or visible radiation) is different from that needed to influence molecular rotation (microwave).

2.1.2 Boltzmann Distribution Law

Before discussing particular spectroscopic methods, we should take a brief look at a law that underlies all forms of spectroscopy. Given a set of molecular energy levels (Figure 2.3), we might expect that a molecule would prefer to rest in the lowest possible energy level. However, if the energy gap between the lowest energy level and that immediately above is small enough, it would not take much effort, perhaps simple absorption of thermal energy from the environment, to move it to a higher energy state. In other words, although there is a high probability that a molecule will exist in the lowest available energy state, there is a finite probability that it may rest in a higher energy state. The relative populations of these two states will depend on how much extra energy is required to reach the higher energy level. This is the basis for the *Boltzmann distribution law*. At a given temperature (T) the relative populations of an upper and lower state are related to the energy difference between these two states. Given a set

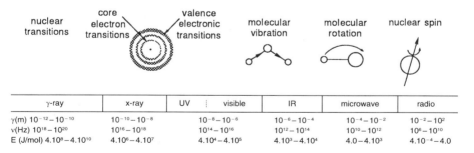

	core		valence			
nuclear transitions	electronic transitions		electronic transitions	molecular vibration	molecular rotation	nuclear spin

γ-ray	x-ray	UV	visible	IR	microwave	radio
γ(m) $10^{-12}-10^{-10}$	$10^{-10}-10^{-8}$	$10^{-8}-10^{-6}$		$10^{-6}-10^{-4}$	$10^{-4}-10^{-2}$	$10^{-2}-10^{2}$
ν(Hz) $10^{18}-10^{20}$	$10^{16}-10^{18}$	$10^{14}-10^{16}$		$10^{12}-10^{14}$	$10^{10}-10^{12}$	$10^{6}-10^{10}$
E (J/mol) $4.10^{8}-4.10^{10}$	$4.10^{6}-4.10^{7}$	$4.10^{4}-4.10^{5}$		$4.10^{3}-4.10^{4}$	$4.0-4.10^{3}$	$4.10^{-4}-4.0$

Figure 2.2 Spectrum of electromagnetic radiation. The boundaries between domains are not sharp, and each region is not shown to scale. The visible domain actually forms an extremely small part (~350–750 nm) of the electromagnetic spectrum.

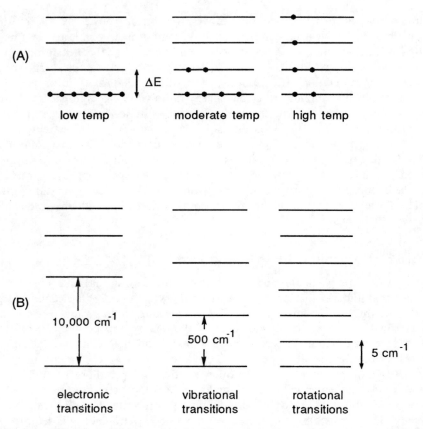

Figure 2.3 Relative population of molecular energy levels or energy states. (A) Higher-energy levels may be populated with increasing temperature (thermal energy). (B) In electronic and vibrational spectroscopy, only the ground state is populated under normal conditions since the excited states lie at much higher energies. However, several rotational states may be populated.

of molecules (N), with a choice of populating one of two states separated by an energy (ΔE), the relative number of molecules in the lower (N_1) and upper (N_u) states at a temperature T is given by equation (2.3).

$$N_u/N_1 = \exp(-\Delta E/kT) \tag{2.3}$$

If ΔE is large relative to thermal energies (kT),[2] only the lower state will be significantly populated. This equation underlies many fundamental aspects of molecular spectroscopy and statistical thermodynamics. We shall see shortly that it can account for the relative sensitivities of various spectroscopic methods.

2.2 Optical Spectroscopy

The region of the electromagnetic spectrum that includes visible light, and the near-ultraviolet and near-infrared to either side, is appropriate for the stimulation of transitions between electronic states. Various spectroscopic methods have been developed that probe these electronic transitions and provide much detailed information on the electronic and magnetic properties of a metal ion or ligand.

2.2.1 Electronic Absorption

Both the absolute energy and intensity of the absorption bands provide useful information on the types of ligand surrounding a metal ion and their geometry. Absorption of light is accompanied by rearrangement of electrons over the orbitals of the metal and the surrounding ligands. Common electronic transitions are summarized in the energy-level diagram shown in Figure 2.4.

The absorbance at a wavelength λ (A_λ) is a measure of the relative intensity of a beam of light before (I_λ°) and after (I_λ) it passes through a solution containing the absorbing molecule,

$$A_\lambda = \log(I_\lambda^\circ/I_\lambda) = \varepsilon_\lambda cl \tag{2.4}$$

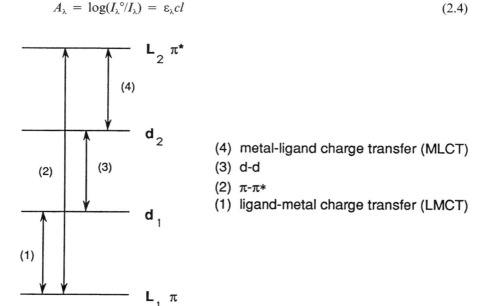

(4) metal-ligand charge transfer (MLCT)
(3) d-d
(2) π-π*
(1) ligand-metal charge transfer (LMCT)

Figure 2.4 Electronic transitions commonly observed in metal complexes. Only d–d absorption bands are formally symmetry forbidden. The symbols d_1, d_2, L_1, and L_2 represent metal centered d-orbitals and ligand π-orbitals, respectively. Typical extinction coefficients are noted in Table 2.1.

where c is the concentration (M), l is the pathlength of the cell (cm), and ε_λ is the extinction coefficient (cm^{-1} M^{-1}). Extinction coefficients are constants that reflect the intrinsic absorbance of a sample at a given wavelength λ. They are, therefore, extremely useful for comparative purposes, since they do not depend on the quantity of the sample being studied.

Selection rules are a simplifying factor in the interpretation of spectroscopic data. It turns out that not all transitions between energy states are allowed, since a physical mechanism to transform a molecule from one electronic, structural, or dynamic state to another may not exist.[3] Typically, ligand–metal charge-transfer (LMCT), metal–ligand charge-transfer (MLCT), and ligand π–π^* absorptions are high-energy allowed transitions that are found in the UV and blue region of the visible spectrum ($\varepsilon > 1000$ M^{-1} cm^{-1}). However, d–d absorptions (corresponding to electronic rearrangements within the d-orbital set) are formally symmetry forbidden for octahedral complexes, and such transitions are typically weak ($\varepsilon \sim 10 - 100$ M^{-1} cm^{-1}). The terms *symmetry allowed* and *symmetry forbidden* have evolved from the language of group theory.[3] The details need not concern us here. Suffice it to say that simple inspection of an equation defining a transformation between a ground and excited state can tell us if the probability of the transformation is zero or finite. Under certain circumstances, forbidden transitions can be observed, although with much reduced intensity, since there may be other less favorable mechanisms that provide a pathway for the transition to occur. For example, vibrations of metal–ligand

(M–L) bonds may distort the complex and remove the O_h symmetry. This provides a mechanism for a d–d transition by coupling electronic rearrangements to molecular vibrations (vibronic coupling). For certain metal ions (e.g., high-spin Mn^{2+} and Fe^{3+}), d–d transitions are also spin forbidden, and so the absorbance may be even more difficult to detect ($\varepsilon < 0.01$ M^{-1} cm^{-1}). Consider high-spin d^5 ions in an O_h ligand field. Any electronic rearrangement must pair off two spins. This requires a change in spin state. However, the electric dipole moment of light cannot act on spin (since the latter is a magnetic parameter), and so there is no mechanism to effect this transition. In reality, light does have a magnetic field perpendicular to the electric field. However, this is extremely weak and provides a very poor coupling mechanism. The d–d bands of most first-row transition-metal ions are normally found in the visible range (350–750 nm). For tetrahedral coordination, d–d transitions are symmetry allowed, and the average extinction coefficient is correspondingly larger ($\varepsilon \sim 100$ to 1000 M^{-1} cm^{-1}). Table 2.1 summarizes this data.

2.2.2 Circular Dichroism

Figure 2.5 illustrates how a chiral (or optically active) center may discriminate between left and right circularly polarized light (lcp and rcp, respectively). The differential absorption of lcp and rcp light (reflected in the different extinction coefficients, $\Delta\varepsilon = \varepsilon_l - \varepsilon_r$) is termed *circular dichroism* (CD). Since biological macromolecules are constructed from chiral building blocks and may adopt a chiral global structure, it should come as no surprise that metal complexes may exhibit circular dichroism when bound to these molecules. Circular dichroism is a particularly useful technique when studying a protein that contains several absorbing centers (ions, clusters, or organic chromophores). The absorption bands from these sites will often overlap, making analysis of the individual prosthetic centers difficult. Even a single chromophore can give rise to several overlapping bands. Figure 2.6 illustrates how circular dichroism spectroscopy will often resolve such overlapping transitions, since these may have opposite signs ($+$ or $-$). A correlation of circular dichroism bands, when comparing spectra from metalloproteins that contain similar prosthetic centers, often suggests a common structural environment for the chromophore.

2.2.3 Raman Spectroscopy

Raman spectra provide information on molecular vibrations. For biological applications the Raman experiment is superior to infrared, since spectra can be obtained in aqueous solutions. IR spectra of aqueous solutions are dominated

Table 2.1 Summary of Extinction Coefficients for Metal and Ligand Transitions in First-Row, Transition-Metal Complexes

	Type of Electronic Transition	ε (M^{-1} cm^{-1})
Ligand	MLCT, LMCT, π–π^*	$10^3 - 10^4$
Metal (d–d)	Tetrahedral (allowed)	$10^2 - 10^3$
	Octahedral (forbidden)	(M^{2+}) ~ 10, (M^{3+}) ~ 50
	Spin forbidden[a]	$10^{-3} - 10^{-2}$

Common examples include high-spin Mn^{2+} (d^5), high-spin Fe^{3+} (d^5), and low-spin Co^{3+} (d^6).

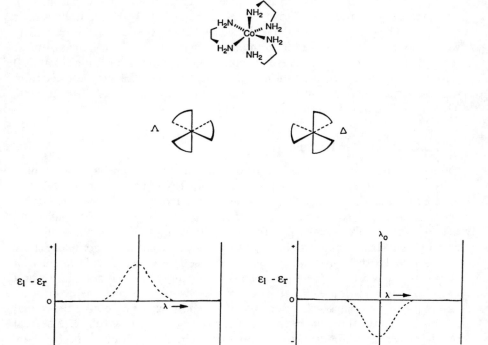

Figure 2.5 Interaction of circularly polarized light with the chiral complex Co(en)$_3^{3+}$ (Δ isomer shown). The circular dichroism spectrum is given by the absorption difference $A_{lcp} - A_{rcp}$. Spectra on the left and right are for Λ and Δ isomers, respectively.

by vibrational frequencies from 55.5 M H_2O. The Raman experiment depends on inelastic scattering of light, that is, the frequency of the scattered light is different from that of the incident radiation, which is typically in the UV or visible region. In ordinary Raman spectroscopy, the molecule is raised to a higher electronic energy state called a *virtual state*. In a formal sense this is not a true excited electronic state, although in resonance Raman the virtual state closely corresponds to an excited electronic state (Figure 2.7). The molecule loses this excess energy by returning to the ground electronic state. If it does so by emitting light of the same frequency (elastic scattering), the term *Rayleigh scattering* is used. If a bending or stretching mode of a molecule is influenced by the absorption of this energy, it will change the vibrational state of the molecule, and so we observe *Stokes* lines (lower energy than the incident radiation) or *anti-Stokes* lines (higher energy than the incident radiation), according to whether the molecule finally occupies a higher or lower vibrational state, respectively. These processes are summarized in Figure 2.7. As a result of the Boltzmann distribution law, anti-Stokes lines are much weaker than Stokes lines

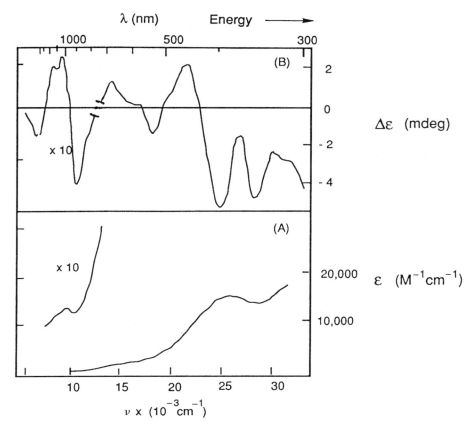

Figure 2.6 Absorption (A) and circular dichroism (B) spectra of reduced *Chromatium vinosum* high-potential iron protein (HiPIP). This low-molecular-weight redox protein (mol wt ~ 9650 Da) contains a [Fe$_4$S$_4$] cluster (see Chapter 5). There is a striking difference in the number of resolved features when comparing (A) and (B).

because of the differing populations of the ground vibrational states. The Rayleigh line is more intense than the others but arises through elastic scattering and provides no information on vibrational levels; therefore, the Stokes lines are the experimental observables. The probability of the scattering process is increased if the exciting light corresponds to the energy difference between two electronic states. However, only those vibrations that are coupled to the electronic transition being probed are resonance enhanced. Resonance Raman spectroscopy is a sensitive method that can be used to selectively probe vibrations that are coupled to a particular chromophore; it can be readily applied to prosthetic centers with strong absorption bands in electronic spectra (e.g., hemes, iron, and copper proteins, and pigment molecules such as chlorophylls). Note that the resonance Raman intensity is proportional to ε^2, where ε is the extinction coefficient of the absorption band. In this way we can avoid the spectral

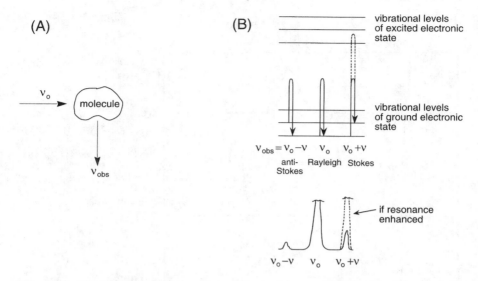

Figure 2.7   The Raman experiment depends on inelastic scattering of light (A) that gives rise to frequencies of lower (Stokes) or higher (anti-Stokes) energy than the incident radiation (B). In resonance Raman, the virtual state overlaps with an excited electronic state, which greatly enhances the intensity of Raman transitions.

complexity that would result if all possible vibrations were detected. Specific examples will appear in later chapters.

2.2.4   Luminescence Spectroscopy

Luminescence covers a range of emission processes that includes fluorescence (singlet to singlet transition) and phosphorescence (triplet to singlet transition). From the emission wavelengths, we can ascertain the energies of excited singlet and triplet states relative to the ground state, and determination of decay lifetimes for each state provides information on the rates of processes that compete with natural decay pathways. Figure 2.8 defines the electronic states involved and typical lifetimes for each. Full arrows ($\rightarrow$) indicate absorption or emission of a photon. Wavy lines ($\rightsquigarrow$) imply loss of energy by nonradiative pathways, typically involving heat loss through molecular vibrations. Singlet excited states lie at higher energy than triplet states.[4] Since ground states are normally singlets, fluorescence is formally allowed, decay is rapid (rate constant k_f), and the excited singlet state has a short lifetime. Phosphorescence from the triplet state is a spin-forbidden transition; therefore, the decay rate (k_p) of this excited state is slower, and the lifetime is longer. Rapid intersystem crossing (k_{isc} or k_i) between the singlet and triplet states occurs by nonradiative pathways. Assuming that all decay pathways are first-order processes, the overall rate constant for depopulation of an excited state can be written as the sum of the individual rate constants for all radiative and nonradiative (k_i)

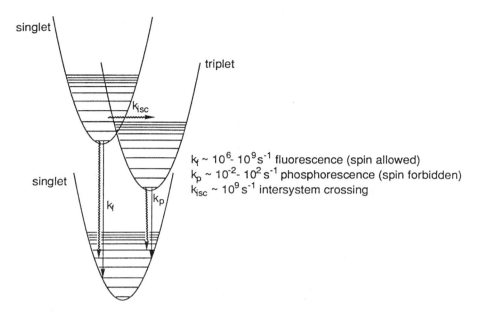

singlet

triplet

$k_f \sim 10^6 \text{-} 10^9 \text{s}^{-1}$ fluorescence (spin allowed)
$k_p \sim 10^{-2} \text{-} 10^2 \text{s}^{-1}$ phosphorescence (spin forbidden)
$k_{isc} \sim 10^9 \text{s}^{-1}$ intersystem crossing

singlet

Figure 2.8   Excited electronic states and their radiative ($\rightarrow$) and nonradiative ($\rightsquigarrow$) decay pathways. The terms *fluorescence* and *phosphorescence* normally apply to singlet $\rightarrow$ singlet and triplet $\rightarrow$ singlet radiative decay, respectively. Luminescence is used as a general term to cover emissive processes between any pair of states. Nonradiative pathways typically involve energy loss via vibrational deactivation with a rate of $\sim 10^9 \text{ s}^{-1}$.

pathways. For example, in the case of the excited singlet state, the overall decay rate (k_{tot}) is given by equation (2.5).

$$k_{tot} = k_f + \Sigma k_i \tag{2.5}$$

The lifetime or relaxation time (τ) of the decay is given by the reciprocal of the rate constant (equation [2.6]).

$$\tau = 1/k = 1/(k_f + \Sigma k_i) \tag{2.6}$$

The fraction of the excited molecules that decay by a particular pathway is defined as the quantum yield (ϕ) for that pathway. Continuing with the example outlined earlier, the quantum yield for fluorescence (ϕ_f) is given by equation (2.7).

$$\phi_f = k_f/(k_f + \Sigma k_i) = \tau/\tau_f \tag{2.7}$$

Summary of Sections 2.1–2.2

1. The Boltzmann equation [$N_u/N_l = \exp(-\Delta E/kT)$] describes the relative populations of energy levels that differ in energy ΔE at a temperature T.

2. Optical spectroscopic techniques depend on the absorption or emission of energy between electronic states. The intensity of bands and their energies provide information on the coordination environment or oxidation level of the metal center or ligand chromophore. In later sections we shall see that these optical transitions can be used as indirect probes to monitor ligand binding, redox changes, and a host of other chemical events. In emission spectroscopy, measurements of excited-state lifetimes provide information on the rates of processes that compete with natural decay pathways (e.g., electron transfer, ligand binding, distance determination through energy transfer).

3. Raman spectroscopy probes molecular vibrations. It is superior to infrared spectroscopy for the study of biological molecules, since there is no interference from background water.

2.3   Magnetic Resonance and Related Properties

2.3.1   Nuclear Magnetic Resonance (NMR)

Of all the physical techniques available to the inorganic biochemist, NMR is perhaps the most generally useful. It provides information on structure, dynamics, kinetics, binding processes, acid-base titrations, electronic structure, and magnetic properties, among others. Part of the utility of the method stems from its selective application to specific nuclei. The theory underlying the NMR experiment is closely related to that for electron paramagnetic resonance (EPR). However, the experimental setup for each is very different as a result of technical details relating to the handling of microwave (EPR) and radiofrequency (NMR) radiations. Many nuclei possess a net nuclear spin (I) that can interact with an external applied magnetic field. This can adopt various nondegenerate orientations relative to the applied magnetic field. The number of nondegenerate energy levels is determined by the magnitude of I. The simplest system ($I = 1/2$) is illustrated in Figure 2.9. Unlike EPR, where the promotion of molecules from the ground to an excited state is detected by the *absorption* of energy, the NMR experiment monitors the *relaxation* of molecules from an excited state to the ground state (Figure 2.10). This distinction has had important consequences for the development of each type of resonance experiment. For any particular NMR signal, four pieces of information are available: chemical shift (δ); intensity (I); relaxation times (T_1 and T_2); coupling constant (J). The parameters T_1 and T_2 are called the *longitudinal* and *transverse* relaxation times, respectively, and reflect the two distinct ways in which nuclear spin magnetization may decay. T_2 can be determined from the linewidth of a resonance, whereas T_1 is measured from the variation of signal intensity as a function of the time delay between application of a pulse and monitoring the x,y component (see Figure 2.10). Resonances are defined by the chemical shift (δ), which measures the frequency of a resonance (ν_{obs}) relative to a fixed standard (ν_{ref}) [$\delta = 10^6 (\nu_{obs} - \nu_{ref})/\nu_{spec}$],

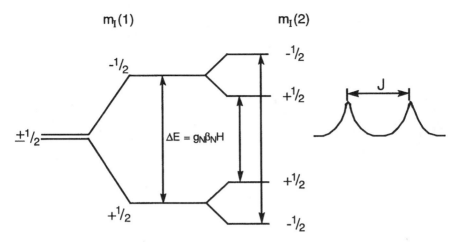

Figure 2.9 Energy level diagram for nuclear spin states $[m_I(1) = \pm\frac{1}{2}]$ showing the splitting of the previously degenerate energy levels and the influence of coupling to a single adjacent nucleus with spin $I = \frac{1}{2}$ $[m_I(2) = \pm\frac{1}{2}]$ The difference in energy of the two resonance peaks is the coupling constant J.

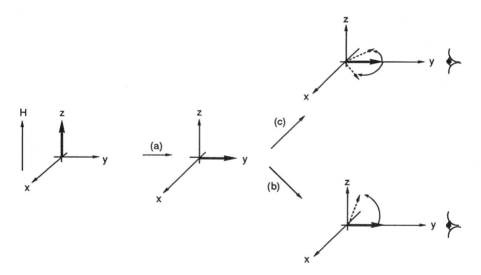

Figure 2.10 NMR as a relaxation phenomenon. In an applied field H aligned along the z axis, one can imagine a net polarization of magnetization along that axis (i.e. a majority of nuclear spins aligned along the z axis). (a) By application of a magnetic field arranged perpendicular to the applied field H (e.g., along the x axis, H_x) these spin vectors can be reoriented along the y axis. A detector monitors the magnetization along the y axis. This magnetization will tend to: (b) decay as the spin vectors realign along the z axis (longitudinal relaxation), or (c) lose coherence and spread out along the xy plane (transverse relaxation). The relaxation rates for these two distinct decay pathways are defined by relaxation times T_1 and T_2, respectively.

where ν_{spec} is the spectrometer frequency for the nucleus under consideration.[5] Discrete resonances are observed for nuclei in chemically distinct environments. Consider as an example the $^{31}P (I = 1/2)$ NMR spectra of ATP and its cobalt complex shown in Figure 2.11. The ATP spectrum shows three discrete resonances corresponding to the α-, β-, and γ-phosphorus sites. Each resonance is split into a multiplet that reflects coupling to neighboring nuclei.[6] The integrated intensities (areas under the bands) of these three sets of resonances are in the ratio 1:1:1 for α:β:γ phosphorus sites, as expected. After forming the Co(III) complex, the resonances shift to very different positions as a result of the change in spin density and the resulting change in local magnetic fields generated at each nucleus.

The energy difference $(g_N \beta_N H_z)$ depends on the properties of the nucleus. For each type of nucleus, the energy difference is sufficiently distinct, relative to perturbations from the environment, that each element can be studied independently. Appendix 5 lists some spectroscopic parameters for important NMR active nuclei. Nuclei with spins $I > 1/2$ are termed *quadrupolar nuclei*, since they possess a quadrupole moment. This provides additional relaxation mechanisms, and so NMR resonances from nuclei with $I > 1/2$ are often broad. The NMR frequency of each nucleus refers to the relative energy splitting of the nuclear spin levels in a defined magnetic field. It is the distinctive NMR frequency that allows each type of nucleus to be monitored independently. Only certain isotopes of each element may possess a nuclear spin (I). The term *natural abundance*

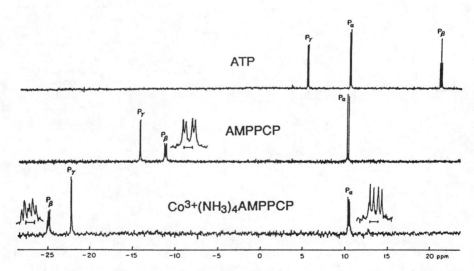

Figure 2.11   NMR spectra and parameters for ATP and AMPCP. Binding of Co³⁺ leads to significant shifts in δ. The molecules ATP [Ad-OP$_\alpha$(O)$_2$OP$_\beta$(O)$_2$OP$_\gamma$O$_3$] and AMPCP [Ad-OP(O)$_2$OP(O)$_2$CH$_2$PO$_3$] are structurally related. In the latter, the oxygen connecting P$_\beta$ and P$_\gamma$ is replaced by CH$_2$. [Adapted from J. Granot et al., *Biochemistry, 19,* 3537–3543 (1980).]

refers to the percentage of a particular NMR active isotope of an element in nature. Since many isotopes may be purchased in an enriched form, the problem of low natural abundance may be avoided. The ability to study selectively the NMR properties of a multitude of nuclei has given rise to the term *multinuclear NMR.*

In this text we can neither delve into the detailed background of the method nor dwell on its many applications. The interested reader may find excellent accounts of theory and applications elsewhere. Of the many manifestations of the NMR experiment, we shall discuss only three examples that are of particular relevance to inorganic biochemistry. This is not meant to underestimate the relevance of other applications, and many key references are given at the end of this chapter for recommended further reading.

Binding and Exchange Phenomena

Many important applications of NMR are concerned with binding events (e.g., ligand binding to a metal center or proton uptake or release from an ionizable residue). Nuclear magnetic resonance can be used to study these chemical processes since the NMR response (both chemical shift and the relaxation times T_1 or T_2) is different in the presence or absence of the bound species. In the limit of fast exchange, an NMR resonance is the average of contributions from the bound and free forms. For example,

$$\delta_{obs} = p_{free} \cdot \delta_{free} + p_{bound} \cdot \delta_{bound}$$

where p is the fractional population of either the bound or free form. Similarly,

$$(1/T_1)_{obs} = p_{free} \cdot (1/T_1)_{free} + p_{bound} \cdot (1/T_1)_{bound}$$

For binding or exchange processes (e.g., ligand exchange, conformational changes, and so on) that are slow on the NMR time scale (millisecond range), two sets of resonances may be observed that correspond to the bound and free forms, where the relative intensities of each resonance reflect the fraction of each form. In the intermediate exchange regime between the slow and fast limits, the two sets of signals begin to coalesce into a single set of resonances (see the following diagram). The effect of chemical exchange between two chemically distinct environments (A and B) is illustrated in Figure 2.12A. In the example shown, site A is preferentially populated (75% relative to 25% for B), since the binding affinity is greater for A. At fast exchange, the spectrum tends toward the chemical shift for A. The signals from A and B have linewidths of 10 Hz and are separated by 100 Hz. By carrying out systematic titration or variable-temperature experiments, much useful information can be obtained on the thermodynamic and kinetic parameters associated with binding phenomena and conformational or structural changes. Figure 2.12 also illustrates some related applications with the relevant equations.

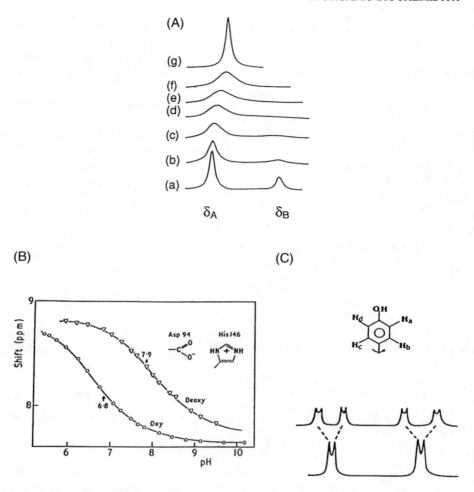

Figure 2.12 Analyzing ligand binding or exchange, or conformational changes induced by metal binding. (A) The appearance of the NMR spectrum depends on the rate of exchange between the two chemically distinct environments, illustrated here for (a) 0 s^{-1}, (b) 10^2 s^{-1}, (c) $2 \times 10^2 \text{ s}^{-1}$, (d) $3 \times 10^2 \text{ s}^{-1}$, (e) $5 \times 10^2 \text{ s}^{-1}$, (f) 10^3 s^{-1}, (g) $\geq 10^4 \text{ s}^{-1}$, from slow exchange (a, where each site can be distinguished) to fast exchange (g, where the chemical shift reflects the weighted average of the two limits). The rate can be varied by changing the temperature of the solution. In this way, kinetic parameters can be evaluated. [Adapted from *Biological Spectroscopy* by I. D. Campbell and R. Dwek, Benjamin-Cummings, London, 1984.] (B) Titration of His 146 in hemoglobin in the deoxy and oxy forms, monitored by ^1H NMR experiments. The acidity constants pK_a were obtained by fitting to a standard equation [$\delta(\text{pH}) = \delta_{HA}[H^+]/(K_a + [H^+]) + \delta_A K_a/(K_a + [H^+])$], where δ_{HA} is the limiting shift of the protonated His, δ_A is the limiting shift of the deprotonated His, and K_a is the acidity constant. [Adapted from I. D. Campbell, in *NMR in Biology* (eds. R. A. Dwek et al.,), Academic Press, London, 1977, p. 33.] (C) In an asymmetric protein environment, each of the four protons on a tyrosine ring may be distinct. Rotation of the ring will reduce the observed NMR spectrum to two resonances since the chemical shifts of the two interchanging pairs are averaged. At low temperature, however, the rotational motion can be frozen out. By studying the line shape, the rotation rate can be determined. [Refer to R. A. Dwek, ed., *NMR in Biology*, Academic Press, London, 1977, for more details on these types of experiments.]

2D Experiments

The increasing use of two-dimensional NMR demands some comment. As with most aspects of NMR spectroscopy, the basic ideas can be summarized in a few lines. A thorough review would require several chapters! Figure 2.9 illustrates the interaction (or coupling) of two nuclear moments. This suggests that the spins of nuclei that are strongly coupled through a small number of chemical bonds (contact coupling) can communicate with each other (i.e., the spins are *correlated*). By use of appropriate pulse sequences, this correlation between nuclear spins can be demonstrated in a two-dimensional NMR experiment. Correlated spectroscopy is abbreviated to *COSY* when nuclei interact through chemical bonds.[7,8] A related experiment based on two-dimensional nuclear Overhauser effect spectroscopy is called *NOESY* and provides information on nuclei that are coupled by through-space (dipolar) interactions. Both experiments are illustrated in Figure 2.13, and these lay the foundation for determination of solution structures of small molecules and biological macromolecules.

NMR of Paramagnetic Molecules

Paramagnetic chromophores influence NMR spectra in two ways: (1) by providing additional mechanisms for extending the range of chemical shifts and (2) by enhancing nuclear relaxation, which is often accompanied by a significant increase in line width. The former results in a shift of important resonances, from active site residues or ligands neighboring the paramagnetic center, away from spectrally congested regions of the spectrum. Paramagnetically shifted resonances also provide valuable data on the magnetic properties of metal centers (Figure 2.14).

Paramagnetic centers provide efficient relaxation mechanisms that influence T_1 and T_2. The paramagnetic contribution to the longitudinal relaxation time T_1 $[(1/T_1)_{\text{paramagnetic}} - (1/T_1)_{\text{diamagnetic}}]$ is particularly important, since the change in relaxation is inversely proportional to the sixth power of the distance between the nucleus and paramagnetic centers [equation (2.12)] and can, therefore, provide distance information.

$$[(1/T_1)_{\text{paramagnetic}} - (1/T_1)_{\text{diamagnetic}}] \; \alpha \; 1/r^6 \qquad (2.12)$$

Figure 2.15 illustrates the application of this method to establish the geometry of a manganese nucleotide complex in an enzyme pocket.

2.3.2   Electron Paramagnetic Resonance (EPR)

Electron paramagnetic resonance spectroscopy (EPR) is a particularly powerful method for studying molecular species that possess unpaired electrons. The basis for the technique is illustrated in Figure 2.16 for a simple organic radical containing one unpaired electron with spin $S = 1/2$. In the presence of an external magnetic field (MF) H applied along the z axis, the two possible ori-

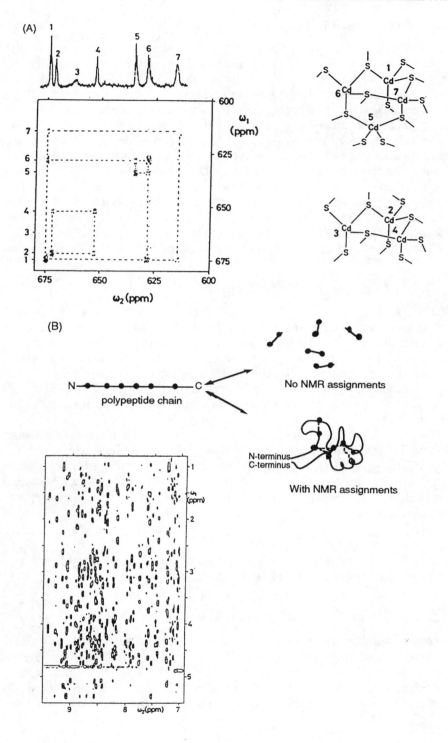

Figure 2.13 (A) Schematic illustration of a ^{113}Cd-COSY spectrum of ^{113}Cd$_7$-metallothionein, which contains a Cd$_4$ and a Cd$_3$ cluster. The off-diagonal peaks correlate peaks in the 1D spectrum that are related by through-bond coupling. Note that certain cross-peaks do not show in the 2D spectrum. Additional connectivities were established by analysis of coupling constants in regular 1D spectra. [Adapted from M. H. Frey et al., *J. Am. Chem. Soc., 107*, 6847–6851 (1985).] (B) In a ^{1}H-NOESY experiment the off-diagonal cross-peaks demonstrate that these nuclei are coupled by through-space interactions, thereby providing structural information. It is necessary to first assign all the resonances in the NMR spectrum to specific protons from residues in the protein, otherwise the data is meaningless. A portion of the 2D-NOESY map for rabbit metallothionein is shown. Each cross-peak indicates a close proximity between the protons represented by the coupled resonances. [Adapted from A. Arseniev et al., *J. Mol. Biol., 201*, 637 (1988).]

entations of the spin vector ($m_s = \pm 1/2$) are no longer equivalent, relative to the field **H**. Both the energy (E) of each level and the difference in energy (ΔE) can be calculated from equation (2.8);[9] thus, we have a basis for a form of spectroscopy where an electron is excited from the lower to the upper energy state.

$$\Delta E = E_2 - E_1 = g_e\beta_e[m_s(\alpha) - m_s(\beta)]H = g_e\beta_e H \tag{2.8}$$

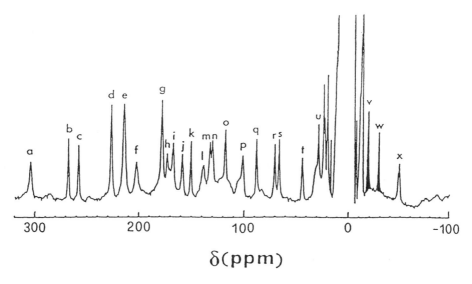

$\delta(ppm)$

Figure 2.14 NMR spectrum of a paramagnetic metalloprotein; the 90-MHz ^{1}H NMR spectrum of Co$_7$-metallothionein. Note how the magnetic fields from the paramagnetic cobalt ions shift the ^{1}H resonances outside of the "diamagnetic" region (0–12 ppm). The shaded resonances correspond to exchangeable protons and disappear in D$_2$O. [From I. Bertini et al., *J. Am. Chem. Soc., 111*, 7296 (1989).]

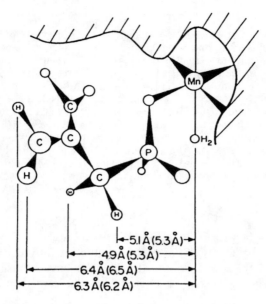

Figure 2.15 Determination of internuclear distances in an enzyme-bound Mn–organophosphate complex by measurement of relaxation times. [Adapted from T. L. James and M. Cohn, *J. Biol. Chem.*, *249*, 3519 (1975).]

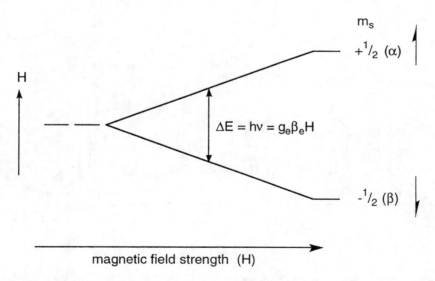

Figure 2.16 Zeeman splitting of two degenerate electron spin levels for an $S = \frac{1}{2}$ ($m_s = \pm \frac{1}{2}$) system in an applied magnetic field $\boldsymbol{H}$. The energy difference $\Delta E = g_e \beta_e \boldsymbol{H}$ varies with the strength of the applied field. The two possible orientations of the magnetic spin vector are shown beside the energy states to which they belong.

The energy difference ΔE depends on the magnitude of the applied field H, and so there are two ways in which the experiment might be performed: (1) by keeping H constant and varying the frequency v of an applied field until it corresponds to $\Delta E(= hv)$, or (2) by keeping the frequency of the oscillatory field constant and varying the magnitude of the external field H. When H is such that equation (2.9) is satisfied, energy is absorbed and a transition occurs.

$$H = \Delta E/g_e\beta_e = hv/g_e\beta_e \tag{2.9}$$

In practice it is the latter method that is employed. Figures 2.16 and 2.17 summarize the basic principles of the EPR experiment. Although the experiment monitors the absorption of energy, note that data are displayed in the form of a derivative (dA/dH) spectrum that more clearly defines peak positions and linewidths.

Application to transition-metal ions is considerably more complex (at least at the level of theory) than simple organic radicals. The reasons for this are twofold: (1) The electronic structures of transition metals are more elaborate, and (2) the unpaired electron density is generally not localized in symmetric s-orbitals. The latter point results in anisotropy of the measurements, since the x, y, and z axes may be inequivalent, resulting in distinct spectral responses (see Table 2.2, for example), while coupling between the spin (S) and orbital (L) angular momenta gives rise to additional mechanisms for removing the degeneracy of orbitals. In essence, the orbital angular momentum generates a magnetic field that acts upon the electron spin (Figure 2.18).[10] Fortunately, the electronic structures of transition-metal ions are very well understood. It is therefore possible to rationalize the complexity of EPR spectra and to derive many important physical parameters that relate to the electronic structure of the metal

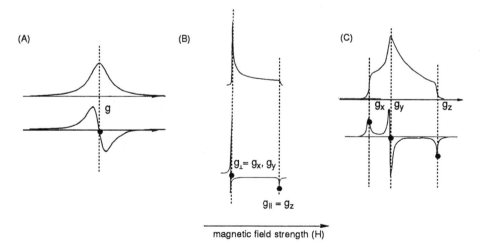

Figure 2.17 Characteristic absorption (upper) and derivative (lower) spectra for (a) isotropic, (b) axial, and (c) rhombic spin systems (see text).

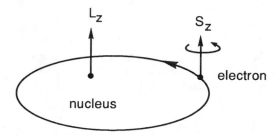

Figure 2.18 The motion of the electron around the nucleus generates an orbital angular momentum (L) that adds an additional component to the magnetic field experienced by the electron spin (S).

ion. In keeping with the format adopted in the rest of this chapter, we will aim to reach an understanding of the principles of the method and how it may be usefully employed, without necessarily focusing on every little detail.

Any EPR spectrum contains four sources of information: intensity, line-width, g-value, and multiplet structure (defined by a coupling constant A). These give information on concentration, dynamic processes and spin-spin interactions, energies of spin states, and interactions with neighboring nuclei, respectively. Both the g-value and coupling constants reflect the coordination environment of the metal center. For studies in the solid state or frozen solution, g and A are normally anisotropic (i.e., the paramagnetic site does not possess spherical symmetry, and so its interaction with the applied magnetic field will depend on the orientation relative to that field). The terms noted in Table 2.2 are commonly employed. Two factors that arise in discussions of transition-metal ions, which have lesser importance for organic radicals, are rapid relaxation rates and zero-field splitting (ZFS). The former results from the effects of spin-orbit coupling that gives rise to additional relaxation mechanisms (Figure 2.18). In effect, the orbital angular momentum generates a fluctuating magnetic field that can turn the spin back to its ground-state orientation. Enhanced relaxation results in line broadening, and so low temperatures that minimize other relaxation pathways are often required to observe signals from transition metals. Zero-field splitting results from interactions between electron spins in metal ions

Table 2.2 Terms Commonly Used to Define EPR Spectral Characteristics

Environment	Site Symmetry	g-Value	Hyperfine Coupling Constant
Isotropic	$x = y = z$	$g_x = g_y = g_z$	$A_{xx} = A_{yy} = A_{zz}$
Axial	$x = y \neq z$	$g_x = g_y \neq g_x$	$A_{xx} = A_{yy} \neq A_{zz}$
Rhombic	$x \neq y \neq z$	$g_x \neq g_y \neq g_z$	$A_{xx} \neq A_{yy} \neq A_{zz}$

possessing two or more d-electrons. These interactions lead to a splitting of the energy levels, even in the absence of an applied magnetic field (zero field). Imagine each electron as a source of a magnetic field, with the splitting resulting from the effect of internal rather than external applied fields.

Much like the chemical shift (δ) in NMR, the g-value defines the field position of an EPR signal. It depends on the electronic environment of the paramagnetic center and has characteristic values that may be interpreted by comparison with known standards. This is the approach that most inorganic biochemists use in the analysis of EPR spectra of metalloproteins. Although it is possible to interpret spectra in quantitative detail, much useful information can be derived by simple inspection and comparison of data. The g-value is defined by equation (2.10)

$$g = h\nu/\beta_e H_o \tag{2.10}$$

where ν is the frequency of the electromagnetic radiation applied perpendicular to the external field H_o (typically $\nu \sim 9$ GHz for X-band spectrometers and $H_o \sim 0.3$–1.2 T), β_e is the electronic Bohr magneton (0.92×10^{-23} JT), and h is Planck's constant (6.636×10^{-34} Js). Molecules are randomly oriented in solution, and so the g-values are spread over a range of values. The absorption spectrum appears as an envelope encompassing the range of molecular absorptions. The shape of this envelope reflects the symmetry of the paramagnetic center. Figure 2.17 illustrates the situation for isotropic, axial, and rhombic symmetry.

For transition metals, the g-values may vary widely. This arises from the effects of spin-orbit coupling. The electron spins about its axis and orbits the nucleus, giving rise to spin and orbital angular momenta, respectively (Figure 2.18). These angular moments produce magnetic dipoles that interact and generate local anisotropic magnetic fields at the electron. The electron, therefore, experiences a local magnetic field that depends on the electronic structure of the metal center, and so anisotropic g-values that differ from the free-spin value (g_e) may result. Table 2.3 notes the metals most commonly studied by EPR and lists the available spin states for a variety of oxidation levels.

Transition-metal nuclei may have nuclear spins (e.g., Cu^{2+} $I = 3/2$) that couple to unpaired electron spins and produce multiplet splitting in EPR spectra. Such electron–nuclear interactions are termed *hyperfine splittings*. Divalent copper ($I = 3/2$) provides the classic example (Figure 2.19). The additional splitting from interaction with the nuclear moment I is defined by equation (2.11).[11]

$$H = g_e\beta_e S_z H_z - g_N\beta_N I_z H_z + AS_z I_z \tag{2.11}$$

A nuclear moment I can adopt $2I + 1$ orientations relative to a defined axis. These are represented by distinct M_I values. The interaction of I and S is dependent on their relative orientations, and so a variety of energy levels are generated. If ligands bound to the paramagnetic metal also have nuclear spin (e.g., ^{14}N has $I = 1$), these ligand nuclei may also couple to the electron. These

Table 2.3   Spin States for Transition Metal Ions Commonly Studied by EPR

Metal Ion (d^n)	Geometry	Electronic Configuration	Spin (S)
Mo^{5+} ($4d^1$)	O_h	$t_{2g}^1 e_g$	1/2
	T_d	$e^1 t_2$	1/2
Fe^{3+} ($3d^5$)	O_h	$t_{2g}^3 e_g^2$ HS	5/2
		$t_{2g}^5 e_g$ LS	1/2
		$t_{2g}^4 e_g^1$ intermediate	3/2
	T_d	$e^2 t_2^3$ HS	5/2
		$e^4 t_2^1$ LS	1/2
		$e^3 t_2^2$ intermediate	3/2
Mn^{2+} ($3d^5$)	O_h	$t_{2g}^3 e_g^2$ HS	5/2
	T_d	$e^1 t_2$ HS	5/2
Fe^{2+} ($3d^6$)	O_h	$t_{2g}^4 e_g^2$ HS	2
		$t_{2g}^6 e_g$ LS	0
	T_d	$e^3 t_2^3$ HS	2
		$e^4 t_2^2$ LS	1
Co^{2+} ($3d^7$)	O_h	$t_{2g}^5 e_g^2$ HS	3/2
		$t_{2g}^6 e_g^1$ LS	1/2
	T_d	$e^4 t_2^3$	3/2
Cu^{2+} ($3d^9$)	O_h	$t_{2g}^6 e_g^2$	1/2
	T_d	$e^4 t_2^5$	1/2

super-hyperfine splittings are much smaller and may not be resolved (although the linewidth of a signal will increase). The number of lines produced by coupling to a specific nucleus is given by $2nI + 1$, where n is the number of equivalent nuclei of a given isotope.

We have seen that the effect of a magnetic field is to remove the degeneracy of electron spin levels. These energy levels may also lose their degeneracy in the absence of an applied magnetic field as a result of local magnetic fields generated in multielectron systems. This zero-field splitting can result in additional signals in the EPR spectrum. Figure 2.20 illustrates the situation for Fe^{3+} (d^5, high spin) in the event of (1) no ZFS, (2) a small ZFS, and (3) a large ZFS. An important theorem (Kramer's theorem) states that ions with a nonintegral spin (odd number of unpaired electrons) must have doubly degenerate energy levels in the absence of an applied magnetic field. This clearly holds for the example in Figure 2.20. It may be impossible to observe any transitions either when the number of unpaired electrons is even and the Kramer's degeneracy is lost or when the zero-field splitting is large. The classic high-spin ferric heme spectrum from myoglobin is considered in detail in Figure 2.21. Figure 2.22 illustrates some common signature spectra from other important prosthetic centers. With experience, one can often identify a prosthetic center by its characteristic EPR signature. Although it is desirable to understand the principles underlying the appearance of the spectrum, this is not always necessary. As with most physical

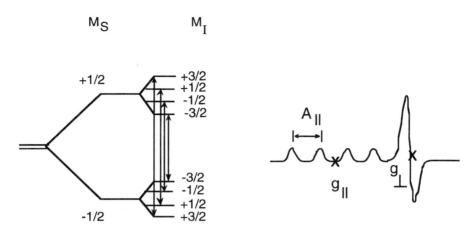

Figure 2.19 Hyperfine coupling in Cu^{2+}. The axial spectrum shows discrete parallel ($\parallel$) and perpendicular ($\perp$) features. Typically, the hyperfine coupling constant $A_\perp$ is poorly resolved; however, $A_\parallel$ can be correlated with the coordination symmetry of the copper center [$A_\parallel$(tet or sp) > $A_\parallel(T_d)$]. The four transitions indicated in the energy-level diagram on the left give rise to the four components of the $g_\parallel$ signal in the spectrum on the right. An energy-level diagram could also be drawn for the $g_\perp$ transitions. However, the difference in energy between the M_I levels is very much smaller and the coupling is not resolved.

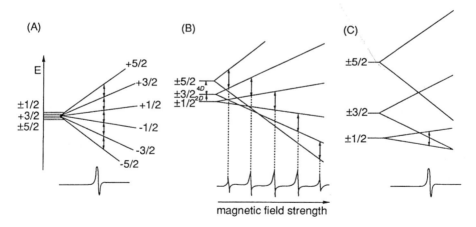

Figure 2.20 Influence of zero-field splitting on the appearance of an EPR spectrum for an $S = \frac{5}{2}$ spin system (e.g., Fe^{3+} d^5). (A) no ZFS, complete degeneracy; (B) small ZFS between the degenerate Kramer's doublets in a sextet ($S = \frac{5}{2}$) state; (C) large ZFS; only the $-\frac{1}{2} \rightarrow +\frac{1}{2}$ transition can be observed. High-spin ferric heme spectra tend to fall in the latter category (Figure 2.21).

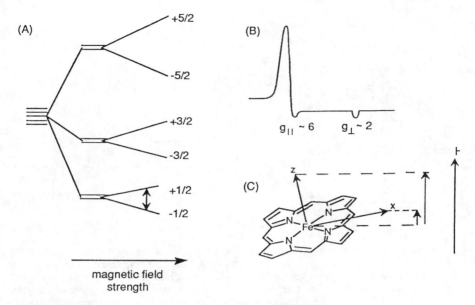

Figure 2.21 (A) Energy-level splitting for high-spin d^5 ions Fe^{3+} and Mn^{2+}. (B) Only the $-\frac{1}{2} \rightarrow +\frac{1}{2}$ transition can be observed and is often referred to as the "fictitious" $S = \frac{1}{2}$ spectrum (see Figure 2.20). A common example is the spectrum of a heme with axial symmetry ($g_{\perp} \equiv g_x = g_y, g_{\parallel} \equiv g_z$). (C) Two components of the signal are observed since either the z component or x,y components of the spin vector, respectively, are projected onto the applied magnetic field $\mathbf{H}$. The reader can refer to Figure 2.17b to see how the absorption envelope of a signal can give rise to two components.

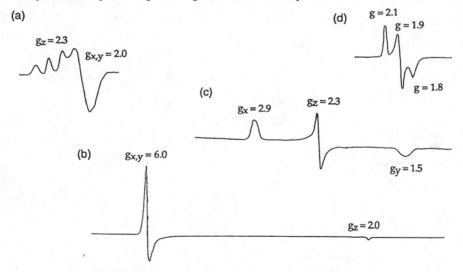

Figure 2.22 Representative EPR spectra for a variety of common paramagnetic centers in metalloproteins. Typical g-values are indicated. (a) Cu^{2+} in azurin, (b) high-spin heme in myoglobin, (c) low-spin heme in cytochrome c, (d) $[Fe_4S_4]^+$ in ferredoxin.

methods, the amount of time one spends mastering the details of the technique very much depends on the information level required.

2.3.3 Coupling of Magnetic Moments

There are many examples of bi-, tri-, or multinuclear metal centers as prosthetic sites in metalloproteins or enzymes. Several of these will be discussed in Chapter 4. If each of the two metal centers possesses a net spin (S_1 and S_2), then, it is possible for these two spins to couple together to form a resultant spin (S), which may be the sum ($S = S_1 + S_2$) or the difference ($S = S_1 - S_2$) of each spin. The extent of coupling is defined by an exchange coupling constant (J) defined by the following equation:

$$H = H_o - 2JS_1 \cdot S_2 + \text{other terms}$$

where H is a Hamiltonian operator. J may be measured by magnetic suscepti-bility experiments[12] or by monitoring the temperature dependence of either EPR spectral intensities or the chemical shifts of paramagnetically shifted resonances. If J is positive, $S_1 + S_2$ is lower in energy than $S_1 - S_2$ and the term *ferromag-netic coupling* is used. If J is negative, $S_1 + S_2$ is higher in energy than $S_1 - S_2$ and the term *antiferromagnetic coupling* is used. In essence, the extent of fer-romagnetic or antiferromagnetic coupling between spins S_1 and S_2 is described by the Hamiltonian term $H = -2JS_1 \cdot S_2$. When $S_1 = S_2$, the energy difference between the spin-paired and spin-free states is $2J$. This is summarized in the following scheme.

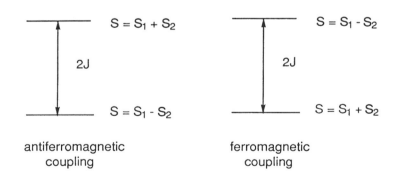

This kind of interaction is important for understanding the magnetic properties of a number of binuclear iron and copper centers that we shall come across later in the text.

Summary of Section 2.3

1. In a magnetic field, both electron and nuclear spins adopt preferred orientations. The immediate electronic environment modulates the relative energies of these spin levels, and so the frequency of NMR or EPR transitions provides a probe of chemically important parameters (oxidation level, neighboring atoms, coordination geometry).

2. The absolute energy of the transition (g-value or chemical shift δ) reflects the immediate environment of the spin center. Coupling constants provide information on ligand atoms that are chemically bonded to the spin center. Relaxation times reflect dynamic changes in the environment at the spin center.

3. As in the case of optical experiments, these spectral features can be used as direct, or, more commonly, as indirect probes of structure, dynamics, kinetics, binding processes, acid/base titrations, and electronic structure.

Review Questions

• In what ways might NMR studies of RNA or DNA molecules differ from similar studies on proteins? Think in particular of the nuclei used and the application of experiments such as COSY or NOESY to evaluate structural details.
[Wuthrich, K. *NMR of Proteins and Nucleic Acids,* Wiley-Interscience, 1986]

• An EPR signature has been observed from the Mo center in xanthine oxidase. (a) What is the oxidation state of the molybdenum center? (b) Sketch the appearance of the spectrum if $S = 1/2$, and if $S = 5/2$. (c) You expect either a Mo–C or Mo–H intermediate to be formed during catalysis. Suggest experiments to obtain support for (or refute) such a proposal.

2.4   Solution Methods Dependent on X-Radiation and γ-Radiation

2.4.1   EXAFS (Extended X-Ray Absorption Fine Structure)

When an atom absorbs a photon of X-radiation ($h\nu'$), sufficient energy may be available to remove an inner core electron (ionization energy, IE). The remaining energy appears as the kinetic energy (KE) of the ejected electron [equation (2.13) and Figure 2.23].

$$KE = h\nu = hc/\lambda = h\nu' - IE \tag{2.13}$$

To understand the principles of the method, it is best to consider this ejected electron as a wave with an associated wavelength λ. Figure 2.24 shows a molybdenum center and sulfur ligand atoms. X-rays of appropriate frequency (ν') have been used to ionize the Mo center. The ejected electron, which is shown in the form of a wave, may interact with the electron clouds of the ligand atoms.

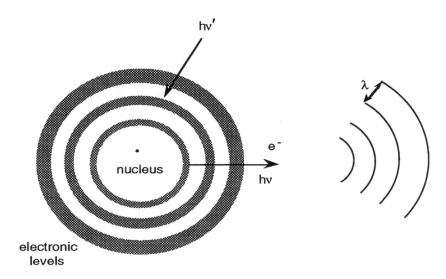

Figure 2.23 Absorption of X-rays ($E = h\nu'$) by an atom results in the loss of a core electron (shown as a wave packet) with a resulting kinetic energy $h\nu$ defined by equation (2.13).

The electron wave is scattered from these atoms, and the scattered waves subsequently interfere with the original wave packet to give regions of constructive and destructive interference. A detector at a fixed position (P) may be used to monitor the intensity of the final waveform. If X-rays of distinct frequency (ν') are employed, the wavelength of the emitted electron wave packet (λ_2) will change. The interference patterns with the scattered waves will now give a different intensity (I_2) at P. By monitoring the intensity as a function of λ [$h\nu'$ and IE are known; see equation (2.13)], a graph similar to that shown in Figure 2.25 may be plotted. The interference effects depend on the distance between the metal center and the scattering ligand, and so Fourier transformation (FT) yields a plot of intensity versus distance. Fourier transformation is a mathematical procedure that is commonly used to convert data from one physical dimension to another related dimension (e.g., time and frequency, or wavelength and distance). The EXAFS method offers no insight on the relative geometry of the ligand atoms. However, the scattered waves depend on the electronic structure of these atoms, and so their identity and number may be deduced from the intensity (Figure 2.24). From EXAFS data three important facts may be determined:

1. Identity of ligand atoms. Inasmuch as the scattering depends on the electron density of the surrounding ligands, it is difficult to distinguish atoms or ions with similar electronic configurations (e.g., N and O, or S and Cl).
2. Distance of ligand atoms from the metal center.
3. Number of scattering ligand atoms of a given type.

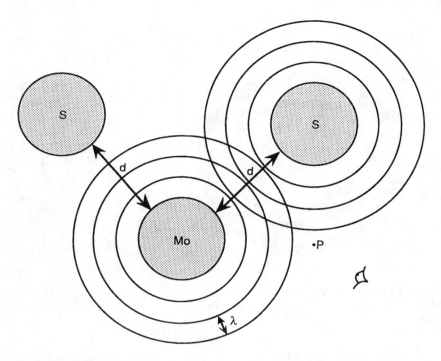

Figure 2.24   The wave from the ejected electron is scattered by neighboring atoms. The intensity and relative phase of the scattered wave depends on the electron density and distance to the scattering atom. The intensity of the resulting wave (following constructive or destructive interference) at any observation point P will vary according to the chemical nature and position of the scattering ligand atom.

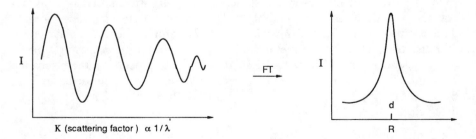

Figure 2.25   EXAFS patterns from a $(MoS_4)^{2-}$ species with four identical Mo–S bonds. Fourier transformation (FT) takes the data from the wavelength-dependent interference pattern to provide both distance information from the peak position and an estimate of the number of scattering atoms from the peak intensity.

Note that, although distances may be determined fairly accurately (± 0.1 Å), the number of scattering ligands is subject to larger errors ($\pm 30\%$) because of the dramatic effect of phase incoherence on the signal amplitudes.

2.4.2 Mossbauer Spectroscopy

Mossbauer spectroscopy probes high-energy transitions within the nucleus. The only appropriate energy source for these experiments is the γ-radiation emitted from an excited nucleus of the same isotope during a natural radioactive decay pathway. Figure 2.26 illustrates the formation of ^{57}Fe from ^{57}Co. Inasmuch as available γ-sources are restricted to select nuclei, this limits the applicability of the method. Almost without exception, studies on biological molecules focus on ^{57}Fe, since many other viable nuclei are not biologically relevant (e.g., ^{119}Sr, ^{127}I, ^{129}Xe).

The idea underlying the Mossbauer experiment is straightforward. The 14.41-keV transition shown in Figure 2.27 is a convenient energy to use, and so the energy (E_t) emitted by the γ-source is well defined. The energy difference between the nuclear energy shells is influenced by the charge on the metal ion, the inner sphere ligands and their geometry, and the effects of nuclear quadrupole moments that may result in a loss of degeneracy of the energy levels for these shells (Figure 2.27). Since the energy of the source radiation is fixed and may not match the energy difference between two nuclear states, it must be varied by use of the Doppler effect (Figure 2.28). By moving the source either toward or away from the target, the energy (E) of the radiation from the source may be increased or decreased, respectively, by an amount equivalent to the kinetic energy of motion. When the energy of the radiation corresponds to the energy

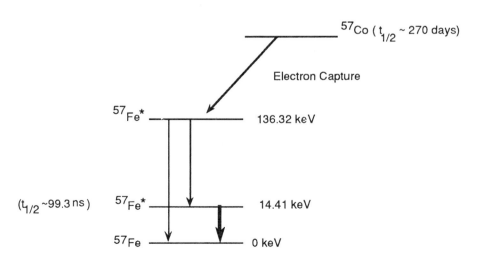

Figure 2.26   Source of the 14.41 keV γ-radiation used in Mossbauer experiments.

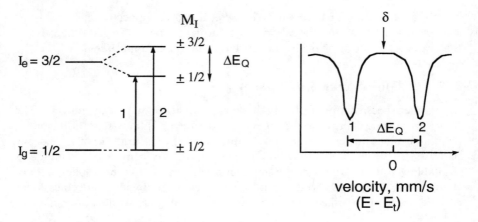

Figure 2.27 Absorption of γ-radiation promotes nuclear transitions. I_g and I_e are the nuclear spin quantum numbers in the ground and excited state, respectively. The $\pm\frac{3}{2}$ and $\pm\frac{1}{2}$ levels of the excited state lose their degeneracy if an electric field gradient exists at the nucleus that can interact with the quadrupole moment. However, each level remains doubly degenerate in the absence of an applied magnetic field. In the case shown here, two transitions are observed, separated by an energy dependent on the quadrupole coupling constant ΔE_Q. The average position between these two transitions is the isomer shift (δ) given by $\delta = E - E_t$.

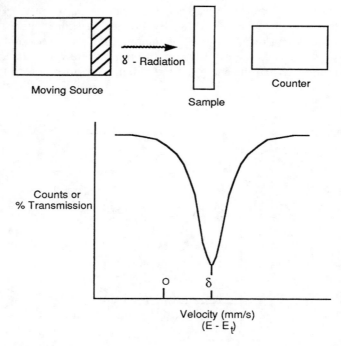

Figure 2.28 Schematic illustration of an experimental apparatus for obtaining Mossbauer spectra. The isomer shift δ is determined relative to the energy of the source radiation (E_t).

of a nuclear transition, the γ-ray is absorbed, and the number of counts monitored on the detector decreases. The energy of the transition is measured as a shift relative to the energy of the source (E_s) and is termed the isomer shift (δ). The splitting produced by the quadrupole moment may also be measured and is termed the quadrupole coupling constant (ΔE_Q). These parameters are illustrated in Figure 2.27. Both have units of mm/s.

The Mossbauer signal is influenced by the charge on the nucleus, the identity of the surrounding ligands, and the symmetry of the ligand field. The charge density at the nucleus is a function of the oxidation state of the metal and the identity of the ligand atoms. These features influence the isomer shift (δ). The quadrupole coupling constant (ΔE_Q) is dependent on electric field gradients at the nucleus ($\Delta E_Q = eQq/h$), where e is the electronic charge, Q is the quadrupole moment, q is the electric field gradient generated by asymmetry in the ligand environment (either geometric and/or charge distribution), and h is Planck's constant. These principles are best illustrated through the data listed in Table 2.4.

Further information can be obtained by variation of external conditions. Since ΔE_Q is determined by the symmetry of the ligand environment at a metal site, the variation of ΔE_Q with temperature reflects conformational flexibility in metal–ligand binding. Mossbauer can also be used to study magnetic properties of the nucleus. In the presence of an applied magnetic field, the degeneracy of nuclear energy levels is removed and additional signals appear in the spectrum.

Table 2.4   Representative Mossbauer Parameters for High-Spin Ferrous and Ferric Centers in a Variety of Ligand Environments

Complex (Coordination)		δ_{Fe} (mm/s)	$|\Delta E_Q|$ (mm/s)
Fe^{3+}-O	(6)	0.35–0.60	<0.9
	(4)	0.30–0.4	<0.9
Fe^{3+}-S	(6)	0.2–0.45	<0.8
	(4)	0.15–0.25	<0.5
Fe^{2+}-O	(6)	0.9–1.3	<4
	(4)	0.8–0.95	<2.5
Fe^{2+}-S	(6)	0.8–1.0	—
	(4)	0.5–0.7	<3.2

A few common trends should be noted: (1) $\delta(Fe^{2+}) > \delta(Fe^{3+})$ since there is more electron density at the divalent ion. (2) $|\Delta E_Q|(Fe^{2+}) > |\Delta E_Q|(Fe^{3+})$ since high-spin Fe^{3+} is a half-shell d^5 ion with spherical symmetry, whereas d^6

has an additional electron (see diagram). Note that nitrogen ligation has not been included in this table since this often leads to low-spin configurations. In general, however, $|\Delta E_Q|$(low-spin Fe^{3+}) > $|\Delta E_Q|$(high-spin Fe^{3+}) as a result of the loss of spherical symmetry (again, refer to the diagram).

Summary of Section 2.4

1. EXAFS depends on the ejection of core electrons from a metal ion by an incident γ-ray and the scattering of the resulting electron wave by the surrounding ligands.
2. The information available from EXAFS data includes (1) identification of ligand atoms, (2) determination of the number of ligand atoms of a given type, and (3) determination of metal–ligand bond distances.
3. Mossbauer spectroscopy probes nuclear energy states. The energy of these states depends on the oxidation level of the metal center and the identity of surrounding ligands.

2.5 Electrochemical Methods

Electron transfer between redox prosthetic centers underlies much of the chemistry of respiratory pathways in living cells and plays an important role in enzymatic oxidation or reduction of organic functional groups. To understand the probability of such reactions proceeding in one direction or another requires some knowledge of the relative affinities of redox centers for electrons (i.e., whether the center prefers to be oxidized or reduced). The following half-reaction concerns a metal center M; however, similar equations can be written for organic redox cofactors.

$$M^{n+} + e^- \rightleftharpoons M^{(n-1)+}$$

The potential $E°$ refers to the solution potential, where $[M^{n+}] = [M^{(n-1)+}]$.[13] Typically, potentials are determined relative to the normal hydrogen electrode (NHE).[14] This potential may be generated either by an electrode in solution or by addition of chemical oxidants or reductants. Both will be discussed later. It is convenient to distinguish electrochemical methods where electron exchange occurs directly at the electrode from those techniques where exchange occurs in bulk solution via an electron mediator. These will be termed *direct methods* and *indirect methods,* respectively. At this time the latter techniques are more commonly applied in protein studies.

2.5.1 Direct Methods

Although many electrochemical methods may yield information on the kinetics of electron exchange between an electrode surface and the molecule undergoing oxidation or reduction, it is the half-reaction potential ($E°$) that is of most interest to the inorganic biochemist. In part this reflects the experimental difficulties inherent in protein electrochemistry and the lack of information on the orientation and binding of the protein to the electrode surface. As a general rule, the larger the protein, the more difficult it is to study. Cyclic voltammetry and differential pulse polarography are two commonly used direct methods. Both require efficient electron exchange at an electrode surface. Appendix 6 lists commonly used electrodes and their range of application in aqueous solution before electrolysis of H_2O becomes limiting. Promoters can be used to enhance the interaction of the protein and the electrode surface. These are molecules [e.g., 4,4'-bipyridine, bis(4-pyridyl)disulfide, (Cys-Glu)$_2$, 2-aminoethane thiol] that bind to the electrode surface and promote both surface binding and electron exchange by the redox protein (Figure 2.29).[15]

Cyclic Voltammetry

The voltage applied to the working electrode is scanned linearly from an initial value E_i to a predetermined limit E_f, at which point the direction of the scan is reversed. The cathodic and anodic currents and peak potentials are shown in Figure 2.30. As the potential is scanned in the negative direction, there is an increase in current flow that results from an exchange of electrons between the electrode and electroactive material around the electrode. Electrolysis depletes the local concentration of this material, and the current decays. The reverse holds true for the return scan. Assuming Nernstian behavior for reversible electron transfer, the following relationships hold:

$$i_a \sim i_c \tag{2.14}$$
$$(E_c - E_a) \sim (59/n) \text{ mV at 298 K}$$
$$E^\circ = E_{1/2} = (E_a + E_c)/2$$

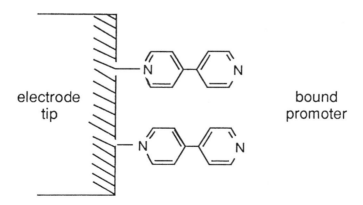

electrode
tip

bound
promoter

Figure 2.29 Promoters such as 1,4-bipyridine facilitate the interaction of the redox protein and the electrode surface.

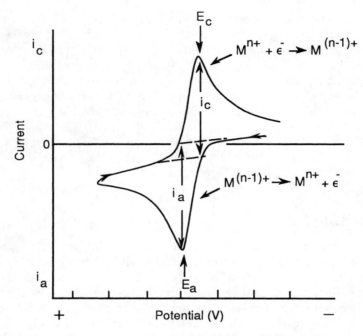

Figure 2.30   A typical cyclic voltammogram showing reversible electron transfer. Anodic and cathodic currents (i_a, i_c) and potentials (E_a, E_c) are indicated.

where i_a and i_c are the anodic and cathodic currents, and E_a and E_c are the anodic and cathodic peak potentials, respectively.

Differential Pulse Polarography

In differential pulse polarography, a uniform square voltage pulse is applied to a linear voltage ramp. This ramp and the current response are shown in Figure 2.31. The polarographic method offers better resolution of signals from redox

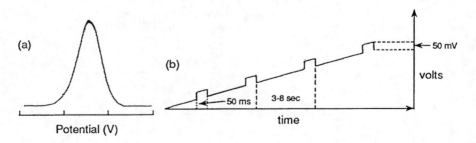

Figure 2.31   (a) Differential pulse polarogram and (b) potential time sequence.

centers with comparable $E°$s. However, cyclic voltammetry provides more information on the kinetics of electron exchange with the electrode.

2.5.2 Indirect Methods

In this category of experiment, the electrochemical potential of the solution is poised at a series of values and the fraction of oxidized and/or reduced molecule is determined at each potential. The solution potential is controlled by use of a low molecular weight organic or inorganic complex (termed a *mediator*) that is able to exchange electrons with the prosthetic center in the metalloredox protein and has an $E°$ close to that being measured. Appendix 7 provides details of some common mediators. Inasmuch as the potential ($E°$) of the redox protein may not be known, it is usually necessary to test several redox mediators, either individually or as a mixture (a "cocktail"). The potential is established by addition of a suitable mixture of oxidized and reduced mediator[16] or by bulk electrolysis of a solution of the mediator at a predefined potential. Unlike the situation described earlier, there is no direct electron exchange between the protein and the electrode surface (Figure 2.32). Having poised the solution potential at a defined value, the ratio of oxidized to reduced protein can be determined by use of EPR or optical spectroscopy, as explained in the following sections.

Reduction Potentials from Electron Paramagnetic Resonance Experiments

If one of the redox states possesses a characteristic EPR signature, this may be used to determine $E°$. The percentage of oxidized (or reduced) protein is plotted as a function of the solution potential ($E_{applied}$) (Figure 2.33). By use of the Nernst equation (2.15), it is clear that the reduction potential is defined at the equilibrium point $[c_{ox}]/[c_{red}] = 1$.

$$E = E° - (RT/nF) \log [c_{red}/c_{ox}] \tag{2.15}$$

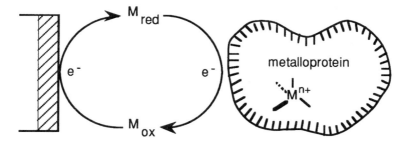

Figure 2.32   The redox mediator acts as an electron carrier *to* or *from* the electrode surface and metalloprotein. No direct contact between the protein and electrode surface is required. Unlike the redox promoters described earlier, electron exchange does not take place by way of an electrode-bound mediator molecule.

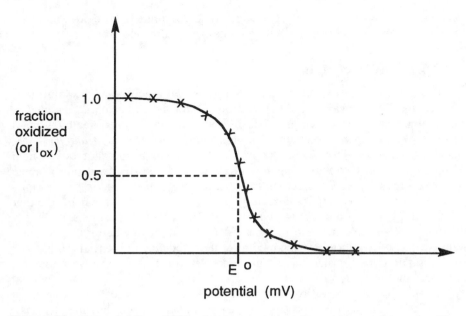

Figure 2.33 Determination of $E°$ by electron paramagnetic resonance spectroscopy. The solution potential is altered by addition of chemical oxidants or reductants. The fraction of the EPR active oxidized (or reduced) protein is determined from the intensity of the EPR signal, which in this case derives from the oxidized species (I_{ox}). As the solution potential is made more positive, the metalloprotein is reduced to a diamagnetic form. The fractional loss of oxidized protein is plotted against solution potential. When [ox] = [red], the solution potential is $E°$.

In equation (2.15), we define E as the solution potential, $E°$ is the standard half-reaction potential, R is Boltzmann's constant ($8.314 \, JK^{-1}$), F is Faraday's constant ($96,487 \, C \, mole^{-1}$), T is the temperature (K), n is the number of electrons that are transferred, c_{red} is the concentration of reduced center, and c_{ox} is the concentration of oxidized center.

Reduction Potentials from Optical Spectroscopy

The ratio of oxidized and reduced protein may be monitored by a variety of optical methods. Electronic absorption is most convenient, but circular dichroism can be used to resolve overlapping absorption bands from distinct redox centers. Normally, the wavelength that results in the largest difference between the absorbance of the oxidized (A_{ox}) and reduced (A_{red}) forms is chosen. Possible interference from absorption bands of the mediator should also be taken into account. All these methods ultimately depend on the Nernst equation (2.15). Equation (2.16) can be readily derived from (2.15) and (2.4),

$$E = E° + (RT/nF) \log[(A_{red} - A)/(A - A_{ox})] \tag{2.16}$$

where A_{red} is the absorbance of the fully reduced center, A_{ox} is the absorbance of the fully oxidized center, and A is the absorbance at the solution potential E. Figure 2.34 shows a typical data set. From the Nernst equation, when $c_{ox}/c_{red} = (A_{red} - A)/(A - A_{ox}) = 1$, we find that $E = E°$.

2.5.3   Useful Thermodynamic Relationships and Applications of Reduction Potentials

Having determined $E°$ for a variety of half-reactions, we now have both a powerful method to evaluate the thermodynamics of biological redox processes and a tool to probe the immediate environment of the prosthetic center. Table 2.5 lists some important potentials in inorganic biochemistry. If the reduction potentials ($E°$) for two half-reactions are known, $\Delta E°$ may be readily calculated. Equations (2.17) make the important connection with solution redox chemistry,

$$\Delta G° = \Delta H° - T \Delta S° = -RT \ln K = -nF \Delta E° \tag{2.17}$$

where $\Delta G°$, $\Delta H°$, and $\Delta S°$ are the standard free energy, enthalpy, and entropy for the reaction under consideration; and K is the equilibrium constant for the redox reaction. For example, consider the reaction of spinach ferredoxin and cytochrome c. The more positive half-reaction will tend to proceed as written, and the less positive half-reaction will go in reverse. Using the expressions in (2.17), free energies and equilibrium constants under standard conditions can be readily evaluated: $\Delta E° = +0.684$ V, $\Delta G° = -66$ kJ mole^{-1}, $K = 3.1 \times 10^{11}$.

ferredoxin (red) $\rightleftharpoons$ ferredoxin (ox) + e^-	0.430 V
cytochrome c(ox) + e^- $\rightleftharpoons$ cytochrome c(red)	0.254 V
Fd(red) + cyt c(ox) $\rightleftharpoons$ Fd(ox) + cyt c (red)	0.684 V

From equations (2.17), it is clear that the enthalpic and entropic contributions to the free energy of a half-reaction may be determined from a variable temperature experiment.

$$E° = (\Delta S°/nF)T - \Delta H°/nF$$

Plotting $E°$ as a function of temperature gives $\Delta S°$ from the slope and $\Delta H°$ from the intercept (assuming $\Delta H°$ and $\Delta S°$ do not vary with temperature).

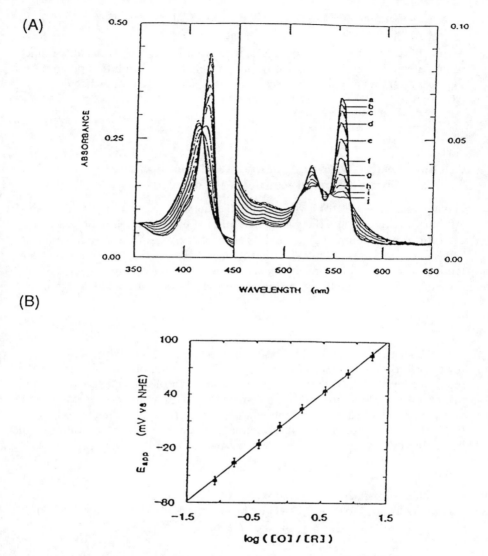

Figure 2.34 Determination of the reduction potential for the heme center in cytochrome b_5 by spectroelectrochemistry. A family of spectra obtained at different applied potentials (E_{app} vs. NHE) using $Ru(NH_3)_6^{2+/3+}$ as mediator is shown in (A): (a) -255.6 mV, (b) -55.6 mV, (c) -35.6 mV, (d) -15.6 mV, (e) 4.4 mV, (f) 24.4 mV, (g) 44.4 mV, (h) 64.4 mV, (i) 84.4 mV, (j) 244.4 mV. The resulting Nernst plot is shown in (B) using ΔA data obtained at 556 nm. The midpoint potential and slope are 5.1 ± 0.6 mV and 59.7 ± 0.4 mV, respectively. [Reprinted with permission from L. S. Reid et al., *J. Am. Chem. Soc.*, *104*, 7516; copyright (1982) American Chemical Society.]

Table 2.5   Reduction Potentials for Some Important Biochemical Redox Couples

Reduction Half-Reaction	$E°$(V) vs. NHE
$\frac{1}{4}O_2 + H^+ + e^- \to \frac{1}{2}H_2O$	0.816
$Fe^{3+} + e^- \to Fe^{2+}$	0.771
Photosystem P700	0.430
$NO_3^- + e^- \to NO_2^-$	0.421
Cytochrome $f(Fe^{3+}) + e^- \to$ cytochrome $f(Fe^{2+})$	0.365
Cytochrome $a_3(Fe^{3+}) + e^- \to$ cytochrome $a_3(Fe^{2+})$	0.35
Cytochrome $a(Fe^{3+}) + e^- \to$ cytochrome $a(Fe^{2+})$	0.290
Rieske Fe–S$(Fe^{3+}) + e^- \to$ Fe–S(Fe^{2+})	0.28
Cytochrome $c(Fe^{3+}) + e^- \to$ cytochrome $c(Fe^{2+})$	0.254
Cytochrome $c_1(Fe^{3+}) + e^- \to$ cytochrome $c_1(Fe^{2+})$	0.220
$UQH· + H^+ + e^- \to UQH_2$	0.11
$\frac{1}{2}UQ + H^+ + e^- \to \frac{1}{2}UQH_2$	0.100
Cytochrome b (mitochondrial) $(Fe^{3+}) + e^- \to$ cytochrome b (Fe^{2+})	0.077
$\frac{1}{2}$ fumarate $+ H^+ + e^- \to \frac{1}{2}$ succinate	0.031
cytochrome b_5 (microsomal) $(Fe^{3+}) + e^- \to$ cytochrome b_5 (Fe^{2+})	0.020
$\frac{1}{2}$ oxaloacetate $+ H^+ + e^- \to \frac{1}{2}$ malate	-0.166
$\frac{1}{2}$ pyruvate $+ H^+ + e^- \to \frac{1}{2}$ lactate	-0.185
$\frac{1}{2}$ acetaldehyde $+ H + e^- \to \frac{1}{2}$ ethanol	-0.197
$\frac{1}{2}$ FMN $+ H^+ + e^- \to \frac{1}{2}$ FMNH_2	-0.219
$\frac{1}{2}$ FAD $+ H^+ + e^- \to \frac{1}{2}$ FADH_2	-0.219
$\frac{1}{2}$ glutathione (oxidized) $+ H^+ + e^- \to \frac{1}{2}$ reduced glutathione	-0.230
$\frac{1}{2}$ lipoic acid $+ H^+ + e^- \to \frac{1}{2}$ dihydrolipoic acid	-0.290
$\frac{1}{2}$ NAD$^+ + \frac{1}{2}$H$^+ + e^- \to \frac{1}{2}$ NADH	-0.320
$\frac{1}{2}$ NADP$^+ + \frac{1}{2}$H$^+ + e^- \to \frac{1}{2}$ NADPH	-0.320
$H^+ + e^- \to \frac{1}{2}H_2$	-0.421
Ferredoxin (spinach), $Fe^{3+} + e^- \to Fe^{2+}$	-0.430
$\frac{1}{2}$ Fe$^{2+} + e^- \to \frac{1}{2}$Fe	-0.44
$\frac{1}{2}$ succinate $+ \frac{1}{2}$ CO$_2$ $+ H^+ + e^- \to \frac{1}{2}$ α-ketoglutarate $+ \frac{1}{2}$ H$_2$O	-0.67

In some texts these reactions may be written as two-electron reductions. When handling equations where different numbers of electrons are involved, it is safer to convert $E°$'s to $\Delta G°$ values, since in a formal sense $E°$ refers to a one-electron reduction. Consider the following examples:

(Example 1).

$$FMN + 2H^+ + 2e^- \rightleftharpoons FMNH_2 \quad \Delta G° = -2 \times F \times (-0.219) = 0.438F$$
$$2(\text{cyt } c, Fe^{2+} \rightleftharpoons \text{cyt } c, Fe^{3+} + e^-) \quad \Delta G° = 2 \times (-1)F \times (-0.254) = 0.508F$$

$$FMN + 2H^+ + 2 \text{ Cyt } c, Fe^{2+} \quad\quad \Delta G° = 0.946F$$
$$\rightleftharpoons FMNH_2 + 2Cyt c, Fe^{3+}$$

and so $E° = -\Delta G°/nF = -0.946F/2F = -0.473$ V.

(Example 2).

$$Fe^{2+} + 2e^- \rightleftharpoons Fe \quad \Delta G° = -2 \times F \times (-0.44)$$
$$Fe^{3+} + e^- \rightleftharpoons Fe^{2+} \quad \Delta G° = -1 \times F \times (-0.77)$$

$$Fe^{3+} + 3e^- \rightleftharpoons Fe \quad \Delta G° = 0.11F$$

and so $E° = -\Delta G°/nF = -0.11F/3F = -0.04$ V

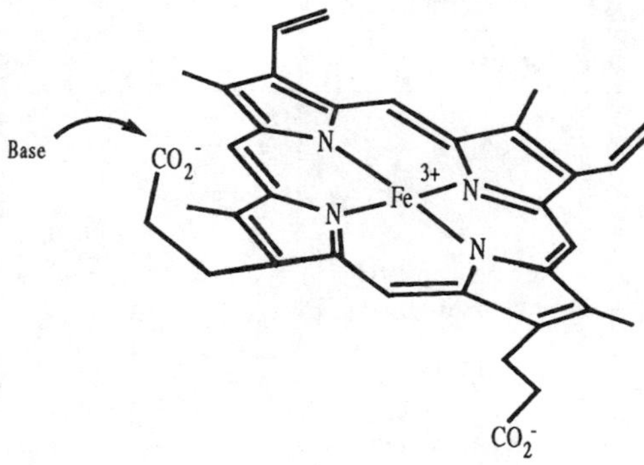

If the redox center is adjacent to an ionizable residue or ligand, the reduction potential will vary with the state of ionization, since protonation will stabilize the reduced form and lead to a more positive reduction potential. By plotting $E°$ as a function of pH, the pK_a of the ligand may be estimated (Figure 2.35).

Summary of Section 2.5

1. Electrochemical methods provide information on the affinities of metal ions for electrons [i.e., $M^{n+} + e^- \rightleftharpoons M^{(n-1)+}$] and the kinetics of such processes. The half-reaction potential $E°$ refers to the solution potential where $[M^{n+}] = [M^{(n-1)+}]$.
2. Potentials can be determined by monitoring electron exchange directly at an electrode surface (cyclic-voltammetry or polarography) or indirectly by use of the Nernst equation, $E = E° - (RT/nF) \log[c_{red}/c_{ox}]$, and measuring the relative concentrations of oxidized and reduced species by optical methods or EPR.
3. Key equations relating redox measurements to solution chemistry include

$$\Delta G° = \Delta H° - T \Delta S° = -RT \ln K = -nF \Delta E°$$

Review Question

• The E°s for some cobalt complexes are given. Provide an explanation for these trends.

$Co^{3+/2+}$	+1.88 V	$[Co(bipy)_3]^{3+/2+}$	+0.31
$[Co(ox)_2(H_2O)_2]^{1-/2-}$	+0.78	$[Co(en)_3]^{3+/2+}$	+0.18
$[Co(ox)_3]^{3-/4-}$	+0.57	$[Co(NH_3)_6]^{3+/2+}$	+0.11
$[Co(edta)]^{1-/2-}$	+0.37	$[Co(CN)_6]^{3-}/[Co(CN)_5]^{3-}, CN^-$	-0.31

ox = oxalate, edta (see Appendix 1), en = ethylenediamine, bipy = bipyridine

(A)

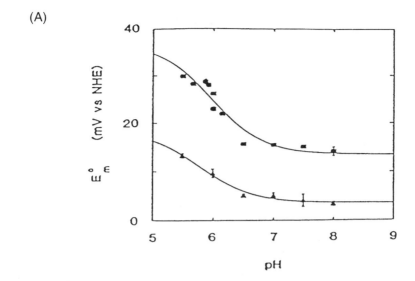

(B)

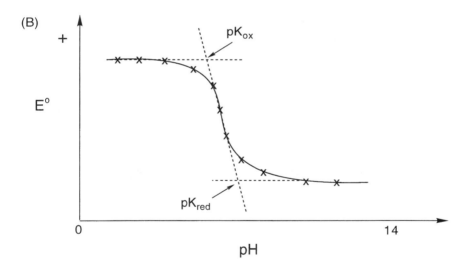

Figure 2.35 (A) Variation of $E°$ with pH for cytochrome b_5 at two ionic strengths [μ = 0.1 M (Δ), μ = 0.5 M ($\square$)]. The data is fit to an equation $E = E° + RT/(nF) \ln(K_{red} + [H^+]/(K_{ox} + [H^+])$. The p$K_a$ of a neighboring ionizable functional group is dependent on the oxidation state of the redox center. In this case, the ionizable functionality is believed to be a heme propionate that folds back to stabilize the ferric center. For μ = 0.1 M, pK_{ox} = 5.7, pK_{red} = 5.9, $E°$ = 11 mV; μ = 0.5 M, pK_{ox} = 5.8, pK_{red} = 6.2, $E°$ = 25 mV. [Reprinted with permission from L. S. Reid et al., *J. Am. Chem. Soc., 104,* 7516; copyright (1982) American Chemical Society.] (B) An idealized pH-dependence plot.

PART B.   BIOCHEMICAL METHODS

2.6   Enzyme Kinetics

2.6.1   Steady-State Kinetics

Many metalloenzymes catalyze chemical reactions where a substrate molecule S is converted into a product molecule P. Detailed kinetics studies can provide significant insight into the reaction mechanism. In this section we shall review some useful kinetic models.

At the turn of the century, it was experimentally demonstrated that there exists a hyperbolic relationship between the initial velocity ($v_0 = -d[\text{substrate}]/dt$ at $t = 0$), and the initial substrate concentration $[S]_0$, for many single-substrate enzyme-catalyzed reactions (Figure 2.36). The maximum reaction velocity for a given enzyme concentration is V_{max}, and the constant b is the initial substrate concentration necessary to give an initial reaction velocity $v_0 = 1/2\ V_{max}$. An equation of the same form [equation (2.18)] may be derived from the reaction scheme (2.19) by assuming steady-state conditions.[17] Steady-state conditions asssume that the intermediate [ES] breaks down as rapidly as it is formed (i.e., $k_1[E][S] = (k_{-1} + k_2)[ES]$). Substrate binds to enzyme and the resulting enzyme–substrate complex breaks down to yield the products P. By looking at *initial* velocities, the back-reaction k_{-2} may be neglected, since [P] is low. The Mi-

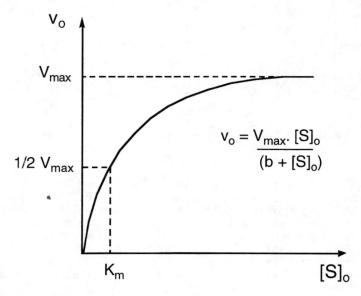

Figure 2.36   Dependence of initial reaction velocity (v_0) on substrate concentration [S] in an enzyme-catalyzed reaction.

chaelis constant (K_m) and turnover number (k_{cat}), are defined by relationships (2.20). Equation (2.18) is usually rewritten in the form of (2.21) or (2.22),

$$v = k_{cat}[E]_0[S]_0/([S]_0 + K_m) \tag{2.18}$$

$$E + S \underset{k_{-1}}{\overset{k_1}{\rightleftharpoons}} ES \overset{k_2}{\rightarrow} E + P \tag{2.19}$$

$$K_m = (k_{-1} + k_2)/k_1 \qquad k_{cat} = V_{max}/[E]_0 \tag{2.20}$$

$$1/v_0 = 1/V_{max} + K_m/V_{max} [S]_0 \tag{2.21}$$

$$v_0 = V_{max} - K_m v_0/[S]_0 \tag{2.22}$$

which give straight-line plots called the Lineweaver–Burk plot and Eadie–Hofstee plot, respectively. V_{max} (or k_{cat}) and K_m may be evaluated by noting the slope and intercept on the y axis (Figure 2.37). Within the simple model outlined earlier, $k_{cat} = k_2$ corresponds to the first-order rate constant for conversion of enzyme-bound substrate to product (ES → E + P, or ES → EP). For more complex reactions, k_{cat} is a function of many first-order rate constants. It should be noted that the equilibrium constant K_m is equal to the enzyme-substrate dissociation constant ($K_d = k_1/k_{-1}$) only when $k_{-1} \gg k_2$. The ratio k_{cat}/K_m is referred to as the specificity constant, since it reflects the specificity of an enzyme for competing substrates. It can best be thought of as a second-order rate constant for the enzyme and substrate.

When the kinetics of a reaction are under investigation, it can be useful to examine the influence of molecules that are structurally similar to the substrate, bind reversibly and competitively to the substrate binding site, but do not react.[18] These molecules are classed as competitive inhibitors. By applying the steady-state approximation to the following reaction scheme, a Michaelis–Menten-type equation (2.23) may be derived:

$$EI \underset{-I}{\overset{+I}{\rightleftharpoons}} E + S \underset{k_{-1}}{\overset{k_1}{\rightleftharpoons}} ES \overset{k_2}{\rightarrow} E + P$$

$$v_0 = V_{max} [S]_0/\{[S]_0 + K_m(1 + [I]_0/K_i)\} \tag{2.23}$$

In effect K_m is increased by a factor $(K_i + [I]_0)/K_i$. The value of K_i, which is a

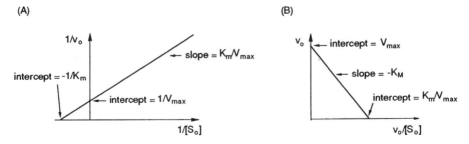

Figure 2.37   (A) Lineweaver–Burk and (B) Eadie–Hofstee plots.

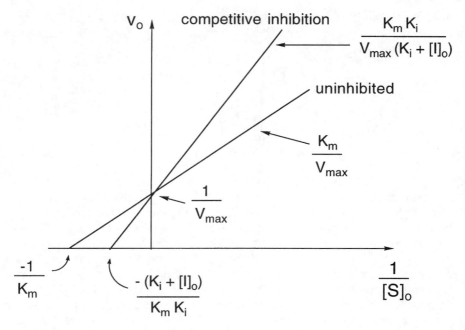

Figure 2.38 Lineweaver–Burk plot demonstrating competitive inhibition for inhibitor concentration $[I]_0$. The inhibitor constant K_i may be readily determined. Parameters for the uninhibited plot were noted in Figure 2.37.

true dissociation constant, may be derived from Lineweaver-Burk plots (Figure 2.38).

The binding energy for the substrate can be used to lower the activation energy for the second order reaction, $\Delta G_t^* = \Delta G^* - \Delta G_d$ (defined by k_{cat}/K_m). It is best to consider the ratio k_{cat}/K_m because this encompasses both the overall activation energy ΔG^* and the binding energy ΔG_d (Figure 2.39). In the context

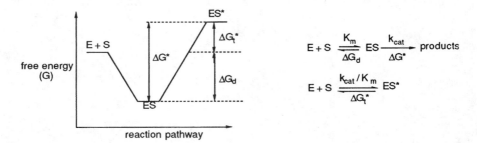

Figure 2.39 Free energy diagram outlining the free energy changes accompanying substrate binding to an enzyme (ΔG_d) and the contributions from ground and transition states to the activation energy (ΔG^*) for the enzyme–substrate complex (ES).

of simple Michaelis–Menten kinetics (assuming $K_m \sim K_d$), variable temperature kinetics studies can be used to evaluate whether the contribution of substrate binding to the activation barrier (ΔG^*) for the reaction of enzyme and substrate, is utilized in ground-state or transition-state stabilization. The former is bad for catalysis because this increases ΔG^*, whereas the latter is favorable since it results in a decrease in ΔG^*. Equation (2.24) can be derived from simple transition-state theory by considering the reactions of ES and E + S with effective rate constants given by k_{cat} and k_{cat}/K_m, respectively (Figure 2-39).

$$RT \ln(k_{cat}) = RT \ln(kT/h) - \Delta G^* \tag{2.24a}$$

$$RT \ln(k_{cat}/K_m) = RT \ln(kT/h) - \Delta G^* - \Delta G_d \tag{2.24b}$$

These equations offer a powerful method for evaluating mutant enzymes or modified substrates. If mutation of a residue affects predominantly ΔG_t^* or ΔG_d, this reflects a specific contribution of that active site residue to either ground-state or transition-state stabilization, respectively. The same considerations apply to the analysis of substrate analogues.

2.6.2 Transient Kinetics

The steady-state parameters (k_{cat} and K_m) described in the previous section are complex functions of all the reactions that occur when reactant molecules are turned into products. For example, the simple reaction defined by equation (2.19) gives rise to expressions for k_{cat} and K_m defined by equations (2.20). However, (2.25) shows a more realistic mechanism, and equations (2.26) describe the appropriate steady-state parameters. These simplify to the earlier expressions ($k_{cat} \sim k_2$ and eq. 2.20) when $k_3 \gg k_2$.

$$E + S \underset{k_{-1}}{\overset{k_1}{\rightleftharpoons}} E{\cdot}S \underset{k_{-2}}{\overset{k_2}{\rightleftharpoons}} E{\cdot}P \underset{k_{-3}}{\overset{k_3}{\rightleftharpoons}} E + P \tag{2.25}$$

$$k_{cat} = \frac{k_2 k_3}{k_2 + k_{-2} + k_3} \tag{2.26a}$$

$$K_m = \frac{k_2 k_3 + k_{-1} k_{-2} + k_{-1} k_3}{k_1(k_2 + k_{-2} + k_3)} \tag{2.26b}$$

For a reaction pathway with a single intermediate X (2.27), the equations defining the steady-state parameters become even more complex (2.28).

$$E + S \underset{k_{-1}}{\overset{k_1}{\rightleftharpoons}} E{\cdot}S \underset{k_{-2}}{\overset{k_2}{\rightleftharpoons}} E{\cdot}X \underset{k_{-3}}{\overset{k_3}{\rightleftharpoons}} E{\cdot}P \underset{k_{-4}}{\overset{k_4}{\rightleftharpoons}} E + P \tag{2.27}$$

$$k_{cat} = \frac{k_2 k_3 k_4}{(k_2 + k_{-2})(k_{-3} + k_4) + k_2 k_3 + k_3 k_4} \tag{2.28a}$$

$$K_m = \frac{k_{-1}[k_{-2}(k_{-3} + k_4) + k_3 k_4] + k_2 k_3 k_4}{k_1[(k_2 + k_{-2})(k_{-3} + k_4) + k_2 k_3 + k_3 k_4]} \tag{2.28b}$$

Clearly, if we were able to determine each of the rate constants k_1, k_{-1}, k_2, k_{-2}, etc., we would know a great deal about the enzyme-catalyzed reaction. Detailed mechanistic studies are therefore concerned with characterization of the enzyme-bound substrate, and intermediate and product states by spectroscopic, chemical, or other physical methods and also with a thorough kinetic analysis that elucidates as many of these individual reaction rate constants as possible. Sections, 1.7–1.9, and 2.6.1 demonstrated how the evaluation of rate constants affords much information on chemical reactivity, permits identification of rate-limiting steps, and allows one to determine how the binding energy of substrates and intermediates is used to define enzyme specificity and facilitate catalysis, the main function of an enzyme. Recall from Figure 2.39 that this involves lowering the transition-state energy or decreasing the binding energy to reduce ΔG^*.

To carry out transient kinetic studies, one must utilize a method that is appropriate for the time regime of the experiment. Two common methods involve stopped-flow and quenching techniques. Both are limited by the mixing time (typically $\sim$ 1 to 4 ms) required to bring together two or more solutions. This is equivalent to a rate constant of ca. 250 to 1000 s^{-1}. Obviously, we cannot measure a reaction that occurs with a (pseudo) first-order rate constant significantly greater than this. The reaction will be completed during the mixing process before reaching the detection unit. Fortunately, very few enzymatic reaction steps occur with rates that are faster than the ms timescale, and so stopped-flow or rapid-quench methods are convenient to use.

After mixing the reactants, one must follow the progress of the ensuing chemical reactions. There are many methods that can be employed to monitor this chemistry. All that is needed is an observable change in at least one physical or chemical parameter that reflects the chemical process under investigation. These may reflect the optical, magnetic, or electrochemical properties of reactants, intermediates, or products. Other methods of detection include viscosity measurements or the use of radiolabeled reactant molecules. Several illustrations are included in the text.

Stopped-Flow Measurements (ms timescale): Two solutions, held in syringes, are rapidly mixed and passed into an observation cell. Usually, an optical property is monitored (absorbance, fluorescence, circular dichroism).

Rapid-Quench (ms timescale): Two or more solutions of reacting species are mixed. Subsequently the reaction mixture is allowed to age. The reaction mixture is quenched at various time points after mixing, either by addition of a chemical that stops the reaction (chemical-quench), or by freezing (freeze-quench). A property of the solution is monitored for each time point. This may be the product distribution of a radioisotope or the EPR signature of a reactant or product.

Flash Photolysis (ns-ms timescale): Here reactions are initiated by an intense burst of light from a laser or flash lamp. The radiation initiates chemistry, which is typically monitored by absorption or emission from reactant or product species. The time scale of the experiment that can be followed depends on the instrument used.

NMR (μs-ms timescale): Section 2.3.1 earlier described the use of NMR to follow dynamic processes in solution.

Although we began by introducing these methods through a discussion of enzyme kinetics, they are more generally applicable to studies of binding chemistry, transport processes, or structural rearrangements. The key difference between steady-state and transient kinetic studies, especially for enzyme studies, is that one monitors the overall limiting rate of turnover in a steady-state experiment, where it is assumed that the enzyme is catalyzing multiple turnovers with a constant equilibrium concentration of reaction intermediates. In contrast, transient studies assume a single turnover event. Such studies use an excess of enzyme over substrate.

Assuming that one of the reactant or product species for a particular step of a reaction can be monitored, the rate profile can be fit to obtain the relevant rate constants. The power of modern computers has greatly improved the ease of analyzing kinetic data. Table 2.6 summarizes some of the common rate laws that are used in studies of transient kinetics. These are derived similarly to simple first-order or pseudo-first-order processes, reviewed in Section 1.9.

Summary of Section 2.6

1. Kinetic studies provide extensive information on rates of individual steps in complex reactions, quantitation of energy barriers, and insight into reaction mechanism.
2. In enzymology, the Michaelis–Menten equation is often adopted for analysis of steady-state kinetics [$v = k_{cat} [E]_0 [S]_0/([S]_0 + K_m)$].
3. Transient kinetics differs from steady-state kinetics inasmuch as the experimental system has not attained a dynamic equilibrium. Individual steps in a complex reaction pathway may be monitored by a variety of chemical and spectroscopic techniques.

Table 2.6   Rate Laws in Common Use in Pre-Steady-State Chemical Kinetics

Description of Rate Law	Differential Form	Integrated Form	Comments
1. first-order	$\dfrac{d[Y]}{dt} = -k_1[Y]$	$Y_t = Y_o \exp(-kt) + c$	
2. second-order (pseudo-first-order)	$\dfrac{d[Y]}{dt} = k_2[X][Y]$	$Y_t = Y_o \exp(-k_2[X]_o t) + c$	where $[X] \gg [Y]$
3. second-order (one species)	$\dfrac{d[Y]}{dt} = -k_2[Y]^2$	$Y_t = \dfrac{Y_o}{(1 + k\,Y_o\,t)}$	
4. consecutive (pseudo-first-order) $A \rightarrow Y \rightarrow C$	$\dfrac{d[Y]}{dt} = k_1[Y] - k_2[Y]$	$Y_t = Y_o\,[\exp(-k_2 t) - \exp(-k_1 t)] + c$	

4. Evaluation of activation parameters and the influence of added inhibitors provides information on the bonding interactions that underlie catalysis and the structures of intermediates and transition-state complexes.

2.7 Measuring the Molecular Mass of a Protein

2.7.1 SDS-PAGE

Gel electrophoresis involves the movement of charged macromolecules through a defined matrix material under the influence of an applied electric field. Treatment with sodium dodecyl sulfate (SDS) both disrupts the structure of a protein and saturates it with the negatively charged detergent molecules. The protein, then, migrates toward the positively charged anode, and the electrophoretic separation depends only on its molecular weight (mol wt). By referencing to proteins of known molecular weight (Figure 2.40), that of the protein of interest may be determined, since mobility is proportional to log (mol wt).

2.7.2 Gel Filtration Chromatography

Column chromatography is an alternative method for the determination of molecular weight. The column material typically consists of beads of a cross-linked support matrix (e.g., *Sephadex*) that separate proteins by molecular size. Larger molecules do not interact with the pores in the gel beads and, thus, move through the column more rapidly. There is a linear relationship between log (mol wt) and elution volume (the volume of aqueous buffer required to remove the protein from the column). By comparing the elution volume of a given

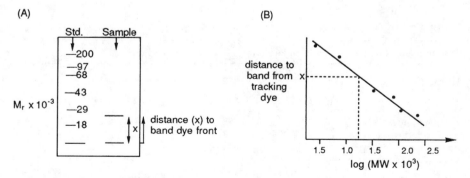

Figure 2.40 Determination of protein molecular weight by SDS-polyacrylamide gel electrophoresis. (A) A sample is run against a series of molecular-weight markers (proteins of known molecular weight and the migration distance measured relative to a tracking dye. (B) The molecular weight of the unknown may be quickly deduced from a calibration plot. For the unknown, we can determine a mol. wt. ~23,000.

protein with a calibration plot for proteins of known molecular weight, that of the unknown may be determined.

Gel filtration chromatography can be viewed as a method complementary to SDS-PAGE. Denaturing conditions are employed in running SDS-PAGE gels, and so a multimeric protein will be dissociated into several subunits. Although the mass of each subunit may be determined, the relative numbers of each and the total mass of the protein will not be clear. Gel filtration is carried out under nondenaturing conditions and gives the overall molecular mass. For example, an $\alpha_2\beta_4$ hexamer gives two bands on a denaturing gel of molecular mass 10 kDa and 50 kDa for the α and β subunits, respectively. By gel filtration, it is known that $M_r \sim 220,000$.

2.8   Measuring the Molecular Mass and Length of Polynucleotides

The ideas that we have developed for determination of protein molecular weights are also applicable to oligo- and polynucleotides. Gel electrophoresis is commonly used. Agarose is employed as the gel matrix for polynucleotides that are of the order of 100s to 1000s of bases in length. Data is again referenced to nucleic acids of known length (defined by the number of nucleotide units or base pairs in the DNA or RNA molecule of interest). For smaller oligonucleotides (10 to several hundred base units), a polyacrylamide gel is used. These DNA molecules may be visualized in several ways. However, the two most common methods use radiolabeling and visualization on X-ray film or staining with ethidium bromide solution. When ethidium binds to DNA, it fluoresces under UV light and a photograph of the bands may be taken.

2.9   Measurement of Macromolecule–Ligand Binding Affinities

Later we shall see that many biological macromolecules (commonly proteins or enzymes) bind specific ligands with high affinities. The dependence of mobility on molecular size may be used to determine the binding affinity of the ligand to the macromolecule. Consider the following equilibrium reaction between a protein P and a ligand molecule L.

$$P + L \rightleftharpoons PL \qquad K_a = [PL]/[P][L] \tag{2.29}$$

It can be shown easily that the binding constant K_a is related to the protein concentration that results in 50 percent of the ligand forming a complex with the protein, where $[L] = [PL]$, and so equation (2.29) takes the following form.

$$K_a = [PL]/[P][L] = 1/[P]$$

The fraction of bound and free ligand may be readily determined by gel electrophoresis, since the bound and free ligand molecules have distinct molecular
weights and therefore different mobilities. After staining, the relative amounts
of L and PL may be estimated by densitometry.[19] By varying the concentration
of protein, the equilibrium is shifted, and a plot of the fraction of free (or bound)
ligand versus [P] will give the binding constant. Figure 2.41 shows data obtained
for binding a protein to a short oligonucleotide.

2.10 Protein Isolation and Purification

It is obvious that the study of metalloproteins, metalloenzymes, and nucleic
acids requires an adequate supply of pure material to support investigations.
No protein, no measurements! Some molecules may be commercially available,
but many are not. Most biological molecules that are of interest must be isolated

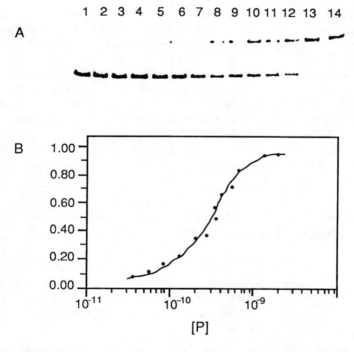

Figure 2.41 Determination of equilibrium dissociation constant ($K_d = 1/K_a$) for a protein–DNA complex. (A) Solution contains 8.3×10^{-12} M DNA and increasing protein
concentrations from none in lane 1 to 6.2×10^{-10} M in lane 14. (C) The apparent K_d is
obtained by plotting the fraction of DNA bound (obtained from the relative intensities
of the bands for free DNA and bound DNA) vs. log [protein] and fitting an equation
derived from equation (2.29) in the text. By simple manipulation, one can derive $\log([P])$
$= \log K_d - \log([D]/[PD])$. The apparent K_d is 5.2×10^{-10} M.

from the natural source (plant or animal tissue, bacteria, fungae) and purified. We close this chapter by briefly commenting on the type of procedures commonly used to obtain samples of pure proteins or enzymes from a common bacterial source. The ideology for the isolation of proteins from plant or animal tissue is very similar.

Figure 2.42 summarizes a typical isolation procedure. The bacterium of interest is grown in a medium that contains nutrients and minerals.[20] After a day or two, the cells are harvested by centrifugation. Isolation protocols are normally performed at 4 °C, since degradative enzymes that are present during the initial stages of isolation are less active at low temperatures. The cell contents can be released by breaking open the membrane walls by one of several techniques (sonication, French press, chemical lysis, or grinding with a pestle and mortar). Cell debris and large genomic DNA are removed by centrifugation. At this point, the solution contains many other proteins and metabolites in addition to the one of interest. Differences in molecular weight (size), surface charge, affinity for ligands, or solubility properties can be used to effect a separation. Materials that discriminate on the basis of these parameters are available commercially. Appendix 8 outlines some of the commonly used chromatography materials and their basis for separation. It is extremely rare that one method alone will suffice. When running chromatography columns, the appearance of

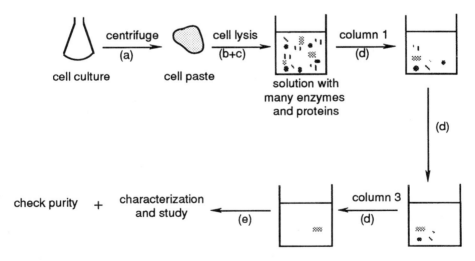

Figure 2.42 Summary of a typical "protein prep." (a) Centrifugation to harvest the cell paste. (b) Cell lysis. (c) Centrifugation to remove cell debris. (d) The supernatant is applied to a series of chromatography columns. After each column, the fractions containing the protein of interest are collected, concentrated, and applied to the next column. At some stage, an ammonium sulfate precipitation may be included that uses the differing solubilities of proteins in a solution of high salt concentration. If the desired protein is precipitated, it must later be redissolved. (e) The purity of the protein can be tested at each stage by an electrophoretic technique, such as SDS-PAGE.

specific proteins or enzymes can be monitored by following a characteristic absorption in the optical spectrum, catalytic activity, reaction with a specific antibody, and so on. Typically, several of these separation procedures will be combined, one after the other, to achieve a final purification. Many criteria of purity exist. However, SDS-PAGE (see Section 2.7) is commonly used to determine the number of protein components in a sample.[21]

Summary of Sections 2.7–2.10

1. The molecular weights of proteins and nucleic acids can be evaluated by gel electrophoresis experiments.
2. Binding constants of proteins to DNA sequences are readily evaluated by gel-shift experiments.
3. In most cases, interesting biological molecules must be isolated from their natural source. This often involves harvesting the source, isolation of a crude extract of the molecule of interest, and purification by chromatographic methods.
4. Differences in molecular weight (size), surface charge, affinity for ligands, or solubility properties can be used to effect a separation and purification.

Notes

1. For example, the wavenumber ν' has units of cm^{-1} and is defined as $\nu' = 1/\lambda = \nu/c$, where λ is in units of cm.
2. An energy packet kT corresponds to approximately 2.5 kJ mole^{-1}, 0.6 kcal mole^{-1}, 208 cm^{-1}, or 0.026 eV at a temperature of 300 K (27°C).
3. By application of a mathematical discipline known as group theory (or symmetry theory) the allowed transitions, and therefore selection rules, may be deduced.
4. Singlet states have no net electron spin, since all electrons are paired ($\downarrow\uparrow$). Triplet states have a net spin ($S = 1$), since two electrons are uncoupled ($\uparrow\uparrow$). The latter is a more stable arrangement.
5. The electron density at a given nucleus induces local magnetic fields that influence the effective field at the nucleus ($H_{eff} = H_{appl} - H_{induced}$). The chemical shift reflects electronic differences in the environment of the nucleus, and so the resonance frequency $\nu = (g_N\beta_N)H_{eff}/h$ is sensitive to the environment.
6. In this case, the nuclei are the same, but this need not always be true. To a first approximation, a resonance is split into $2nI + 1$ lines, where n is the number of adjacent chemically equivalent nuclei possessing a nuclear spin I. For example, in Figure 2.19, α-P is adjacent to β-P and appears as a doublet; β-P is adjacent to α-P and γ-P and appears as a triplet; γ-P is adjacent to β-P and appears as a doublet.
7. Pulse programs form the backbone of NMR methodology. Readers who are interested in learning more should refer to one of a large number of excellent texts in the area. For example, *Modern NMR Spectroscopy*, J. K. M. Sanders, B. K. Hunter, Oxford (1987); *Modern NMR Techniques for Chemistry Research*, A. E. Derome, Pergamon (1987).
8. Experiments that correlate one nucleus (e.g., ^{1}H–^{1}H) are termed homonuclear (e.g., HOMO–COSY, where COSY = correlated spectroscopy). Experiments that correlate two distinct nuclei (e.g., ^{1}H–^{13}C) are called heteronuclear (e.g., HET–COSY). COSY experiments provide information on nuclei that are connected and separated by one or more chemical bonds. NOESY

experiments (2D nuclear Overhauser effect spectroscopy) correlate nuclei that are adjacent in space but not connected by chemical bonds. Examples of each type of experiment are described in the text. See especially Figure 2.13.

9. Equation (2.8) refers to the simplest situation of a $S = 1/2$ system in, for example, an organic radical: g_e is the free electron g-value ($g_e = 2.00232$), β_e is the electronic Bohr magneton ($\beta_e = 0.92 \times 10^{-23}\ JT$), $m_s(\alpha)$ and $m_s(\beta)$ are the m_s values $+1/2$ and $-1/2$, respectively, and H is the magnitude of the magnetic field (in Tesla, T).

10. In multielectron ions, the orbital and spin angular momenta are defined by the electronic arrangement of all the electrons. Such an *electronic state* is defined by a *term symbol* that defines the overall spin (S), orbital (L), and total (J) angular momenta.

11. Equation (2.11) shows an electron Zeeman component S_zH_z, a nuclear Zeeman component I_zH_z, and an electron-nucleus coupling S_zI_z.

12. The magnetic susceptibility (χ_m), a measure of the inherent magnetism of a compound, is related to the magnetic moment (μ), where $\mu = e^{-J/kT}$. A plot of χ_m versus $1/T$ can be used to determine J.

13. Reduction potentials are referenced to the half-reaction, $H^+ + e^- \rightleftharpoons 1/2\ H_2$, $E^\circ = 0.0$ V, where $[H^+] = 1$ M, partial pressure of $H_2 = 1$ atm, temperature $= 25°C$.

14. By definition, reduction pathways that are favored over the standard half-cell reaction for H^+ reduction (see note 13) have positive E° values. Solution potentials may also be referenced to the standard calomel electrode (SCE, Hg_2Cl_2/Hg), or silver/silver chloride electrode (Ag/AgCl). To obtain the potential relative to NHE, the following constants, 0.244 V and 0.199 V, may be subtracted from potentials measured relative to SCE and Ag/AgCl, respectively, assuming data is taken under standard conditions at 25°C.

15. Refer to Guo and Hill, *Adv. in Inorg. Chem.* 36, 342.

16. This potential may be measured by placing the working and reference electrodes in the solution.

17. Equation (2.18) is commonly referred to as the Michaelis–Menten equation, although this form contains key contributions from Briggs and Haldane. The equation assumes a single-substrate, enzyme-catalyzed reaction with one substrate binding site on the enzyme and formation of a single intermediate complex. However, pseudo-single-substrate reactions, and enzymes with two or more noninteracting reaction sites are also covered.

18. Several types of competitive and noncompetitive inhibitions are recognized. Michaelis–Menten-type equations have been derived, each having a characteristic Lineweaver–Burk plot from which K_i may be determined. Excellent summaries of enzyme inhibition and general introductions to enzyme structure, kinetics, and mechanisms can be found in *Understanding Enzymes,* by T. Palmer (Ellis Harwood) or from *Enzyme Structure and Mechanism,* by A. Fersht (Freeman).

19. Densitometry is a technique that measures the intensities of bands on a photographic negative of a stained gel by monitoring the amount of light transmitted through the band.

20. The American Type Culture Collection (ATCC) is a good source of bacterial strains, yeasts, fungae, and so on, and of recommendations for growth conditions.

21. *Protein Purification* by R. Scopes (Springer-Verlag) 1982 is a valuable guide to the methods and strategies employed in protein purification, and the various criteria of purity.

Further Reading
General Introduction to Biophysical Techniques

Banwell, C. N. *Fundamentals of Molecular Spectroscopy*, McGraw–Hill, 1972.
Campbell, I. D., and R. A. Dwek, *Biological Spectroscopy*, Benjamin–Cummings, 1980.
Cantor C. R., and P. R. Schimmel. *Biophysical Methods* (Parts I, II, and III), Freeman, 1980.
Stout, G. H. and L. H. Jensen. *X-Ray Structure Determination,* 2nd ed., Wiley-Interscience, 1989.

Magnetic Resonance

Carrington, A., and A. D. McLachlan. *Introduction to Magnetic Resonance*, Chapman & Hall, 1984.
Sanders, J. K. M., and B. K. Hunter, *Modern NMR Spectroscopy*, Oxford, 1987.
Symons, M. *Chemical and Biochemical Aspects of Electron-Spin Resonance Spectroscopy*, Van Nostrand Reinhold, 1978.
Wertz, J. E. and J. R. Bolton. *Electron Spin Resonance: Elementary Theory and Practical Applications*, Chapman & Hall, 1986.

NMR Applications in Protein Chemistry

Berliner, L. J., and J. Reuben, eds. *Biological Magnetic Resonance*, Vols. 1–13, Plenum.
Bertini, I. and C. Luchinat. *NMR of Paramagnetic Molecules in Biological Systems*, Addison–Wesley, 1986.
Dwek, R. A. *Nuclear Magnetic Resonance in Biochemistry*, Oxford, 1975.
Wuthrich, K. *NMR of Proteins and Nucleic Acids*, Wiley-Interscience, 1986.

Physical Inorganic Methods

Drago, R. S. *Physical Methods for Chemistry*, 2nd ed., Saunders, 1992.
Gray, H. B., and A. B. P. Lever. *Iron Porphyrins (Parts II & III), Physical Bioinorganic Chemistry*, Addison-Wesley, 1983 and VCH, 1989.
Huynh, B. H. Mossbauer spectroscopy in the study of cytochrome cd1 from Thiobacillus denitrificans, desulfoviridin, and iron hydrogenase. *Methods Enzymol. 243*, 523–43 (1994).
Johnson, M. K., A. E. Robinson, and A. J. Thomson. Low-temperature MCD in Fe-S proteins, in *Iron-Sulfur Proteins,* Wiley-Interscience, pp. 367–406, 1982.
Lever, A. B. P. *Inorganic Electronic Spectroscopy*, 2nd ed., Elsevier, 1984.
Nakamoto, K. *Infrared and Raman Spectra of Inorganic Coordination Compounds*, 4th ed., Wiley-Interscience, 1986.
Sawyer, D. T., and J. L. Roberts. *Experimental Electrochemistry for Chemists*, Wiley-Interscience, 1974.
Wilkins, R. G. *Kinetics and Mechanisms of Reactions of Transition Metal Complexes*, 2nd ed., VCH, 1991.

Biochemical Methods

Cooper, T. G., *The Tools of Biochemistry,* Wiley-Interscience, 1977.
Fersht, A. *Enzyme Structure and Mechanism*, Freeman, 1985.
Johnson, K. A. "Transient Enzyme Kinetics" in *The Enzymes* vol. XX, 1–61, Academic Press, 1992.
Palmer, T. *Understanding Enzymes*, Ellis Harwood, 1985.
Robyt, J. F., and B. J. White. *Biochemical Techniques: Theory and Practice*, Waveland Press, 1990.
Scopes, R. *Protein Purification*, Springer-Verlag, 1982.

Worked Problems

Question 1: In Cu^{2+} EPR spectra, why is $A_\perp$ typically not well resolved, while $A_\parallel$ is well resolved?

Solution 1: In Cu^{2+}, the free electron typically lies in the $d(x^2 - y^2)$ orbital and so the electron density interacts best with the component of the nuclear spin in the x-y plane (designated by $A_\parallel$).

Question 2: Carbon monoxide dehydrogenase catalyzes the reversible oxidation of CO to CO_2 and the synthesis of acetyl-coenzyme A from CO and coenzyme A. Freeze-quench EPR experiments have been used to monitor the kinetics (inset) of the reaction of CO with CoA at 25 °C by plotting the intensity of signal $g = 2.08$ as a function of time.

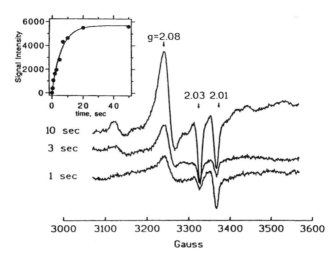

(a) Estimate the exchange rate from the kinetic plot.
(b) Estimate the exchange rate at 55 °C if the activation free energy is 80.6 kJ/mole. It is known that k_{cat} for synthesis of acetyl CoA is ~ 1.1 s⁻¹ at this temperature. Is the species observed in the EPR experiment relevant to the physiological reaction?
(c) For the oxidation of CO to CO_2, k_{cat} is ~ 2000 s⁻¹. Comment on the significance of this observation for acetyl-CoA synthesis.

Solution 2: (a) The maximal intensity is ~ 55000. The half-life for the reaction can be obtained either by fitting the data or can be estimated from the time to reach the half-maximum intensity. We shall use the latter approach and estimate the half-life to be ~ 6 sec. Since rate = (half-life)⁻¹, this is equivalent to a rate of ~ 0.17 s⁻¹.
(b) We can use the relationship $k = (\mathbf{k}T/h) \exp(-\Delta G^*/RT)$ to estimate a rate constant at 55 °C of ~ 1 s⁻¹, and so this species is kinetically competent to serve as an intermediate in the reaction pathway.
(c) The species identified in this pathway for formation of acetyl-CoA cannot be an intermediate for the oxidation of CO to CO_2, since formation of the intermediate is too slow to support a turnover rate of 2000 s⁻¹. It is likely that CO oxidation and synthesis of acetyl-CoA take place at different sites on the enzyme.
[Kumar, M., et. al., *J. Am. Chem. Soc. 115*, 11646 (1993), Figure is reproduced with permission]

Problems

1. A zinc porphyrin shows an absorption band at 410 nm, a fluorescence at 636 nm, and a phosphorescence at 720 nm.
   (a) Construct an energy level diagram that shows the relative energies of the singlet and triplet states and assign the above absorptions and emissions to particular transitions.
   (b) Typically, a singlet lifetime is of the order of 10^{-9} s whereas triplets have a lifetime of $10^{-6} - 10^{-3}$ s. The heme in myoglobin may be replaced with zinc protoporphyrin. An emission is observed at 636 nm with a lifetime of 10^{-5} s. Suggest an explanation for this observation.
   (c) The phosphorescence emission from a myoglobin substituted with a platinum porphyrin is shown (full line). The lifetime of the triplet state is 103 μs. When a Ru(III) label is attached to the protein surface the emission decreases (dashed line) as a result of electron-transfer from the porphyrin. Estimate, by inspection, the electron transfer rate from the photoexcited platinum porphyrin to the surface Ru(III).

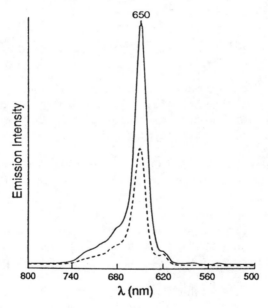

[Cowan, J. A. and H. B. Gray, *Inorg. Chem. 28,* 2074–2078 (1989), Figure is reproduced with permission.]

2. Carbon monoxide dehydrogenase is a complex enzyme that catalyzes the synthesis of acetyl-coenzyme A from coenzyme A, a methyl group, and carbon monoxide. The enzyme contains a nickel center that is bridged to an iron–sulfur cluster. The identity of the binding site can be investigated by resonance Raman experiments and isotope effects. The stretching frequency

between two atoms connected by a bond can be estimated from Hooke's law, $v = 2\pi fc \sqrt{m_1 m_2/(m_1 + m_2)}$, where v is the frequency of the vibration, c is the speed of light, f is the force constant (a measure of bond strength), and m_1 and m_2 are the masses of the two atoms.

(a) If the $^{12}C–O$ and $M–^{12}C$ stretching frequencies for the metal-bound CO are 1993 cm^{-1} and 360 cm^{-1}, respectively, estimate the corresponding stretching frequencies for the ^{13}C-labeled ^{13}CO. Assume here that the molar mass of $M \sim 57$ (an average of the Fe and Ni masses). [Remember that you must use the mass of a single atom in the above equation. Also, it is not necessary to know the values f or c!].

(b) The resonance Raman bands for the metal–carbon stretch for enzyme-bound CO are shown. Spectra **A** show the influence of enrichment with specific metal isotopes relative to natural abundance (NA) levels (the enzyme is isolated from bacteria grown in a minimal medium containing the isotope of choice), whereas spectra **B** show the effect of using $^{13}C^{16}O$ and $^{12}C^{18}O$, relative to $^{12}C^{16}O$.

Which binding mode do these results support: Fe–CO or Ni–CO? Also, why is there a shift for the $C^{18}O$ sample?

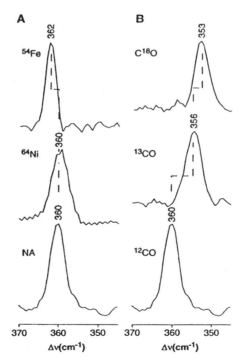

[Qiu, D. et. al., *Science 264,* 817–819 (1994), Figure is reproduced with permission]

3. (a) In a standard 2D COSY experiment, the resonance from a backbone amide NH often shows a cross-peak to other protons as a result of scalar coupling. This cross-peak usually takes the form of **X**. Explain the multiplet structure observed. [Hint: Draw a small section of the backbone].

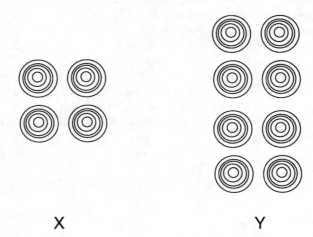

$$\mathbf{X} \qquad\qquad \mathbf{Y}$$

   (b) The NH resonance from a glycine residue has the form shown in **Y**. Provide an explanation.
   (c) What kind of cross-peak would you expect from proline?
   (d) In proteins that contain one or more prolines, the NMR spectrum will sometimes suggest the presence of more than one species. Provide an explanation.

[Chazin, W. J. et al., *Proc. Natl. Acad. Soc. USA 86*, 2195 (1989)]

4. The reaction of [PtCl(dien)]Cl with S-adenosyl-1-homocysteine (SAH) has been followed by ^{1}H and ^{195}Pt NMR in the range $2 < pH < 12$. Three products are formed. No coordination to the adenine ring is observed under these solution conditions. At $pH < 7$, only **1** was formed ($t_{1/2} = 75$ min for 5 mM concentrations). At $pH > 7$, **1** spontaneously isomerizes to **2** ($t_{1/2} = 10$ min). This process can be reversed at $pH < 5$ ($t_{1/2} = 2$ hrs). At $pH > 7$, the following sequence of reactions occurs between 1 equiv of SAH and 2 equiv of [PtCl(dien)]Cl: $SAH \rightarrow 1 \rightarrow 2 \rightarrow 3$. No pathway leading directly from reaction of SAH to **3** could be detected. All three complexes react with sodium diethyl-dithiocarbamate Na(ddtc) forming, eventually, free SAH and [Pt(dien)ddtc]$^+$. Complexes **2** and **3** both consist of a pair of diastereomers. For complex **1**, it could be shown that the interconversion of these isomers was slow on the NMR time scale at 255 K.

   (a) Provide an explanation for the appearance of the chemical shift versus pH profile (upper left), and estimate the observed pK_a's from each resonance. To which functional groups do you attribute these ionizations?
   (b) The chemical shift of the H_α proton of complex **1** is shown as a function of temperature (upper right). Explain the appearance of the data.

(c) ^{125}Pt NMR spectra show the following resonances for **1, 2,** and **3: 1,** δ
− 3358 ppm; **2,** δ − 2935 ppm; **3,** δ − 2938 and − 3362 ppm. Suggest a
reasonable explanation for the trends shown by this data.

(d) Using the ^{1}H NMR data in the summary table and your answers to parts
(a) to (c), suggest structures for complexes **1, 2,** and **3.**

(e) Propose a mechanism for the sequence of reactions leading from SAH to
3 after addition of [PtCl(dien)]Cl.

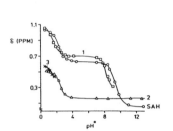
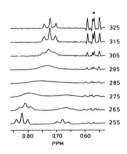

| | | | | chem shift | | | | | |
compd	H8	H2	Hα	Hβ	Hγ	H1′	H2′	H3′	H4′	H5′,5″
free SAH	5.17	5.06	0.62	−1.09	−0.50	2.89	1.69	1.24	1.15	−0.16
1	5.18	5.11	0.68	−0.84		2.96	1.78	1.33	1.41	0.32
2	5.19	5.09	0.16	−1.23	−0.52	2.91	1.71	1.24	1.16	−0.16
3	5.20	5.14		−1.00		2.97	1.78	1.34	1.40	0.30

[Lempers, E. L. M. and J. Reedijk, *Inorg. Chem. 29,* 1880 (1990), Figures and
table are reproduced with permission]

5. Assimilatory nitrate reductase from *Chlorella vulgaris* catalyzes the reduction
of nitrate to nitrite and contains a molybdenum center, a flavin, and a heme
unit.

(a) Use the data in the illustration below to estimate the midpoint potential
for the heme prosthetic group. If E° for the flavin and Mo centers are
− 120 mV and − 70 mV, respectively, comment on the role of each redox
center in the protein.

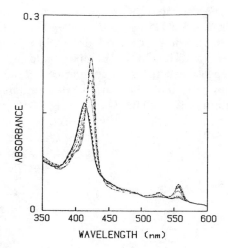

Optical spectra obtained during potentiometric titration of nitrate reductase are shown. Nitrate reductase (2 µM heme) in 50 mM MOPS buffer containing 1 mM EDTA, pH 7.0, in the presence of dye mediators (1.5 µM each mediator), was poised at a series of potentials using MV^+ (20 mM). Spectra were recorded at equilibrium and correspond to the following applied potentials and degree of heme reduction: -125 mV, 11% reduced; -148 mV, 28% reduced; -169 mV, 53% reduced; -190 mV, 75% reduced; -212 mV, 85% reduced; -244 mV, 96% reduced.

(b) Estimate the values of pK_{ox} and pK_{red} from the pH profile below. Suggest which chemical group is being titrated. Explain the shape of the plot.

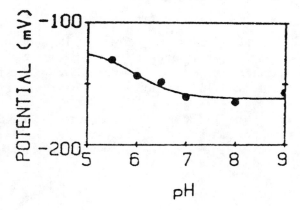

(c) Estimate the enthalpic and entropic components for reduction of the heme center given the following data (temp, $E°$): 12 °C, -142 mV; 18 °C, -148 mV; 25 °C, -152 mV; 32 °C, -163 mV; 38 °C, -171 mV.

[Kay, C. J. et al., *J. Biol. Chem.* **261**, 5799–802 (1986), Figure is reproduced with permission]

6. Show that significant disproportionation occurs if one tries to make a saturated solution of AuCl. $E°(Au^+/Au) = 1.83$ V, $E°(Au^{3+}/Au) = 1.52$ V, $K_{sp}(AuCl) = 2 \times 10^{-13}$.

7. The trace shown was obtained from a stopped-flow experiment to follow the pre-steady-state kinetics of nitrite reduction by a reduced sulfite/nitrite reductase containing a coupled Fe_4S_4–siroheme center. The reaction was monitored at 438 nm.

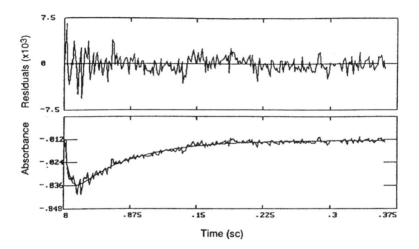

(a) The kinetic trace was fit to a rate law for a consecutive reaction $A \rightarrow Y \rightarrow B$, of the form $Y_t = Y_o [\exp(-k_2 t) - \exp(-k_1 t)] + c$. Rate constants of 14 s^{-1} and 360 s^{-1} were determined. The reaction was carried out under pseudo-first-order conditions with [enzyme] = 30 μM, and [NO$_2^-$] = 100 mM. One reaction step is first-order and the other is second-order. How might you tell which rate constant corresponds to the first- and second-order steps, respectively? Given that 360 s^{-1} is the pseudo-first-order rate constant, calculate the actual second-order rate constant. Be sure to write in the correct units.

(b) Provide an explanation for the absorbance changes observed during the kinetic trace (shown above) that is consistent with the optical spectra for the reduced and oxidized enzyme, and arsenite-bound reduced enzyme (shown over).

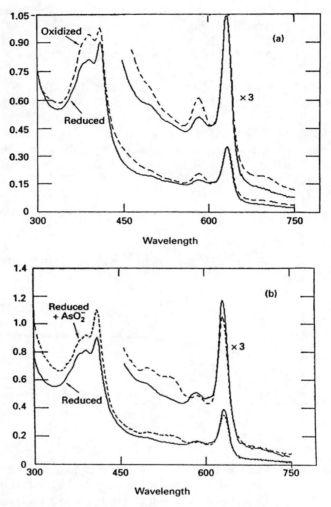

(c) Why is arsenite (AsO_2^-) rather than nitrite (NO_2^-) used to obtain the optical spectrum of a ligand-bound enzyme in part (b)?

(d) Propose a mechanism for the chemical steps that take place after mixing the reduced enzyme with nitrite. Explain how your proposal is consistent with the order of the observed rate constants and the absorbance changes.

(e) At the end of the reaction, the product mixture gives the EPR spectrum shown. This is characteristic of a nitric oxide radical bound to a ferrous heme center. Account for the splitting pattern observed. Also, suggest how this might have been formed.

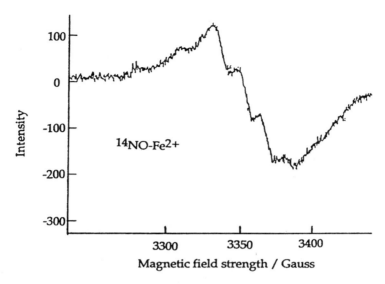

[Lui, S. M. et al., *J. Am. Chem. Soc. 115,* 10483 (1993)]

8. Carbon disulfide is a potent inhibitor of carbon monoxide dehydrogenase (CODH). The Michaelis plots shown in (A) were obtained for CS_2 concentrations of (top to bottom) 0 mM, 0.8 mM, 1.66 mM, 3.3 mM, 6.6 mM, 13.3 mM, and 39 mM. The EPR spectrum of CODH is perturbed by CS_2 binding with the appearance of featrues at g = 2.2, 2.087, 2.017 (g_{av} = 2.15). By monitoring the intensity of these features as a function of CS_2 concentration, indicated in (B), the binding affinity can also be estimated. Evaluate the binding constant of CS_2 for CODH using both the kinetic and spectroscopic data.

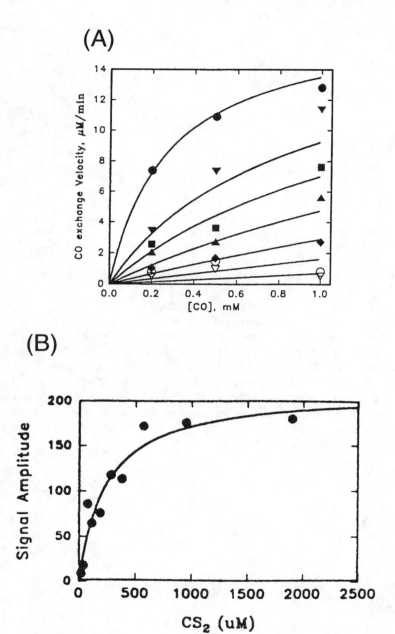

[Kumar, M. et al., *Biochemistry 33*, 9769 (1994), Figures are reproduced with permission]

9. The potential for reduction of hydrogen ion in a solution at ambient temperature and neutral pH is determined to be $E° \sim -0.414$ V versus the NHE. Quantitatively account for this apparent contradiction.
10. The SDS-PAGE gel shows the overexpression of the heavy (H) and light (L) chains of the iron storage protein ferritin. The positions of molecular weight (MW) markers are indicated. Estimate the MW of each of these two overexpressed proteins by use of the equation (described in the text) that relates MW to migration distance. The dye front may be assumed to lie at the bottom of the gel.

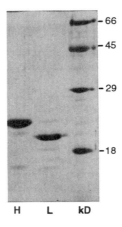

3

Transport and Storage

3.1 Introduction

In Figure 1.1 we introduced the bulk, trace, and ultratrace elements essential for the health of an organism. We define an essential element as one where a deficiency results in either death or severe abnormalities in the development or functioning of an organism. Table 3.1 defines the relevant abundance of these elements in a typical human being.

The *bulk elements* include both main group species (C, H, N, O, P, S) that are used predominantly for construction of the proteins, enzymes, lipids, polysaccharides, and other organic components of a cell, and the mineral ions (Na^+, K^+, Mg^{2+}, Ca^{2+}, Cl^-, and HPO_4^{2-}) that provide an essential balance of background electrolytes. These ions regulate electrical potentials across membranes, mediate the transmission of electrical signals from nerve impulses in the central nervous system, and provide electrostatic stabilization for a large number of proteins, enzymes, and polynucleotides.

Many metal ions are required in only small *or trace* amounts. These ions typically interact with specific biological molecules (often proteins or enzymes) and activate or regulate the function of these molecules by causing a change in molecular structure and/or serving as catalytic centers for chemical reactions. A few of the bulk elements (Mg^{2+}, Ca^{2+}, K^+) can also act in this manner. Some of the chemical roles of these trace or ultratrace elements are noted in Table 3.2.

One should keep in mind that, although these elements are essential for life, an excess will normally result in toxic side effects and, in some instances, cell

Table 3.1 Approximate Elemental Composition of a Typical 70 kg Human

Bulk elements and mineral ions

Oxygen	44 kg	Phosphorus	680 g
Carbon	12.6 kg	Potassium	250 g
Hydrogen	6.6 kg	Chlorine	115 g
Nitrogen	1.8 kg	Sulfur	100 g
Calcium	1.7 kg	Sodium	70 g
		Magnesium	42 g

Trace and ultra trace elements

Iron	5000 mg	Barium	21 mg
Silicon	3000 mg	Molybdenum	14 mg
Zinc	1750 mg	Boron	14 mg
Rubidium	360 mg	Arsenic	~3 mg
Copper	280 mg	Cobalt	~3 mg
Strontium	280 mg	Chromium	~3 mg
Bromine	140 mg	Nickel	~3 mg
Tin	140 mg	Selenium	~2 mg
Manganese	70 mg	Lithium	~2 mg
Iodine	70 mg	Vanadium	~2 mg
Aluminum	35 mg		
Lead	35 mg		

death. In larger concentrations any one of these metal ions may compete for other coordination sites that normally bind a different metal, and so the biological activity of that site may be lost. In Chapter 1 we saw that metal ions can vary widely in their chemistry, and so one metal is not necessarily as good as another.

Table 3.2 Functions of Some Representative Trace Elements

Element	Function
Iron	Electron transport; O_2 transport; catalysis of oxido-reductase reactions
Silicon	Biomineralization of bone and the shells of crustaceans
Zinc	Essential metal cofactor in many hydrolase enzymes and in proteins or enzymes involved in the transcription and translation of DNA and RNA
Copper	Electron transport; O_2 transport; catalysis of oxidation–reduction reactions; synthesis of connective tissue
Manganese	Essential cofactor of a few enzymes in carbohydrate biochemistry; catalase
Iodine	Essential constituent of thyroid hormones
Molybdenum	Catalysis of oxidation–reduction reactions
Cobalt	Vitamin B_{12}
Chromium	Interaction with insulin in carbohydrate metabolism
Nickel	Catalysis of oxidation–reduction chemistry (hydrogenase, F430 cofactor) and hydrolysis (urease)
Selenium	Glutathione peroxidase
Fluorine	Incorporated into the hydroxyphosphate lattice of teeth.

It should now be clear that a balanced distribution of the elements is required for the health of any living cell. To maintain this balance the movement of metal ions *in* and *out* of cells must be strictly regulated and is fundamental to life processes. Transmembrane concentration gradients of ions and neutral species underlie the energetics of cell metabolism, while growth requires a constant influx of essential minerals and nutrients and an efflux of waste and toxic products. Given the subject matter of this text, attention will focus on metal ions and, to some extent, small anionic species. It will become clear, however, that many transport mechanisms for metal ions are inextricably linked to the transport of organic metabolites. Figure 3.1 summarizes four distinct aspects of the problem.

1. Mechanisms must exist for capturing trace quantities of essential mineral ions in the extracellular environment. This may require solubilization of mineral precipitates.
2. These mechanisms must be capable of selective uptake of metal cations.
3. These charged ions must be carried across a hydrophobic membrane.

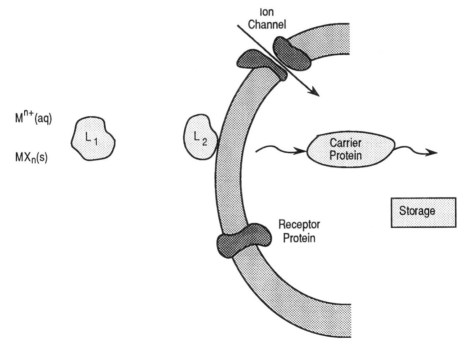

Figure 3.1   Overview of important steps in metal ion uptake by cells. Metal species $M^{n+}(aq)$ may pass into the cell through ion channels. Alternatively, a ligand or protein (L_1 and L_2) must bind and/or solubilize the cation and, then, mediate transfer across the membrane, possibly by a mechanism involving a surface *receptor protein*. Within the cell the metal ion must be transported by a carrier ligand or protein to where it is immediately required or *stored* for later use.

4. Ions must be transported to where they are required within the cell or stored for later use.

Since the essential bulk metal ions Na^+, K^+, Mg^{2+}, and Ca^{2+} are found in relatively high concentrations in the extracellular environment (Table 3.3), the problem of bioaccumulation of these ions is not as acute as that for the essential trace elements, where enrichment factors of 10^5 or greater are common for intracellular relative to extracellular concentrations. Alkali and alkaline earth cations do not bind tightly to common biological ligands, and so the regulation of intracellular levels of these ions is carried out by membrane-bound ion pumps. In contrast, the essential trace transition metals must be taken up by a pathway that involves both extracellular and membrane-bound ligands. The molecular details of uptake, transport, and storage of these metal ions remain for the most part unclear. However, a few examples (iron metabolism in particular) are sufficiently well understood to illustrate the general principles. Figure 3.1 illustrates how cells will either excrete ligands (L_1) that bind tightly to a specific metal ion or the cell surface will possess ligand sites (L_2) that selectively bind trace metal ions. Membrane-bound receptor proteins recognize these metal–ligand complexes and facilitate active transport into the cell. This results in the enrichment of these metals within the cell. Subsequently, distinct proteins or ligands are utilized to transport the metal ion either to a location where it is required for immediate use or to a longer term storage site.

In this chapter we will first consider general mechanisms of transmembrane ion transport mediated by ligands, proteins, or enzymes. Our attention then will focus on mechanisms for iron transport and storage and, to a lesser extent, related mechanisms for copper.

Table 3.3 Ion Concentrations in Sea Water and Extracellular Blood Plasma

Ion	Seawater (mM)	Blood plasma (mM)
Na^+	470	138
Mg^{2+}	50	1
Ca^{2+}	10	3
K^+	10	4
Cl^-	55	100
HPO_4^{2-}	1×10^{-3}	1
SO_4^{2-}	28	1
Fe^{2+}	1×10^{-4}	2×10^{-2}
Zn^{2+}	1×10^{-4}	2×10^{-2}
Cu^{2+}	1×10^{-3}	1.5×10^{-2}
Co^{2+}	3.1×10^{-6}	2×10^{-3}
Ni^{2+}	1×10^{-6}	0

3.2   Metal Ion Uptake and Transmembrane Ion Transport

The question of how a cell "knows" that it is deficient in (or has a surplus of) a particular ion will be considered in Chapters 7 and 8. For the moment, we shall simply assume that a cell is deficient in a specific ion and look at how it goes about selectively accumulating these ions. Nature has evolved a variety of mechanisms to tackle this problem. In the final analysis, these depend on simple considerations of charge, size, and ligand preference of the ion.

The separation of cellular contents from the external environment is defined by the cell membrane (Figure 3.2), which mediates the flow of molecules and ions in and out of the cell. The basic structural unit of the cell membrane is the lipid (Figure 3.3). Although they display considerable structural diversity, all lipids are characteristically amphipathic, that is, they possess a small polar head group and a longer nonpolar tail. Membranes are constructed from two layers of lipids, with the hydrophilic polar head group exposed to solvent and the hydrophobic tail forming the interior. The general problem of transporting charged ions across a membrane relates to the hydrophobicity of these lipid bilayers (Figure 3.3). Gases (O_2, N_2, CO_2, etc.) and ions of low charge density (q^2/r) may move directly through the membrane.[1] Since the charge–radius ratio is smaller for anions, the energy barrier to passage through a hydrophobic environment is lower, and so direct anion transport occurs more readily than for cations. On the other hand, transmembrane movement of metal cations is usually mediated by ligands, proteins, or peptides that bind the cation and wrap it in a hydrophobic shell. The energetics of metal ion translocation and the associated energy-transducing systems will be described in Chapter 6. There are three common mechanisms to effect transmembrane ion transport:

1. Low molecular weight carrier ligands (ionophores) that ferry the ion across the membrane.
2. Membrane-spanning peptides or proteins that define a hydrophilic channel through which the ion can flow.
3. Membrane-bound enzymes that use the energy derived from ATP hydrolysis to pump ions across the membrane.

3.2.1   Ionophores[2]

Ionophores are nonprotein carrier ligands that possess polar functional groups to lend water solubility to the metal-free complex. After binding a metal ion, these functional groups fold inward, leaving a hydrophobic surface that promotes solubility in the lipid membrane. Although ionophores bind cations with a high degree of selectivity, none are currently known for anions. Figure 3.4(a–c) shows some examples of ionophores. A more complete listing with a relative ordering of metal affinities is noted in Table 3.4. Many are macrocycles (closed cyclic molecules) that contain oxygen ligands in the form of carbonyl and, in some cases, ether functionality. In contrast to these cyclic ligands, a family of

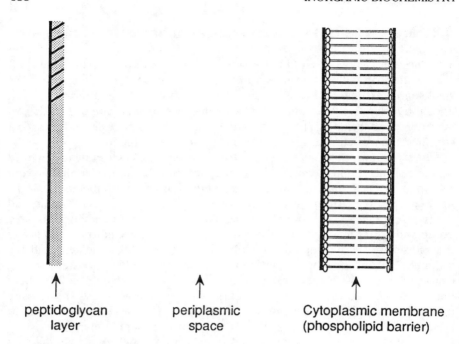

| peptidoglycan | periplasmic | Cytoplasmic membrane |
| layer | space | (phospholipid barrier) |

Figure 3.2 Membrane structure. The inner cytoplasmic membrane contacts the cytosol and is composed of a phospholipid bilayer (see Figure 3.3) in which various transport proteins are embedded. The inner and outer membranes are separated by the periplasmic space. The outer membrane is, in part, formed by a peptidoglycan layer (polysaccharides cross-linked by peptides, illustrated in Figure 6.14) on the inner surface. In gram-positive bacteria, low molecular weight minerals and solutes (mol. wt. < 600) may diffuse directly through this barrier, whereas gram-negative bacteria contain an additional layer of phospholipids that carry proteins (called porins) that allow minerals and solutes to pass freely into the periplasmic space. In all cases the cytoplasmic membrane is the principal barrier to diffusion into the cell.

carboxylate ionophores exists with extended chains of polyether and amine ligands that contain one or two carboxylates. This interesting class of ionophores can be used to couple metal uptake with proton release. Examples include the potassium- and calcium-binding ionophores nigericin and A23187, respectively (see Figure 3.4c). These possess carboxylate ($-CO_2H$) functionality that can exchange H^+ for K^+ or Ca^{2+} to form salts. In Chapter 6 we shall see how such molecules can couple potassium and proton gradients.

Ionophores tend not to show *specificity* (the ability to bind one particular metal ion) but do possess a high degree of *selectivity* (although they can bind to several metal ions, such ligands exhibit a strong preference for one particular ion). Ionophores bind metals predominantly through oxygen ligands, and so selectivity is based primarily on considerations of ion charge and size. The trends noted in Table 3.4 can be understood in terms of a "best-fit" criterion inasmuch as the ligands bind metal ions optimally when the ionic radii best match the "hole-size" in the macrocycle. The antibiotic valinomycin shows high selectivity for K^+ transport. However, this selectivity is not entirely based on "best-fit"

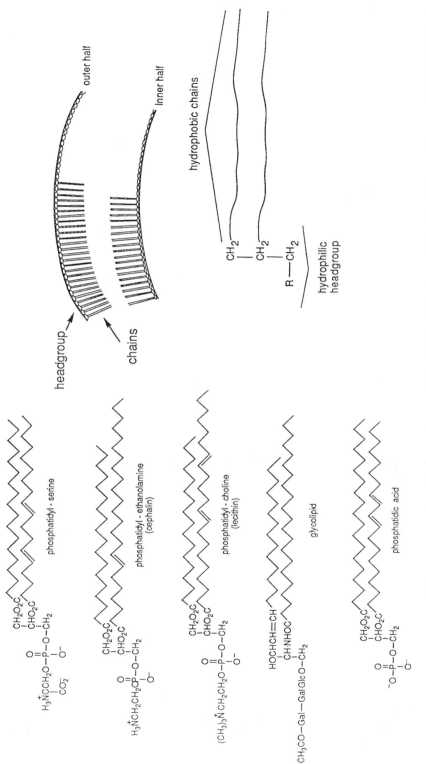

Figure 3.3 Lipids commonly used in the construction of cell membranes. Hydrophobic chains form the interior of the membrane and hydrophilic head groups are solvent exposed. The extended hydrocarbon chains are of variable length and levels of unsaturation, depending on the role and location of the membrane. Extensive unsaturation (more double bonds) produces membranes of greater rigidity. In the case of the glycolipid, the abbreviations Glc = glucose and Gal = galactose have been used. Lipids forming the outer part of a membrane are different from those forming the inner portion. Lipids with smaller head groups are typically found on the inner face of a membrane where the curvature is tighter and may result in steric congestion. Bulky glycolipids form the outer part of the membrane.

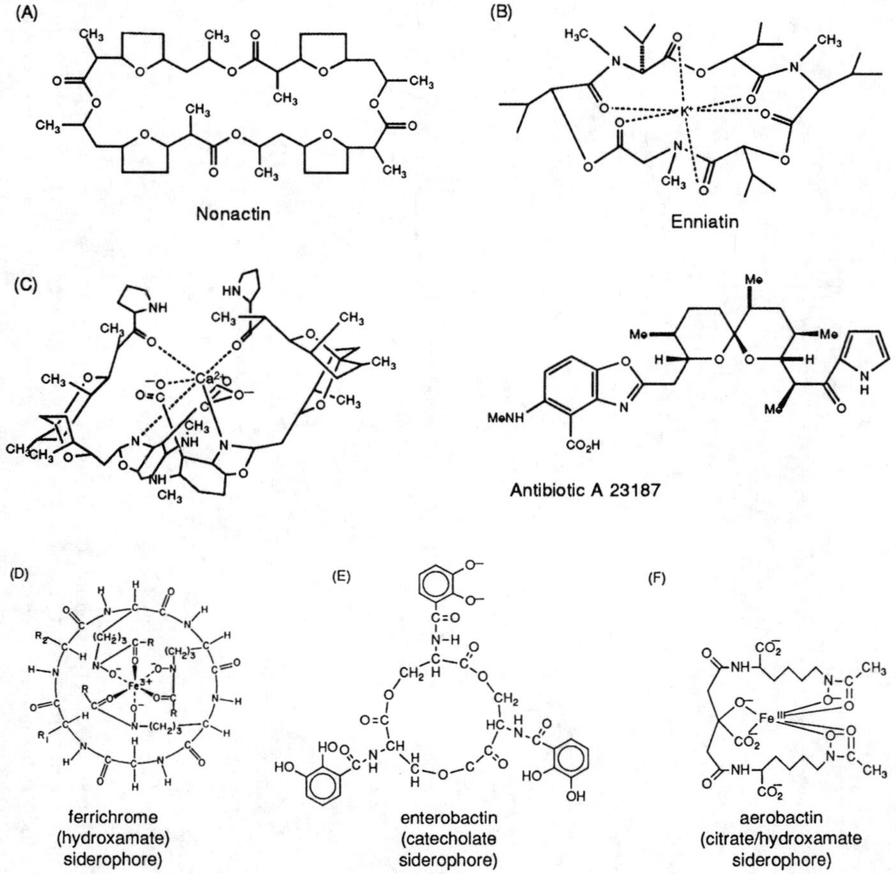

Figure 3.4 Common ionophores. (A) Macrocyclic ethers. (B) Peptides. (C) Carboxylate ionophores. The example shown is a dimeric calcium complex of A23187. The latter group can couple metal ion and H^+ transport. Iron-bearing siderophores are typically derived from (D) hydroxamates, (E) catecholates, or (F) citrates.

criteria, since K^+ and Na^+ can both be accommodated in the central core. Valinomycin also distinguishes these ions on the basis of hydration energies. Both ions must lose their solvation shells before binding to the ligand. The (q^2/r) ratio is higher for the smaller sodium ion (Table 3.5), and so the difference in solvation energy (-71.9 kcal mole^{-1} versus -55.0 kcal mole^{-1} for Na^+ and K^+, respectively) strongly favors the binding of K^+. The energetics of solvent removal is an important aspect of ion-selective uptake (Table 3.5). The ligand atoms must be able to coordinate to the metal center to make up for the loss of hydration energy, and so there is a trade-off between solvent loss and ligand binding. Although most cations lose their entire solvation shell after ligation by the ionophore, a few examples are known where a metal cation binds more

Table 3.4 Metal Selectivity of Ionophores

Ionophore	Specificity	Ionophore	Specificity
Macrocyclic ethers		**Cyclic peptides**	
Nonactin	$K^+ > Rb^+ > Na^+ > Ba^{2+}$	Antamanide	$Na^+ > K^+$
Monactin	$K^+ > Rb^+ > Na^+ > Ba^{2+}$		
Dinactin	$K^+ > Rb^+ > Na^+ > Ba^{2+}$		
		Carboxylates	
		Nigericin	$K^+ > Na^+$
Peptides		Monesin	$Na^+ > K^+$
Valinomycin	$Rb^+ > K^+ > Cs^+ > Ba^{2+}$ $> Na^+$	X-537	$Ba^{2+} > Sr^{2+} > Mg^{2+} > Ca^{2+} >$ $Mn^{2+} > K^+ > Rb^+ > Na^+$
Eniatin B	$K^+ \sim Ba^{2+} > Rb^+ > Na^+$ $> Cs^+$	A23187	$Ca^{2+} > Mg^{2+}$
Beauvericin	$Ca^{2+} >> K^+ \sim Rb^+ Cs^+$ $> Na^+$		

strongly than expected based on comparisons with ions immediately above and below in the periodic table. In these cases, partial solvation can give a larger effective ionic radius.

Ionophores typically bind alkali and alkaline earth metals. The intracellular and extracellular concentrations of these ions are very different, and the maintenance of transmembrane concentration gradients by active transport processes is vital to normal cell metabolism. Briefly, metal ions are pumped against a concentration gradient, and so active transport requires energy (see Figure 3.9 and the discussion in Chapter 6). We have already seen that ionophores relax these concentration gradients, and so most ionophores are active antibiotics (i.e., they result in a breakdown in cell metabolism that leads to cell death). In contrast, a number of specific carrier ligands exist that have been identified as essential for maintaining intracellular levels of iron. These ionophores are given the special name *siderophores* ("iron carriers") and represent one of the most thoroughly investigated ion uptake/transport systems.

Table 3.5 Physicochemical Data for Hydrated Metal Ions

Ion	Li^+	Na^+	K^+	Cs^+	Mg^{2+}	Ca^{2+}	Ba^{2+}
Radius (Å)	0.73	1.16	1.52	1.88	0.86	1.14	1.56
Charge-density $(q^2/r)^a$	1.37	0.86	0.66	0.53	4.65	3.51	2.56
ΔH°_{hyd} (kJ mole^{-1})	-515	-405	-321	-263	-1922	-1592	-1304
Transport number^b	13–22	7–13	4–6	4	12–14	8–12	3–5

^{a} The charge q is taken as the formal valence number (1 or 2).
^{b} Transport numbers estimate the average number of solvent molecules that migrate through solution in close association with an ion. This includes contributions from primary (inner) and secondary (outer) solvation shells and gives a measure of the electrostatic ordering around each metal ion. The large range of values for each metal represents the degree of uncertainty in this type of measurement. However, the general trends are valid.

3.2.2 Siderophores (or Siderochromes)

The name *siderophore* derives from the Greek for "iron carrier." The three main classes of siderophore, which are based on catechol and hydroxamate ligands, are illustrated in Figure 3.4(D–F). Each has a high affinity for ferric ion (Table 3.6). Certain cells and higher plants may, instead, employ citric acid or polyamino acids, respectively, as siderophores. The design of these ligands incorporates features that accommodate the two main functions of this class of molecule. First, they must solubilize iron (the solubility product of ferric hydroxide is 10^{-39} M^4). Second, they must assist in the transport of highly charged ions across a hydrophobic membrane. To effect the solubilization of iron salts, siderophores bind iron tightly. The formation constants for ferric complexes of tris-hydroxamate-type siderophores are approximately 10^{30}, whereas those for catechol-type siderophores are in the range of 10^{45} to 10^{52}. The larger value for the latter reflects the increased negative charge (3- versus 6-) on the ligand. Although there is an obvious need for tight binding of iron ion, at some point the metal must be released to other ligands or proteins within the cell. Clearly, microorganisms must possess special mechanisms for iron release. At one time it was thought that an enzyme (esterase) hydrolyzed the ester bonds in the siderophore backbone, resulting in destruction of the carrier ligand and release of iron. However, structural analogs lacking these ester bonds were fully functional in promoting iron uptake in microorganisms. Current evidence points to a release mechanism involving reduction to Fe^{2+}, which binds weakly ($K \sim 10^8$ M^{-1}) and is readily displaced from the siderophore by protonation. The redox potentials for the hydroxamate ligands noted in Table 3.6 are certainly within the range of biological reductants; however, the potentials for the catecholate complexes are pH dependent and are likely to be moderated by binding to membrane receptors or transport proteins. In this case partial hydrolysis of the ligand backbone is also likely, since this appears to lower $E°$ and bring it within range of biological reductants.

Relative to the other ionophores, these iron-bearing siderophore complexes show greater hydrophilic character, and so membrane translocation must be

Table 3.6 Thermodynamic and Redox Parameters for Hydroxamate Siderophores

Siderophore	log $β^a$	$E°$(mV vs. NHE)
2,3-Dihydroxybenzoate[b]	20	−350
Enterobactin	52	−750
Aerobactin	22.5	−336
Ferrichrome A	32.0	−448
Ferrioxamine B	30.5	−468

[a] The stability constant $β$ refers to the equilibrium where $β = [(Fe^{3+}).L]/[Fe^{3+}][L]$. Note that $E°$ for the biological reductants NADH and FADH$_2$ are −320 mV and −219 mV, respectively.

[b] Three of these ligands bind to form an octahedral ligand set. The log $β$ value listed represents binding of only one ligand (log K_1). Comparison with enterobactin reflects the effect of linking together three such ligand sets.

assisted by a membrane protein. The expression of specific outer membrane receptors and transport proteins to trap and channel solutes (such as vitamin B_{12}, ferric complexes of citrate, enterobactin, aerobactin, and nucleosides) with molecular masses of more than 600 Da from the cell exterior to the cytoplasm is very common. Many of these gene products and their functions have been characterized for *E. coli* (Figure 3.5).

3.2.3 Channels

Ions need not be carried across a membrane by ligands. Rather, proteins or peptides may form *channels* through which the metal ion may pass. Typically, such channels allow about 10^7–10^8 ions per second to pass for a transmembrane potential of approximately 70–100 mV.[3] Figure 3.6 shows gramicidin, a 15-residue helical peptide, which dimerizes to form a membrane-spanning channel that shows high selectivity for K^+. The hole through which the ions pass is called a *pore*. The size of the pore and the ligand atoms that form the inner core form a basis for the selective transport of metal ions, depending on ionic radii and ligand preference. Certain toxins (Figure 3.7) may act by blocking these channels through binding to a component of the channel. Channel selectivity for metal ions derives principally from considerations of ionic radii and the size of the hydrated ion. For example, the antibiotic gramicidin shows a selectivity $Li^+ < Na^+ << K^+ > Rb^+ > Cs^+$. Transport is optimal for K^+. Smaller ions have larger hydration spheres, and so the ions are too big to pass through the channel.

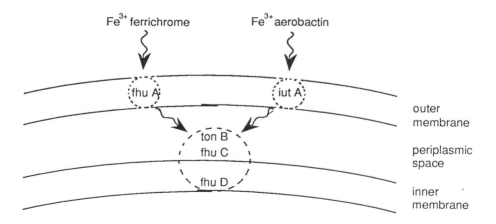

Figure 3.5 Iron transport in *E. coli*. Receptor and transport proteins are often labeled according to the genes that encode them (iut = iron uptake transport, fhu = ferric hydroxamate uptake, ton = transport of iron). Receptors in the outer membrane bind the siderophore complexes and facilitate movement into the cell. Thereafter, a variety of transport proteins (B, C, D) mediate diffusion across the cytoplasmic membrane. The detailed working of these systems is not understood.

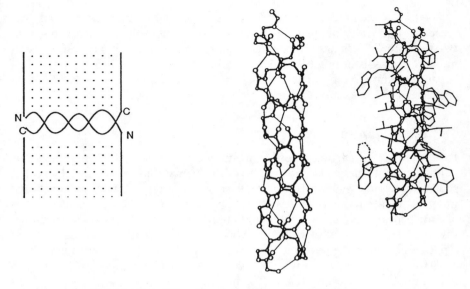

formyl - Val - Gly - Ala - <u>Leu</u> - Ala - <u>Val</u> - Val - <u>Val</u> -
Trp - <u>Leu</u> - Trp - <u>Leu</u> - Trp - <u>Leu</u> - Trp - NHCH₂CH₂OH

gramicidin
peptide sequence

Figure 3.6   The antibiotic gramicidin A is a pentadecapeptide with a C-terminal ethanolamine. It forms antiparallel helical dimers that are approximately 30 Å long with an
internal pore diameter of 5 ± 1 Å. Because the thickness of a typical phospholipid
membrane is approximately 50–60 Å, two of these dimers usually come together in a
head-to-tail fashion to form a functional channel or pore that spans the membrane. A
schematic illustration is shown on the left, and crystallographic models showing the
backbone structure and orientations of side-chains are to the right. The peptide is rich
in nonpolar functional groups that are directed away from the central pore and help to
form a stable complex with the hydrophobic lipid membrane. The residues underlined
in the sequence are uncommon D-amino acids. Positively charged ions migrate through
the interior of the channel by transient coordination to carbonyl O ligand atoms.
[Adapted from Langs, D. A. *Science 241*, 188–191 (1988)].

Since there are minimal binding contacts as the ion passes through the channel,
there is no energetically feasible pathway to remove the solvent molecules.
Larger ions are precluded on the basis of size.

Channel-forming proteins are made up from a number of helical segments.
These helical bundles span the membrane and form the channel (Figure 3.8).
As for ionophores, many of these compounds are antibiotics, since they relax
transmembrane ion gradients, resulting in a breakdown of cell metabolism.
There are, however, a number of channel-forming proteins that are an inherent
part of the cell membrane and that are tightly regulated so that they can be
opened or closed when required. *Voltage-gated* channels are controlled by a
transmembrane electrical potential ψ that inhibits the flow of charged ions from
one side of a membrane to another. When the transmembrane potential changes,

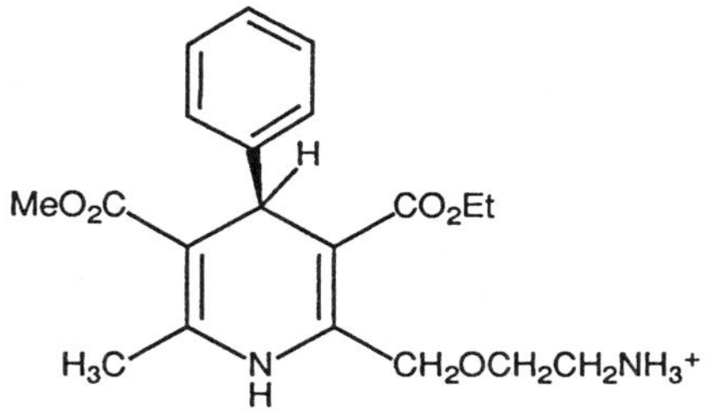

Nifedipine

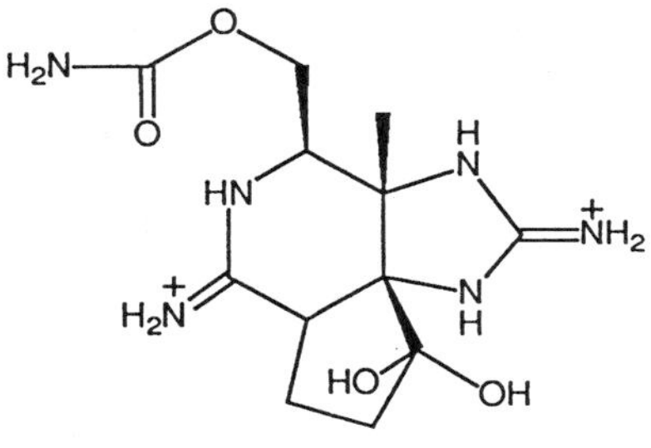

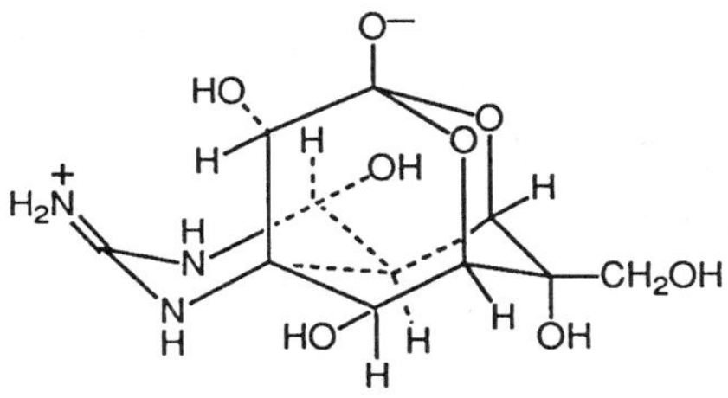

Saxitoxin   Tetrodotoxin

Figure 3.7 Examples of channel-blocking neurotoxins.

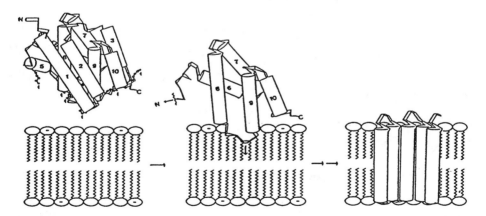

Figure 3.8 Colicin A is an antibiotic protein that forms voltage-dependent channels in membranes. The protein functions by destroying a cell's transmembrane potential. Several distinct structural domains are responsible for the toxicity of the protein. A proteolytic fragment is shown that contains 10 α-helices of the pore-forming domain. Insertion into the membrane produces the voltage-gated channel. [Reprinted by permission from M. W. Parker et al., *Nature, 337,* 93; copyright (1989) Macmillan Magazines Limited.]

ions flow across the membrane to compensate. The energetics of membrane potentials is considered in more detail in Chapter 6. Alternatively, channels may be opened or closed by chemical mechanisms. Normally, an effector molecule will bind to a membrane-bound receptor complex associated with the channel. Chemists may find it easier to think of the effector as a ligand and the receptor as a binding site. The energy from effector binding results in conformational changes that open the channel. The control, or "gating," of ion channels is particularly important in neurochemistry. Electrical signals are transmitted between nerve cells by mechanisms involving ion channels (see Chapter 6).

Both ionophores and channels typically function by what is referred to as a *passive transport* mechanism. This means that ion flow is regulated according to the direction of the concentration gradient of that ion. The alternative, to force an ion against its concentration gradient, is termed *active transport*. Clearly, this requires energy, and so active transport is either coupled to a chemical reaction *or* the transport of another species by an energetically favorable passive route. This provides two additional mechanisms for transporting ions across membranes.

3.2.4   Ion Pumps

One of the best-characterized, although by no means well-understood, examples of a membrane-spanning transport protein is the Na^+/K^+-ATPase (Figure 3.9). Three Na^+ are transported out of the cell for every two K^+ pumped in. In each case, the flow is against the normal concentration gradient for that ion, and energy is provided by the hydrolysis of ATP. Table 3.7 notes a variety of ion pumps and their locations. The direction of ion flow reflects the distribution of the bulk alkali and alkaline earth metals as intracellular or extracellular species and sites of internal metal storage (e.g., Ca^{2+} in the sarcoplasmic reticulum).

Ion transport systems have also been identified for essential trace elements (e.g., Co^{2+}, Mn^{2+}, Ni^{2+}, $MoO_4{}_-$). These may also be used as secondary transport systems for the bulk metal ions (e.g., Mn^{2+} transport proteins can also accommodate Mg^{2+}). Note that molybdenum is transported as the molybdate ion. The requirement for molybdenum is particularly important in nitrogen-reducing bacteria (e.g., *Azotobacter vinelandii* and *Clostridium pasteurianum*), where it is required for biosynthesis of the Fe–Mo cofactor in the enzyme nitrogenase.[4]

Summary of Sections 3.1–3.2

1. Minerals and nutrients must be taken up by cells for growth and development.
2. Ion and metabolite exchange are inextricably linked. In all cases these species must pass through membrane barriers composed of lipids, proteins, and polysaccharides.
3. Three common mechanisms for transmembrane ion transport include (a) specific carrier ligands called ionophores and siderophores that facilitate transfer across the hydrophobic barrier; (b) proteins that form ion channels

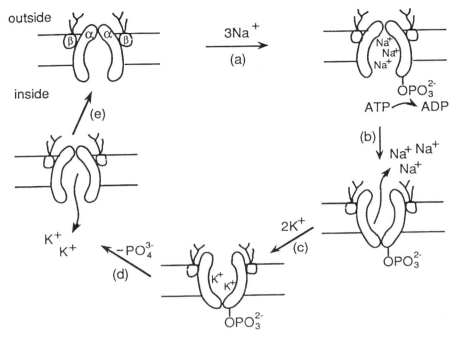

Figure 3.9 Summary of Na^+/K^+–ATPase activity. The enzyme contains two types of subunits (α, β) with the β-subunit bearing oligosaccharide chains. (a) Three intracellular Na^+ ions bind tightly to the enzyme. (b) This promotes phosphorylation of an Asp residue, which mediates a conformational change that weakens Na^+ binding and directs the flow of Na^+ out of the cell. (c) In this conformational state, two K^+ bind tightly. (d) Potassium binding results in dephosphorylation and a return to the original conformation. (e) In this conformational state, K^+ binds weakly, and the ions are released. Overall, there is a net transport of positive charge out of the cell, which helps to establish a negative intracellular potential.

in the cell membrane; (c) membrane-bound enzyme complexes that actively pump ions against concentration gradients by coupling ion movement to ATP hydrolysis.

4. Ion flow in the direction of a concentration gradient is called *passive* transport. Movement against a concentration gradient is called *active* transport.

3.3   Transport and Storage of Metal Ions in Vivo

In multicellular organisms, it may be necessary to transport minerals and nutrients from a point of ingestion to a location where the metal ion will be used or stored. In mammals, for example, minerals and nutrients derived from food and fluids are eventually absorbed into the bloodstream. The alkali (and to some extent the alkaline earth metals) are soluble under biological pH and buffer

Table 3.7 Locations and Functions of Ion Pumps

Transport Protein	Examples of Locations	Function
Na$^+$/K$^+$-ATPase	Mammalian cell	ATP hydrolysis drives Na$^+$ out of and K$^+$ into the cell; both ions are pushed against concentration gradients (see Figure 3.9)
Na$^+$/H$^+$ antiport	*E. coli*	Uses respiratory H$^+$ gradients to generate the transmembrane Na$^+$ gradients that drive Na$^+$-coupled transport of solutes
Na$^+$/glucose symport	Kidney, epithelial cell (blood vessel)	Passive transport of Na$^+$ actively cotransports glucose against unfavorable concentration gradients (see Figure 6.5)
Ca^{2+}-ATPase	Sarcoplasmic reticulum in muscle cells	ATP hydrolysis drives Ca^{2+} into the sarcoplasmic reticulum against unfavorable concentration gradients

conditions and do not interact strongly or irreversibly with ligating groups on cell walls and membranes. These ions can be transported as free ions in the bloodstream (serum) and do not require specific carrier proteins or ligands. This is not the case for the trace transition metals. For example, at physiological O$_2$ concentrations, Fe^{2+} is readily oxidized to Fe^{3+}, which is highly insoluble in aqueous solution at neutral pH ($K_{sp} \sim 10^{-39}$ M^4). The solubility product (K_{sp}) gives a measure of the stability of the salt of a cation and anion. In the case of Fe^{3+} and HO$^-$,

$$Fe^{3+}(aq) + 3HO^-(aq) \rightleftharpoons Fe(OH)_3(s)$$

and $K_{sp} = [Fe^{3+}][OH^-]^3 = 10^{-39}$ M^4. To overcome these serious solubility problems, many trace transition metals require specific carrier proteins to bind and transport the ion in vivo. These prevent precipitation, adventitious complexation to intra- and extracellular proteins and metabolites and to the numerous ligands that line cell walls. Table 3.8 summarizes a selection of carrier proteins

Table 3.8 Protein Carriers in Blood Plasma

Carrier	Ion	Extracellular [M^{n+}]
Transferrin	Fe^{3+}	10^{-16} M
Ceruloplasmin	Cu^{2+}	10^{-14} M
Albumin	Cu^{2+}	10^{-14} M
	Zn^{2+}	10^{-8} M
Phosphoproteins	Ca^{2+}	10^{-3} M
None required	Mg^{2+}, Na$^+$, K$^+$	10^{-3}–10^{-1} M

for trace transition metals commonly found in mammals, although analogous proteins may be found in other eukaryotic and occasionally prokaryotic species. Since iron transport has been most thoroughly investigated, this will again form the initial focus of discussion.

3.3.1 Iron Transport

Transferrin

After entering the mammalian bloodstream by passage through intestinal membranes, free iron is captured by a protein called transferrin (Tf). Similar proteins are also found in milk (lactoferrin) and egg white (ovotransferrin); however, in these cases, transferrin most likely serves a bacteriostatic role rather than functioning as an iron donor to body tissues. Transferrin isolated from blood plasma is occasionally referred to as serotransferrin, whereas ovotransferrin is also called conalbumin. Transferrins are moderate-sized glycoproteins ($M_r \sim 80{,}000$) that bind two ferric ions in similar but structurally distinct sites (termed the N-terminal and C-terminal domains). The structures of human lactoferrin and rabbit serotransferrin have been established by crystallographic analysis (Figure 3.10).

The coordination environment at each iron site is similar and is shown for the N-terminal domain of rabbit transferrin in Figure 3.11. Carbonate ion is

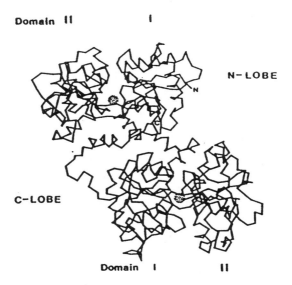

Figure 3.10   Structure of rabbit serotransferrin showing α-helices and β-sheet. The extensive attachment of carbohydrate molecules to the protein surface most likely facilitates interactions with cell membranes. [From R. Serra et al., *Acta Cryst., B46*, 763–771, (1990).]

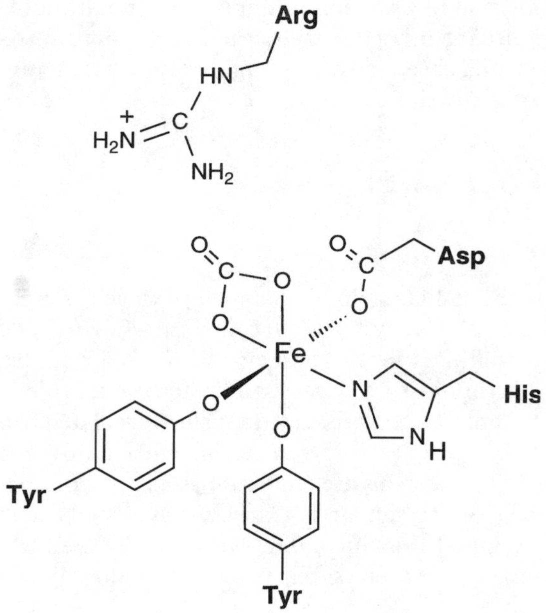

Figure 3.11 The iron at the *N*-terminal iron-binding domain is bound by histidine, two tyrosines, and an aspartic acid. The synergistic anion CO_3^{2-} binds to the ferric center, interacts directly with the neighboring Arg residue, and mediates iron binding by an allosteric mechanism.

required for strong Fe^{3+} coordination. Neither CO_3^{2-} nor Fe^{3+} bind strongly to transferrin on their own, however, together the binding of each is dramatically enhanced. This kind of cooperative behavior is termed *synergism*.[5] The net re-action for iron uptake by transferrin involves the binding of CO_3^{2-} and the release of three protons, which presumably derive from the acidic Tyr and Asp residues.

$$Fe^{3+} + Tf + CO_3^{2-} \rightarrow (Fe^{3+})Tf\,(CO_3^{2-}) + 3H^+$$

Remember, however, that the stepwise mechanism for iron binding involves capture of the more labile ferrous ion $[k_{ex}(Fe^{2+}) \sim 10^7\ s^{-1};\ k_{ex}(Fe^{3+}) \sim 10^3\ s^{-1}]$ followed by oxidation.

$$Fe^{2+} + Tf + CO_3^{2-} \rightleftharpoons (Fe^{2+})Tf(CO_3^{2-}) + xH^+$$

$$(Fe^{2+})Tf\,(CO_3^{2-}) + O_2 \rightarrow (Fe^{3+})Tf\,(CO_3^{2-} + O_2^-)$$

Optical spectroscopy can be a convenient method for monitoring metal bind-ing if there is an accompanying change in absorbance. The red color of the Fe^{3+} protein arises from a tyrosinate $\rightarrow Fe^{3+}$ charge-transfer transition (see Section

2.2.1). There is also an increased absorbance in the ultraviolet region ca. 245 nm and 295 nm from the deprotonated tyrosine. (Proton loss is a necessary prelude for coordination to the metal ion.) By monitoring these absorbance changes in a spectrophotometric titration, the number of ferric ions bound per transferrin molecule may be rapidly established. After all the sites have been occupied, the absorbance no longer changes (Figure 3.12).

Crystallographic studies have generally supported the ligand assignments based on many years of spectroscopic and chemical modification studies.[6] For example, the electronic spectrum of native transferrin in Figure 3.12 shows a strong absorption in the UV at 245 nm that arises from a deprotonated tyrosine. At pH 6 the tyrosine is protonated, whereas even at pH 11 in the absence of ferric ion, the phenolate anion is only partly formed. The presence of ferric ion pushes the equilibrium to the deprotonated form. The optical spectrum also contains a characteristic band (not shown in Figure 3.12) with a maximum at 465 nm and a shoulder at 330 nm (confirmed by CD) that are similar to phenolate–Fe^{3+} charge transfer bands (π to t_{2g}, π to e_g) observed in model phenolate complexes. The energy difference of 11,000 cm^{-1} is also similar to Δ_o for $Fe(H_2O)_6^{3+}$. In electron paramagnetic resonance spectroscopy, coupling of the electron spin (S) to a nuclear spin (I) will result in the splitting of a resonance

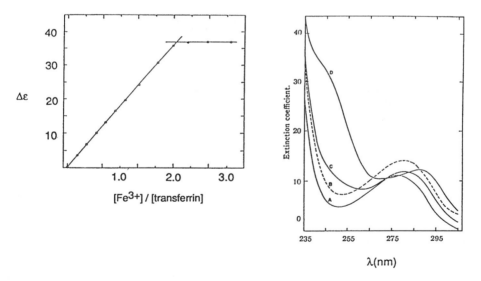

Figure 3.12 (Left) Optical titration of conalbumin (transferrin) with Fe^{3+}. The change in absorbance at 245 nm arises from deprotonation of tyrosine. [Adapted from R. C. Warner and I. Weber, *J. Am. Chem. Soc., 75*, 5094–5101 (1953).] (Right) Absorption spectrum of native transferrin showing the tyrosinate–Fe^{3+} charge transfer band that develops at 245 nm. The spectra were taken under the following conditions: (A) transferrin, pH 6; (B) iron transferrin, pH 6; (C) transferrin, pH 11; (D) iron transferrin, pH 11. At pH 6 the charge-transfer band is lost since the tyrosinate anion is protonated and is released from the ferric ion.

into $2I + 1$ components, and so coordination of histidine is supported by the EPR spectrum of the Cu^{2+} derivative, which gave evidence of coupling to a nitrogen atom through triplet splitting of some of the equatorial components by the ^{14}N ($I = 1$) ligand atom. Metal–carboxylate interactions are generally spectroscopically silent, while octahedral coordination would be expected for an oxygen-rich ligand environment (see Tables 1.3 and 1.4).

Questions relating to the kinetics, thermodynamics, and mechanism of iron binding and release are fundamental to understanding the biological activity of transferrin. The existence of two distinct binding sites leads to complications, however. For example, in the absence of a technique to discriminate iron binding to the N- and C-terminal domains, only the overall thermodynamic binding constants for the following reactions can be evaluated:

$$Fe^{3+} + Tf \rightleftharpoons (Fe^{3+})Tf \qquad (K_1)$$

$$2Fe^{3+} + Tf \rightleftharpoons (Fe^{3+})_2Tf \qquad (K_2)$$

These macroscopic binding constants provide no indication of the distribution of iron between the two sites in the monoferric protein. Figure 3.13 shows how denaturing gel electrophoresis experiments, similar to those described in Section 2.7, give discrete bands for $[(Fe^{3+})Tf]_C$, $[(Fe^{3+})Tf]_N$, and $(Fe^{3+})_2Tf$,[7] and so the individual site (or microscopic) constants for the scheme shown above can be evaluated.[8] The macroscopic (K_1, K_2) and microscopic (k_{1C}, k_{1N}, k_{2C}, k_{2N}) constants are listed in Table 3.9. Perhaps of more significance is the observation that the N-terminal site exhibits greater sensitivity to acid dissociation and does not retain iron below pH 5.7, whereas the C-terminal domain is stable to pH values below 4.8 in the absence of chelating agents. At low pH, there is a conformational transition that favors release of iron from the N-terminal site. The relevance of these results will become clear later in the discussion of intracellular iron release.

In mammals, transferrin is a carrier protein found predominantly in blood serum. However, iron is utilized and stored intracellularly. The mechanism for iron translocation into the cell is summarized in Figure 3.14. First, transferrin binds to a membrane-bound receptor. After binding, part of the cell breaks off to form an internal vesicle in a process known as *endocytosis*. A membrane-bound enzyme (H^+-ATPase) pumps protons into the vesicle and the pH drops to ~5.5. Intracellular versus extracellular pH values can be determined by a method that uses the pH-dependent fluorescence of organic dyes [e.g., by tagging transferrin with fluorescein and comparing the observed fluorescence against a calibration plot (Figure 3.15)]. At 0 °C the $Tf \cdot (Fe^{3+})_2$ complex binds to the surface receptor, endocytosis does not occur, and so the dye emission is determined by the extracellular pH. At 37 °C the complex is taken into the cell, and the lower intracellular pH is reflected in the emission profile. It is likely that this decrease in pH assists in iron release from transferrin, with the free ion then being transported to the cytosol through ion channels. This iron can, then, be used in heme formation, stored in ferritin, or bound by other enzymes. The

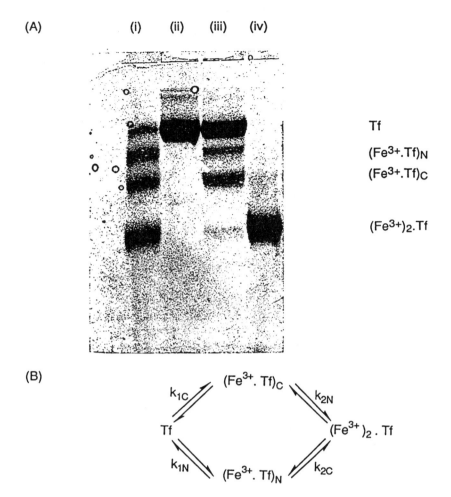

Figure 3.13 (A) SDS-PAGE gels showing discrete bands corresponding to the metal-free protein, Fe^{3+} in the N-terminal domain, Fe^{3+} in the C-terminal domain, and Fe^{3+} in both domains. Columns (i)–(iv) correspond to 60%, 0%, 30%, and 100% iron saturation, respectively. [Adapted from D. G. Makey and U. S. Seal, *Biochem. Biophys. Acta, 453,* 250–276 (1976).] (B) A flow diagram defining the individual microscopic binding constants that can be determined from the relative populations of the N- and C-terminal domains.

vesicle now fuses with the cell membrane and apotransferrin is released from the receptor.

Summary of Section 3.3.1

1. Alkali and alkaline earth metals can move through biological solutions as the free ions, since they interact weakly with other ligands.

Table 3.9 Macroscopic and Microscopic Stability Constants for the Binding of
Iron to Transferrin $(M^{-1})^{a,b}$

pH^c	K_1'	k_{1C}'	k_{1N}'	K_2'	k_{2C}'	k_{2N}'
6.7	3.0×10^{19}	2.9×10^{19}	$<1.4 \times 10^{18}$	2.3×10^{17}	4.8×10^{18}	$<2.4 \times 10^{18}$
7.4	4.7×10^{20}	4.0×10^{20}	6.8×10^{19}	2.4×10^{19}	1.6×10^{20}	2.8×10^{19}

[a] Adapted with permission from D. C. Harris and P. Aisen, in *Iron Carriers and Iron Proteins* (ed. T. M. Loehr),
1989, p. 285.
[b] These are *effective* stability constants determined at constant pH and CO_2 concentration ($K_x' = K_x[HCO_3^-]/$
$[H^+]^3$).
[c] The sharp drop in binding affinity (K_1' and K_2') accompanying a relatively small decrease in pH illustrates how
iron release from transferrin in vivo may be mediated by effecting a local decrease in pH of intracellular vesicles,
as described in the text.

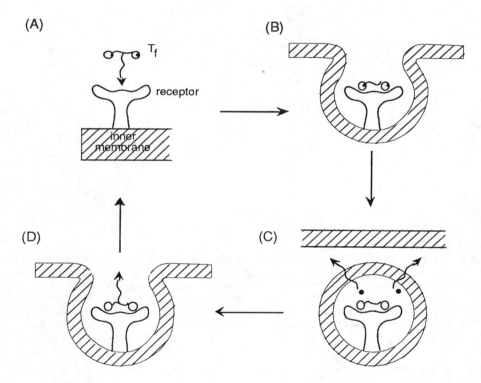

Figure 3.14 Summary of the transferrin-mediated pathway for iron ($\bullet$) translocation
into a cell. (A) Tf binds to the surface receptor. (B) Endocytosis. The receptor–transferrin
complex is carried into the cell where the membrane folds and a vesicle breaks off the
surface. (C) The pH inside the vesicle (or endosome) is lowered by an ATP-driven H^+
pump, accompanied by iron release. (D) Fusion with the cell membrane and release of
apo-Tf.

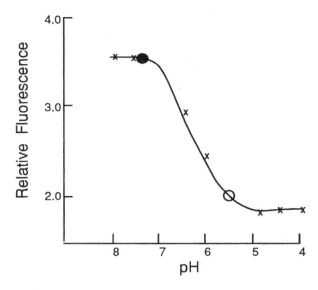

Figure 3.15 Plot of fluorescein fluorescence as a function of pH. The point marked (○) corresponds to the intracellular fluorescein–transferrin complex, and (●) is the same complex bound to a receptor protein on the cell surface. Note the decrease in emission intensity induced by the lower intracellular pH.

2. Transition metals require specific carrier proteins to prevent interactions with other biological ligands and avoid precipitation of insoluble salts. Transferrin is an iron transport protein that binds Fe^{3+} synergistically with CO_3^{2-}. The more labile Fe^{2+} ion is taken up prior to oxidation.
3. Microscopic binding constants for the two iron-binding sites can be determined from denaturing gel electrophoresis.
4. Iron is carried into cells as a complex with the protein transferrin. $Tf \cdot (Fe^{3+})_2$ binds to a receptor protein on the cell surface and is incorporated inside the cell by a process called endocytosis. Release of iron is coupled to a local decrease in intracellular pH.
5. Other transport proteins include ceruloplasmin (Cu^{2+}), albumin (Cu^{2+}, Zn^{2+}), and a number of phosphoproteins (Ca^{2+}).

3.3.2 Metal Storage

Ferritin

In mammals, iron is absorbed and stored in cellular compartments in the liver, spleen, and bone marrow. An average person contains 3–4 g of iron, although most of this is held in storage and only 35 mg (approximately) is actively used in metabolism by enzymes, transport proteins, O_2-binding proteins, and redox proteins. Ferritin is the major repository for iron storage in eukaryotes. This avoids the buildup of large quantities of free iron with accompanying solubility problems and provides a reserve of iron in a stable form that is readily accessible. Horse spleen ferritin is composed of 24 low molecular weight subunits [light (L) $M_r \sim$ 18,500, and heavy (H) $M_r \sim$ 21,000] and may hold up to 4500 ferric centers (Figure 3.16) in a central core that is approximately 60–80 Å in diameter.[9] The structures of horse spleen ferritin and apoferritin have been determined by X-ray crystallography. Figure 3.16 shows a schematic drawing of the ferritin molecule and a diagram of an individual subunit showing the bundle of helices from which it derives its tertiary structure. These subunits form a shell around the hollow core that accommodates the iron ions. Access to this core is by way of a set of funnel-shaped channels that possess a threefold axis of symmetry and are lined with hydrophilic residues (Asp, Glu, Ser) (Figure 3.17a). Another series of channels possesses a fourfold axis of symmetry and are lined with 12 hydrophobic leucine residues. These hydrophobic channels may provide access for organic reductants (NADH or FADH) or chelating agents that are required for release of iron. In common with other iron-binding ligands, ferritin takes up iron as the divalent metal. There are two reasons for this. First, Fe^{3+} is very insoluble at intracellular pHs. Second, Fe^{2+} is kinetically labile relative to Fe^{3+} [$k_{ex}(Fe^{2+}) \sim 10^7$ s^{-1}; $k_{ex}(Fe^{3+}) \sim 10^3$ s^{-1}], which is essential for rapid complexation. Iron, therefore, enters the ferritin core as Fe^{2+}, is subsequently

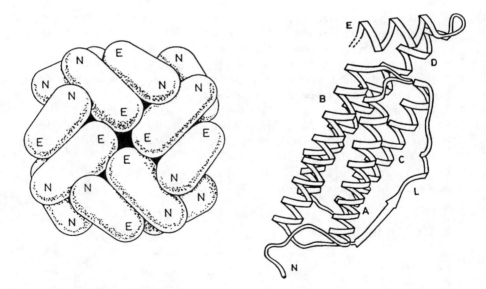

Figure 3.16 Schematic representation of ferritin and a representative subunit showing the bundle of helices that form the major portion of the backbone structure. [From Harrison and Lilley, in *Iron Carriers and Iron Proteins* (T. Loehr, ed.), VCH, 1989, pp. 184–185.]

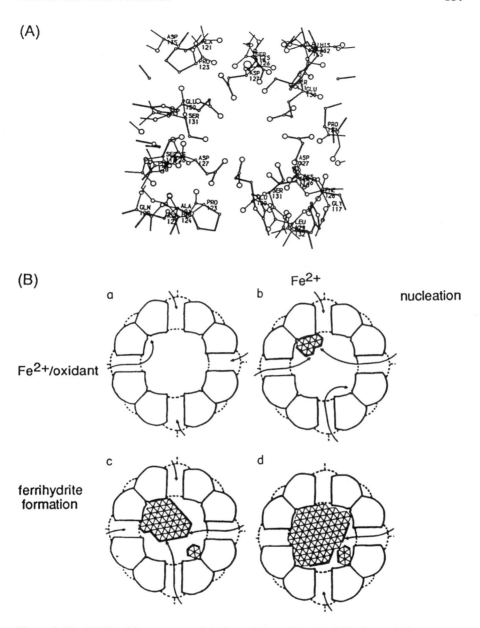

Figure 3.17 (A) Ferritin possesses eight funnel-shaped hydrophilic channels that possess threefold axes of symmetry. These are likely entry ports for ferrous ion. [From Ford et al., *Phil. Trans. Roy. Soc. Lond., B304,* 551 (1984).] (B) Fe^{2+} enters the apoferritin pocket and binds to residues (possibly carboxylate) that favor oxidation to Fe^{3+} and nucleation of the ferrihydrite core. After nucleation, the iron core grows rapidly to form a microcrystalline ferrihydrite phosphate lattice. Additional Fe^{2+} enters the pocket and is oxidized at the mineral surface. [Adapted from Harrison and Lilley, in *Iron Carriers and Iron Proteins* (T. Loehr, ed.), VCH, 1989, pp. 184–85.]

oxidized to Fe^{3+}, and is finally taken up by the growing crystalline lattice within the protein core (Figure 3.17b). Oxidation to the inert ferric ion also stabilizes iron for long-term storage or transport. The ferritin core is composed of a microcrystalline array of ferrihydrite phosphate $[(Fe(O)OH)_8(FeOPO_3H_2) \cdot xH_2PO_4]$ that is formed by oxidation and hydrolysis of Fe^{2+}, and uptake of orthophosphate $(H_2PO_4^-)$. Experiments with $^{18}O_2$ and $H_2^{18}O$ have shown that the oxygen atoms derive from water, and so oxidation and hydrolysis are discrete reactions.[10]

$$4Fe^{2+} + O_2 + 4H^+ \rightleftharpoons 4Fe^{3+} + 2H_2O$$

$$4Fe^{3+} + 8H_2O \rightleftharpoons 4Fe(O)OH + 12H^+$$

The mechanism of biomineralization has also been investigated. In the initial stages of nucleation, binding and oxidation of ferrous iron must take place at sites on the interior surface of the protein. These ideas have been tested by chemical modification studies. Esterification of carboxylate residues in apoferritin prevents formation of the iron core, whereas four carboxylates per subunit are protected from esterification if the core is present. This indicates the presence of four important carboxylate residues within the cavity that most likely form the nucleation sites that initiate mineralization. The initial binding stoichiometry is one iron per two subunits. As the core expands, further stages of oxidation and hydrolysis occur at the ferrihydrite surface of the growing mineral. By use of ^{59}Fe radiolabels, it can be demonstrated that the first iron centers released from ferritin correspond to the last irons to be incorporated in the core, and so the dynamic process of core construction and decomposition takes place in a highly ordered fashion.

The relative amounts of Fe^{2+} and Fe^{3+} within the core and the rate of oxidation of Fe^{2+} can be readily monitored by Mossbauer spectroscopy (Figure 3.18). Remember that typical parameters for Fe^{2+} and Fe^{3+} are Fe^{3+} $\delta = 0.34$–0.89 mm/s, $\Delta E_Q = 0.10$–1.80 mm/s; Fe^{2+} $\delta = 1.19$–1.39 mm/s, $\Delta E_Q = 1.75$–3.00 mm/s. The spectra in Figure 3.18 show (a) 100 percent Fe^{3+}, (b) 50 percent Fe^{3+}, and (c) 25 percent Fe^{3+} in the core. As one might expect, there is evidence for magnetic coupling between these oxygen-bridged ferric centers in the ferrihydrite lattice. For example, the average magnetic moment (μ) of a high-spin d^5 ferric center $(S = 5/2)$ may be estimated from the following simple formula, which gives the magnetic moment for an isolated spin (S) in units of Bohr magnetons.

$$\mu = 2[S(S + 1)]^{1/2} \text{ BM}$$

The spins on neighboring irons tend to be oriented in opposing directions, and the spins cancel. The observed value $(\mu = 3.85$ BM$)$ is therefore lower than the predicted value $(\mu = 5.92$ BM$)$ as a result of the antiferromagnetic coupling between the ferric centers. Cancellation is not complete, however, since there are thermally accessible higher spin states that give rise to a residual magnetism. Remember the Boltzmann population arguments described in Section 2.1. Sim-

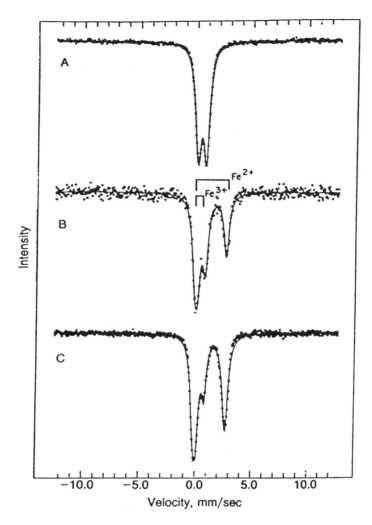

Figure 3.18 Mossbauer spectra showing (A) oxidized ferritin; (B) 50% reduced ferritin; (C) 75% reduced ferritin. [From Watt et al., *Proc. Natl. Acad. Sci. USA, 82,* 3640 (1985).]

ilarly, initial uptake of Fe^{3+} gives a rhombic EPR signal around $g = 4.3$ that is characteristic of HS ferric ion in a low-symmetry environment (see Figure 2.17). As the number of bound ferric ions increases, the intensity of the signal decreases as a result of antiferromagnetic coupling.

The mechanism of iron release from ferritin involves a series of redox reactions that result in the reduction of ferritin-bound iron to Fe^{2+}. The relay of reactions is thought to involve the enzymatic reduction of exogenous FMN by NADH, which then migrates into the hydrophobic channels noted earlier and reduces the ferric centers in the ferritin core. The details are still unclear. Recently, several ferritins isolated from bacterial sources, including *Azotobacter vinelandii* and *E. coli*, were found to possess type b heme bound to the surface. Examples have been crystallographically characterized,[12] and it is proposed that the hemes mediate electron transfer to the iron core.

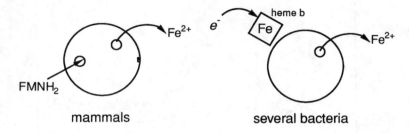

mammals   **several bacteria**

Metallothionein

Metallothioneins are a class of low molecular weight proteins ($M_r \sim 6000–7000$), found in animals and plants, that bind metal ions (typically between four and eight) in cluster arrays. These proteins tend to be rich in cysteine residues, and so metallothioneins selectively bind softer metal ions, such as Cu^{2+}, Zn^{2+}, Cd^{2+}, Hg^+, Pb^{2+}, Ag^+, and Co^{2+}. The structure of the protein has been determined in both the solid state and solution by X-ray crystallography and NMR, respectively (Figure 3.19). Most metal ions appear to be tetrahedrally coordinated by cysteinate ligands. This was previously deduced from analyses of EXAFS, EPR, and electronic absorption data of Co(II)-metallothionein. These show similar features to the Co(II) derivatives of rubredoxin and horse liver alcohol dehydrogenase, which both contain a metal-binding site formed by four cysteine ligands.

In the case of the cadmium derivative of metallothionein, the connectivities of metal ions in the cluster have been evaluated by both X-ray crystallography and NMR. ^{113}Cd has a nuclear spin $I = 1/2$, and so the analysis of spectra is

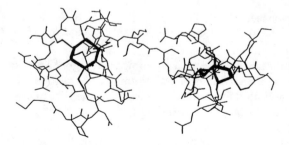

Figure 3.19   Structure of metallothionein determined by solution NMR studies. Strategies and methodologies for determining the structures of the metal clusters and the arrangement of the clusters within the protein were outlined in Figure 2.13.

to some extent similar to ^1H NMR, with which the reader may be more familiar. The derivations of cluster structure and the connectivities to the protein by coordination to cysteine thiolate were described in Section 2.3 (see Figure 2.13). These cluster geometries can vary with the identity of the metal ion and the specific metallothionein under investigation.

Metallothionein most probably serves at least two functions in vivo: first, to promote homeostatic control of zinc and copper levels (i.e., storage)[11] and second, as a detoxification agent to remove cations of the nonessential trace elements (e.g., Cd, Hg, Ag, Au). The ability of a number of these metal ions to induce production of metallothionein by activation of specific genes indicates that metallothionein biosynthesis is part of a tightly regulated system for controlling the levels of such metal ions in vivo. The chemistry of detoxification and gene regulation will be discussed further in Chapters 7 and 10.

Other Storage and Transport Proteins

Many of the best-characterized carrier proteins have been isolated from blood serum. Ceruloplasmin is a copper-binding protein that may also play a role in the detoxification of reactive oxygen species ($O_2^-\cdot$ and O_2^{2-}). Serum albumin binds a variety of metal ions, but principally complexes copper and zinc. Storage and transport proteins that have been characterized for other metal ions are noted in Table 3.8.

Summary of Section 3.3.2

1. Only 35 mg of iron is used in active metabolism by an average person. Another 3–4 g of iron is stored predominantly by the protein ferritin. The inner core can accommodate up to 4500 iron centers in a ferrihydrite phosphate lattice $[(Fe(O)OH)_8(FeOPO_3H_2)\cdot xH_2PO_4]$.
2. Iron is taken up in the labile ferrous form prior to oxidation. Release of iron requires reduction of the core with a biological reductant (NADH or $FADH_2$).
3. Other storage proteins include metallothionein (Cu^{2+}, Zn^{2+}), ceruloplasmin (Cu^{2+}), albumin (Cu^{2+}, Zn^{2+}), and a number of phosphoproteins (Ca^{2+}).

Notes

1. Many physicochemical properties of ions (formation constants, lattice structures, hydrolysis constants) that depend only on electrostatic interactions of the charged ion show systematic variations with the magnitude of q^2/r (see Table 3.5).
2. The term *ionophore* is derived from the Greek words for "ion bearing" or "ion carrying." Later we shall discuss siderophores, or "iron carriers."
3. Transmembrane potentials are discussed in more detail in Sections 6.2 and 6.3.
4. See Chapter 5, Section 5.4.1 for a discussion of nitrogenase chemistry.
5. The word *synergism* is derived from the Greek word *sunergetikos*, meaning "working together."

6. There are, however, many inconsistencies, which highlights the dangers inherent to over-interpretation of spectroscopic or chemical modification data, especially if studies are performed with metal-substituted derivatives where the coordination chemistry may vary from that found in the native protein. Normally, a consensus of results from several approaches provides the clearest picture.

7. In 6 M urea, $(Fe^{3+})_C Tf$ is denatured less than $(Fe^{3+})_N Tf$, reflecting the different numbers of disulfide bridges in the C- and N-terminal domains (11 and 8), respectively, whereas $(Fe^{3+})_2 Tf$ is stable. The differential unfolding forms the basis for the electrophoretic separation.

8. It is relatively easy to prove the following relationships between the macroscopic and microscopic binding constants, $K_1 = k_{1C} + k_{1N}$, $K_2^{-1} = k_{2C}^{-1} + k_{2N}^{-1}$, and $k_{1C} k_{2N} = k_{1N} k_{2C}$.

9. The distribution of H and L chains varies according to the protein origin. Horse spleen ferritin is about 85 percent L chains.

10. On a technical note, $^{18}O_2$ can be produced by heating $Hg^{18}O$.

11. Homeostasis refers to the active maintenance of intracellular concentrations of molecules and ions at physiologically appropriate levels.

12. Frolow, F. et al., Nat. Struct. Biol. 1, 453–460 (1994).

Further Reading

Trace Elements and Bioenergetics

Mertz, W. The essential trace elements. Science 213, 1332-1338 (1981).
Nicholls, D. G. and S. F. Ferguson. Bioenergetics, 2nd ed., Academic, 1992.

Ionophores

Dobbin, P. S. and R. C. Hider. Iron chelation therapy. Chem. Brit. 26, 565-568 (1990).
Neilands, J. B. Siderophores, Adv. Inorg. Biochem., 5, 137–199 (1983).
Raymond, K. N. Complexation of iron by siderophores. A review of their solution and structural chemistry and biological function. Top. Curr. Chem. 123, 49–102 (1984).
Structure and Bonding (Alkali Metal Complexes with Organic Ligands), 16, Springer-Verlag (1973).
Structure and Bonding (Siderophores from Microorganisms and Plants), 58, Springer-Verlag (1984).

Channels and Pores

Langs, D. A. Three-dimensional structure at 0.86 Å of the uncomplexed form of the transmembrane ion channel peptide gramicidin A. Science 241, 188–191 (1988).
Wallace, B. A. and K. Ravikumar. The gramicidin pore: Crystal structure of a cesium complex. Science 241, 182–187 (1988).

Transferrin and Ferritin

Baker, E. N. and Lindley, P. F. New perspectives on the structure and function of transferrins. J. Inorg. Biochem. 47, 147–160 (1992).
Chasteen, N. D. Transferrin: A Perspective, Adv. Inorg. Biochem., 5, 201–233 (1983).
Crichton, R. R. Inorganic Biochemistry of Iron Metabolism. Ellis Horwood, 1991.
Loehr, T., ed., Iron Carriers and Iron Proteins, VCH, 1989.
Welch, S. Transferrin: The Iron Carrier, CRC, 1992.

Metallothionein

Hunziker, P. E., and J. H. R. Kagi. Metallothionein, in Metalloproteins, Part 2 (P. M. Harrison, ed.), Macmillan, 1985, pp. 149–181.

Riordan, J. F., and B. L. Vallee, eds., *Metallobiochemistry: Part B, Metallothionein and Related Molecules,* vol. 205, Academic, 1991.

Problems

1. Vanadate (VO_4^{3-}) is a structural and electronic analogue of phosphate (PO_4^{3-}). In solution the vanadate monomer (V_1) can oligomerize to dimeric (V_2), tetrameric (V_4), or pentameric (V_5) forms.
   (a) Draw likely structures for V_1, V_2, V_4, and V_5.
   (b) Two-dimensional exchange spectroscopy (EXSY) affords a convenient method for identifying species in solution that are in equilibrium. From the ^{51}V EXSY spectrum shown, suggest which species are in equilibrium on the time scale of the experiment. Comment on the relative abundance of each species under the solution conditions of 10 mM vanadate and 0.4 M KCl at pH 8.6.

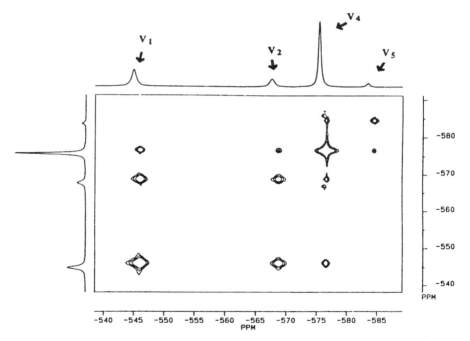

   (c) You are investigating a novel vanadium storage protein that you have isolated from a marine organism. You suspect the protein contains a core of vanadium ions. By analogy with ferritin, suggest possible chemical forms for the core.

[Crans, D. C. *Comments on Inorg. Chem. 16,* 1–33 (1994), Figure is reproduced with permission]

2. (a) Tris catecholate siderophores may exist in enantiomeric forms. Explain with illustration.

(b) Do you expect the affinity for iron to differ for each of these enantiomers? Either way, discuss the implications of your answer.

(c) Describe an experimental approach that might allow you to distinguish which of the two enantiomers is actively transported into the cell. Your answer should briefly outline how you propose to monitor the transport process.

3. Under iron-deficient conditions, the pathogenic bacterium *Pseudomonas aeruginosa* excretes fluorescent peptidic siderophores called pyroverdins. The ligand has six ionizabel sites that form a chelating pocket for ferric ion. The microscopic pK_as range from 12.2 to 4.8. Experimental studies were carried out at $1.8 < pH < 2.3$.

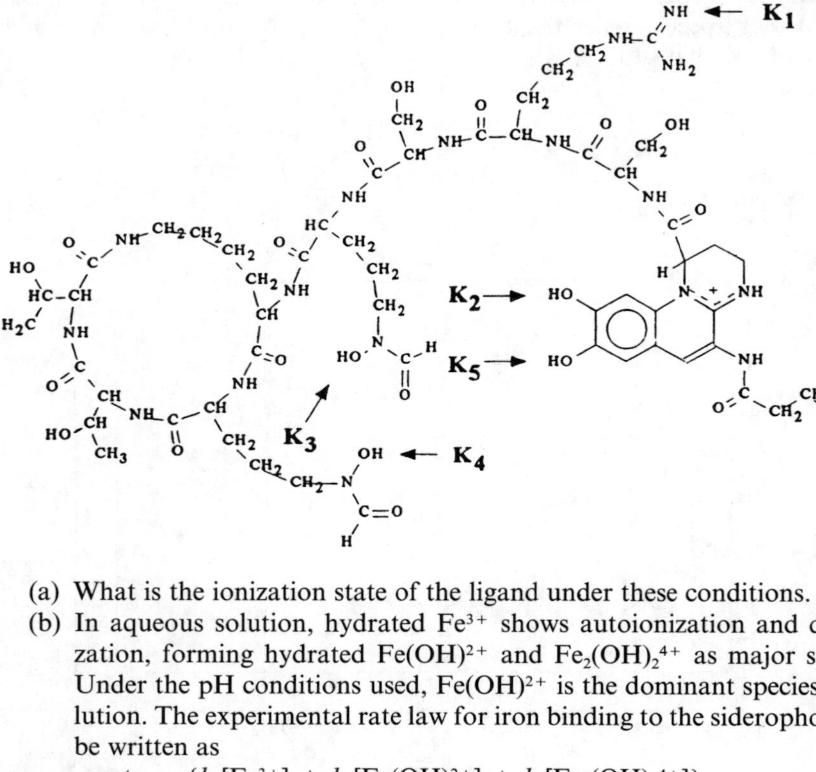

(a) What is the ionization state of the ligand under these conditions.

(b) In aqueous solution, hydrated Fe^{3+} shows autoionization and dimerization, forming hydrated $Fe(OH)^{2+}$ and $Fe_2(OH)_2^{4+}$ as major species. Under the pH conditions used, $Fe(OH)^{2+}$ is the dominant species in solution. The experimental rate law for iron binding to the siderophore can be written as

   rate $= \{k_1[Fe^{3+}] + k_2[Fe(OH)^{2+}] + k_3[Fe_2(OH)_2^{4+}]\}$

   What does this indicate?

(c) From kinetic studies it is found that $k_1 < 80$ M^{-1} s^{-1}, $k_2 = 7.7 \times 10^3$ M^{-1} s^{-1}, and $k_3 = 5 \times 10^4$ M^{-1} s^{-1}. However, under the conditions of the experiment the contribution from $Fe_2(OH)_2^{4+}$ can be neglected. Why?

(d) The $Fe(OH)^{2+}$ pathway has been shown to follow a classical dissociative interchange mechanism with rapid formation of an outer sphere complex (K_{os}, M^{-1}) and a rate-limiting desolvation of the metal (k_{ex}, s^{-1}). Calculate

k_2, assuming $K_{os} \sim 3 \times 10^{-2}$ M^{-1} (determined for a related hydroxamate siderophore).

(e) A series of electronic spectra for the ferric pyroverdin complex were obtained at pH values of 5.1, 4.0, 3.1, 2.6, 2.1, 1.4, and 1.2 for spectra 1 to 7, respectively. If the tris hydroxamate siderophore ferrioximine B has $\lambda_{max} = 425$ nm, $\varepsilon_{425} = 2600$ M^{-1} cm^{-1}, and the tris catecholate siderophore ferric enterobactin has $\lambda_{max} = 495$ nm, $\varepsilon_{495} = 5600$ M^{-1} cm^{-1}, assign some of the absorption bands in the series of spectra shown for ferric pyroverdin to specific chromophores.

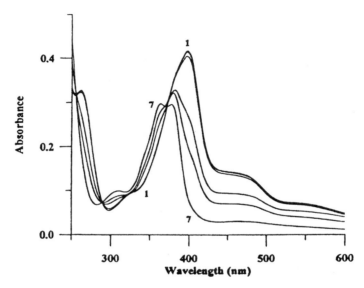

(f) Identify the fluorescent chromophore in ferric pyroverdin.

[Albrecht-Gary, A. M. et al., *Inorg. Chem.* **33**, 6391 (1994), Figures are reproduced with permission]

4. H-type ferritin exhibits fast rates of biomineralization. This has been ascribed to rapid complex formation of a ferrous iron with a specific amino acid side chain and immediate oxidation ($k_{ox} \sim 1000$ s^{-1}). The resulting complex is purple in color with an absorbance at 550 nm that is characteristic of a phenolate–Fe(III) ligand-to-metal, charge-transfer band. These sites are saturated at an Fe/protein ratio of 50. An increase in absorbance is also observed at 420 nm, which corresponds to formation of multinuclear Fe(III)–oxo species. The increase in the 420 nm band is biphasic. After the rapid formation of the 550 nm absorbance band, the 420 nm band continues to increase, although at a reduced rate, and the 550 nm band decreases in intensity. The consumption of oxygen also continues to increase at a rate comparable to that of the 420 nm band. After 30 minutes, it is possible to add an additional aliquot of Fe(II), and a new rapid burst of O$_2$ consumption and increase in absorbance at 550 nm are observed.

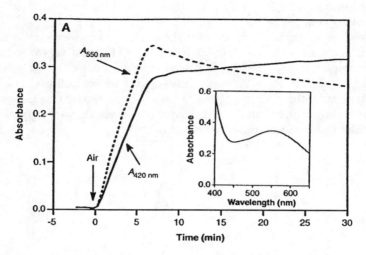

(a) What is the identity of the amino acid to which the iron ion coordinates?
(b) H-subunit ferritin has 24 subunits. Comment on this fact in terms of the number of iron ions taken up during the initial burst phase.
(c) Rationalize these data in terms of a model for iron uptake through the protein coat and mineral formation in the core.

[Waldo, G. S. et al., *Science 259*, 796–798 (1993), Figure is reproduced with permission; Waldo, G. S. and E. C. Thiel, *Biochemistry 32*, 13262–13269 (1993)]

4

Metalloproteins and Metalloenzymes: (I) Oxygen Carriers and Hydrolases

Transition-metal ions have been extensively utilized by nature in the design of metalloproteins and enzymes that catalyze nonredox reactions, store or transfer important cellular substrates, and regulate the activities of other biological macromolecules. This reflects the diverse coordination chemistry (ligand preference, geometry, redox state, kinetic and thermodynamic stabilities and labilities) of the d-block elements, and so it is not surprising that these metals occur in a wide variety of functional roles. As a result of the higher natural abundance of the first-row transition metals, these metals, with just a very few exceptions, are most commonly selected rather than the heavier d- and f-block elements. Molybdenum is a notable exception.

Selection specificity for transition metals cannot be accomplished simply on the basis of ionic radius, as was the case for the alkali and alkaline earth ions. Rather, a combination of ligand preference and geometric constraints is used to differentiate both the metal ion and the oxidation level (Table 1.3). Transition metals are most often associated with oxygen-binding and oxido-reductase proteins and enzymes. The latter function reflects the variable oxidation states available to these cations. However, transition-metal ions may occasionally function as Lewis acids (e.g., Ni^{2+} in urease, Fe^{3+} in acid phosphatases, $[Fe_4S_4]^{2+}$ in hydrolyases), although this is a role that is most commonly filled by magnesium and zinc. The general trend of Lewis acidity follows the order $Ca^{2+} < Mg^{2+} < Mn^{2+} < Fe^{2+} < Co^{2+} < Ni^{2+} < Cu^{2+} > Zn^{2+}$; however, the redox activity of the transition metals makes them less suitable for hydrolysis of sensitive functional groups in proteins and nucleic acids.

In this chapter, we review some examples of oxygen carriers and hydrolytic metalloproteins and metalloenzymes that illustrate the chemical principles underlying the activity of many related systems.

4.1   Oxygen Carriers

Oxygen-binding proteins evolved only when the earth changed from a reducing to an oxidizing environment. Around 2.0–2.5 billion years ago, the proliferation of cyanobacteria resulted in the change from a reducing to an oxidizing atmosphere. These bacteria were equipped with a photosynthetic apparatus capable of decomposing water with the release of O_2. Figure 4.1 and Table 4.1 illustrate the three known O_2-binding systems that are developed around the heme ligand (I) and binuclear iron and copper complexes (II). Related prosthetic centers are discussed in the next chapter, which describes the role of electron carriers and redox catalysts. It is possible that these early oxygen transport and storage proteins evolved from electron-transfer proteins that are found in bacteria and lower organisms since they contain structurally similar prosthetic centers. Binuclear metal sites appeared earlier in the evolutionary time frame than the more complex heme units, since no additional ligand cofactor is required to bind the metal. For example, the binuclear iron carrier may be derived from the simple $[Fe_2S_2]$ ferredoxin-like iron–sulfur proteins that were among the earliest of electron transport proteins. Heme-bearing cytochromes are more highly evolved electron carriers.

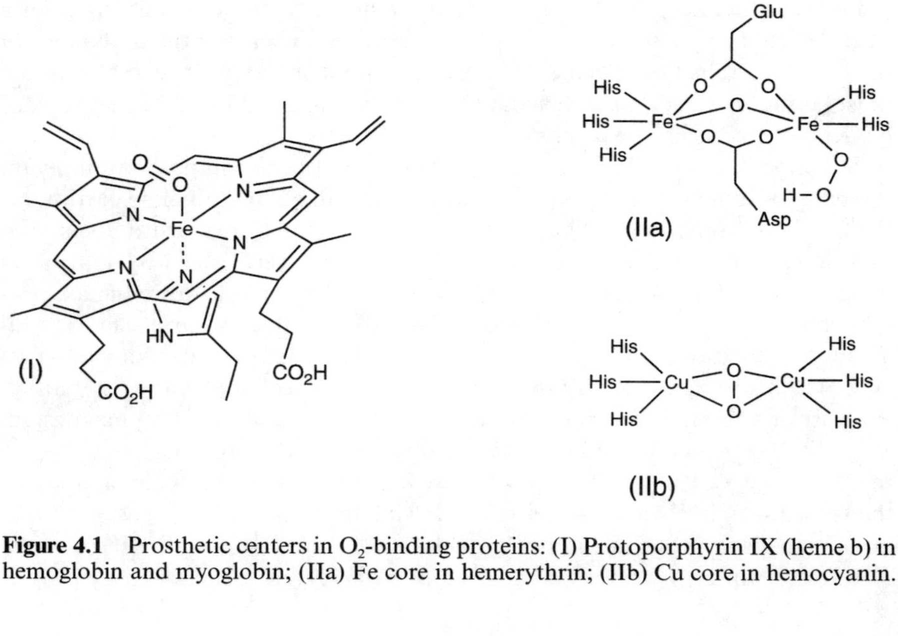

Figure 4.1   Prosthetic centers in O_2-binding proteins: (I) Protoporphyrin IX (heme b) in hemoglobin and myoglobin; (IIa) Fe core in hemerythrin; (IIb) Cu core in hemocyanin.

Table 4.1 Oxygen Transport Proteins

	Hemoglobin	Hemerythrin	Hemocyanin
Source	Higher Animals	Invertebrates	Arthropods, Mollusks
Metal: bound O_2	Fe(heme):O_2	2 Fe(nonheme):O_2	2Cu:O_2
Metal in oxy protein	Fe^{2+}	Fe^{3+}	Cu^{2+}
Metal in deoxy protein	Fe^{2+}	Fe^{2+}	Cu^{+}
Color of oxy form	Red	Burgundy	Blue
Color of deoxy form	Red-purple	Colorless	Colorless

4.1.1 Myoglobin and Hemoglobin

Myoglobin (Mb) and hemoglobin (Hb) were among the first proteins to be structurally characterized (Figure 4.2), and both contain iron protoporphyrin IX (*heme*) as an essential prosthetic center. The function of each protein depends on the reversible binding of O_2; however, their biological roles are different, Mb stores O_2 in cellular tissue, whereas Hb transports O_2 in blood plasma. Mb and Hb may bind up to one and four O_2 molecules, respectively, according to the number of heme ligands.[1] Both Mb and Hb bind O_2 in the reduced state and the terms *oxy-* and *deoxy-Mb* and *Hb* refer to reduced protein, with and without bound oxygen, respectively. The prefix *met-* is used to describe oxidized heme proteins that contain ferric heme centers.

Table 4.2 summarizes some thermodynamic data for the reduction of oxygen and heme centers that is pertinent to a discussion of oxygen-binding functionality. Clearly, O_2 is a powerful oxidant at pH 7 ($E^{O'} = +0.82$ V) and should react readily with reduced heme. In free solution, heme reacts with O_2 according to the following reaction scheme,

$$2\ PFe(II) + O_2 \rightleftharpoons PFe(III)\text{-}O\text{-}O\text{-}Fe(III)P \xrightarrow{2PFe(II)} 2\ PFe(III)\text{-}O\text{-}Fe(III)P$$

and so the stability of heme–O_2 complexes in proteins must originate from *kinetic* rather than *thermodynamic* factors. That is, structural barriers exist that impose high activation energies for the two-electron reduction of O_2. Specifically, the heme prosthetic center is buried in a hydrophobic pocket that is inaccessible to solvent molecules after binding O_2. This serves to inhibit the displacement of O_2 while also preventing μ-oxo dimer formation (as shown earlier, where P = protoporphyrin IX). The stability of the $PFe(II)\cdot O_2$ complex arises in part from the favorable interaction of filled d-orbitals (d_{xy}, d_{yz}) on Fe(II) with π*-orbitals on O_2 (π-back-bonding), and from the overlap of filled σ-orbitals on O_2 with the empty d_{z^2} orbital on Fe(II) (σ-bonding). Figure 4.3 shows that this bonding arrangement favors a bent orientation for O_2 since this geometry maximizes σ-overlap and is facilitated by the steric arrangement of residues within the binding pocket. For many years, however, the structure of the Fe·O_2 complex was a subject of lively debate. Two models (bent and side-on) were pro-

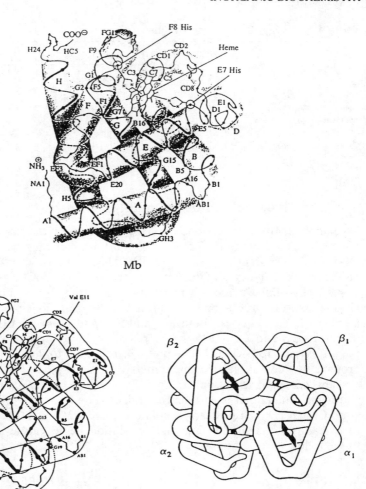

Figure 4.2 Tertiary and quaternary structures of the monomeric and $\alpha_2\beta_2$ tetrameric O_2-binding proteins myoglobin and hemoglobin, respectively.

Table 4.2 Thermodynamic Data for the Reduction of O_2 and Heme

Reaction	$E^{\circ\prime}$ (V vs. NHE)
$4H^+ + O_2 + 4e^- \rightarrow 2H_2O$	$+0.82$
$2H^+ + O_2 + 2e^- \rightarrow H_2O_2$	$+0.27$
$H^+ + O_2 + e^- \rightarrow HO_2$	-0.45
$Hb\cdot(Fe^{3+}) + e^- \rightarrow Hb\cdot(Fe^{2+})$	$+0.17$

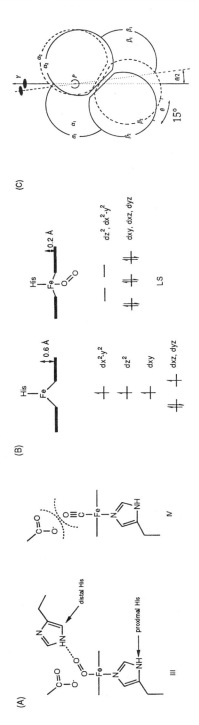

Figure 4.3 (A) O_2 adopts a bent geometry rather than the linear geometry preferred by CO. Bound O_2 is stabilized by hydrogen bond formation to the distal His, whereas CO binding is hindered by steric clashes with neighboring residues. Structural deformations (not shown) of the planar heme also result from ligand binding that favors O_2 coordination (see M. F. Perutz, *Mechanisms of Cooperativity and Allosteric Regulation in Proteins* in Further Reading). (B) O_2 binding results in a change of coordination state (penta- to hexa-coordination) and spin state (HS → LS). The latter is induced by electronic repulsion between the electron density in O_2 and electrons in d-orbitals. By populating the d_{xy}-, d_{yx}-, and d_{xz}-orbitals, the metal ion shrinks in size and is better accommodated in the center of the heme. Population of these orbitals also promotes π-overlap with the π-acceptor ligand His. (C) The overall influence of the change in heme–iron coordination is to effect a structural change in which one $\alpha\beta$ dimer rotates 15° relative to the other. This also disrupts a series of salt bridges (Figure 4.5) that stabilize the deoxy form, thereby promoting O_2 binding. The cooperative effect is enhanced with each heme–O_2 complex that forms. [Adapted from J. Baldwin and C. Chothia, *J. Mol. Biol. 129*, 175–220 (1979).]

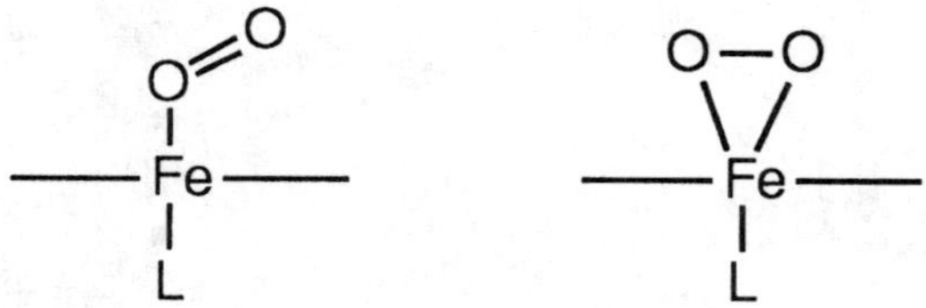

posed. The former was finally established from structural studies on model synthetic compounds, such as Collman's "picket-fence" porphyrin (see structure), and by low-temperature crystallographic studies of oxy-Mb (the instability of oxy-adducts makes such measurements difficult). A close inspection of the model complex shows that this sterically encumbered molecule satisfies the re-

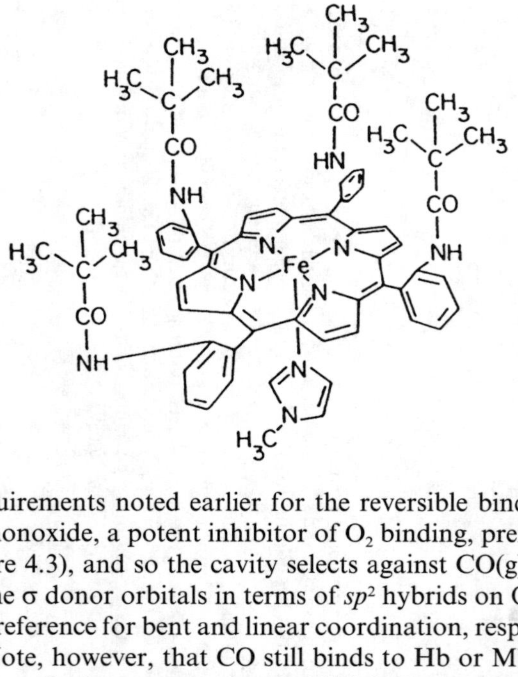

Collman's
"picket-fence"
porphyrin

quirements noted earlier for the reversible binding of O_2. In contrast, carbon monoxide, a potent inhibitor of O_2 binding, prefers a linear binding mode (Figure 4.3), and so the cavity selects against CO(g) coordination. If one thinks of the σ donor orbitals in terms of sp^2 hybrids on O_2 versus sp hybrids for CO, the preference for bent and linear coordination, respectively, is easier to understand. Note, however, that CO still binds to Hb or Mb about 250-fold more strongly than O_2 (as compared with 2000-fold for reduced heme in solution) and is a potent inhibitor of oxygen binding. Since CO is constantly produced in vivo by the decomposition of porphyrins following the lysis of red blood cells (in fact, 1 percent of Hb is bound by CO), the need for strict regulation of O_2 versus CO binding is clear.

The oxygen-binding curves for Mb and Hb are shown in Figure 4.4. The hyperbolic form for Mb is typical of bimolecular binding where initial rapid uptake of O_2 eventually reaches a saturation level. In contrast, the more complex sigmoidal shape for Hb is indicative of cooperativity.[2] The oxy-form of Hb is often referred to as the R-state (relaxed state) and the deoxy-form as the T-state

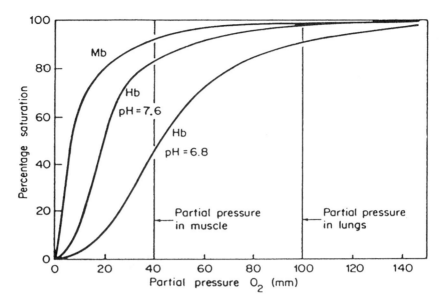

Figure 4.4 Oxygen-binding curves for myoglobin and hemoglobin. Hb takes up O_2 from the lungs and transports it throughout the body. When this oxygen saturated blood passes through tissues (such as muscle), where the O_2 partial pressure is lower, Hb releases some of its O_2 whereas Mb retains >90 percent of its bound O_2. The O_2-binding curves therefore reflect the distinct roles for each: myoglobin (storage) and hemoglobin (transport, and active uptake and release). The reduced O_2 affinity of Hb at lower pH is a manifestation of the Bohr effect. [From F. A. Cotton and G. Wilkinson, *Advanced Inorganic Chemistry*, 5th ed., Wiley, p. 1343.]

(tense state). For both Mb and Hb, oxygen binding results in a transition from high-spin to low-spin iron with accompanying changes in the Fe–N bond lengths and coordination geometry.[3] However, only in Hb do these subtle changes in geometry (Figure 4.3c) ultimately lead to the observed cooperative binding among the four subunits. The changes in quaternary structure accompanying $O_2(g)$ binding are summarized in Figure 4.5. After oxygen ligation to heme iron, there is a 15° reorientation of the $\alpha_2\beta_2$ dimer relative to the $\alpha_1\beta_1$ dimer, and a 0.8 Å translation of $\alpha_2\beta_2$ relative to $\alpha_1\beta_1$. These changes arise from the 0.4 Å movement of Fe into the heme plane after O_2 binding, with simultaneous movement of the proximal His toward the heme plane. In the β subunits, these changes are accompanied by shifts in the positions of other residues in the heme pocket. Specifically, the distal His and Val move away from the O_2-binding site.

Although the oxygen affinity of Mb is independent of pH and ionic strength, that of Hb is lowered by increasing concentrations of H^+, Cl^-, CO_2, and 2,3-D-diphosphoglycerate (DPG), which are present in red blood cells (erythrocytes). The central cavity of deoxy-Hb contains a number of positively–charged ligands to which DPG may bind (Figure 4.5). Since the conformational change accompanying the allosteric transition that is induced by O_2 binding disrupts this site,

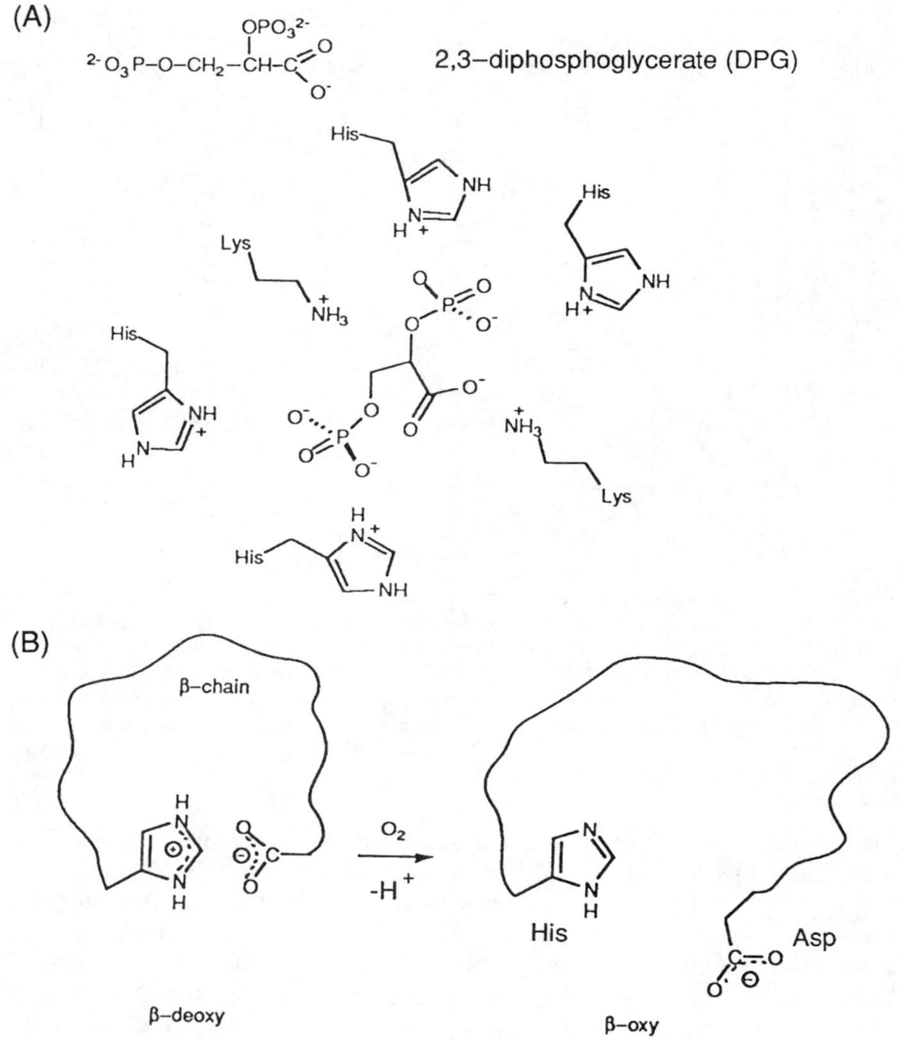

(A)

$$2^- O_3P-O-CH_2-\overset{\overset{\displaystyle OPO_3{}^{2-}}{|}}{CH}-C\overset{\displaystyle =O}{\underset{\displaystyle O^-}{}}$$

2,3–diphosphoglycerate (DPG)

(B)

Figure 4.5 (A) The central cavity of deoxyhemoglobin contains positively–charged residues that form a stable complex with DPG, thereby inhibiting O_2 binding. (B) The Bohr effect is a consequence of salt bridge formation at the C-terminae of the β-chains. O_2 binding disrupts these salt bridges, resulting in the release of H^+.

DPG stabilizes deoxy-Hb. Similarly, ion pairing between subunits is dependent on pH. The conformational change following O_2 uptake results in the breakup of these ion pairs with concomitant release of H^+, and so oxy-Hb is more stable at higher pH values. The pH dependence of oxygen binding to Hb is termed the *Bohr effect*.[4]

In summary, Hb function has two distinct aspects: (1) cooperative O_2 binding arising from heme-driven global structural changes and (2) stabilization of one or the other of the two principal conformational states [oxy-Hb (R-state) and deoxy-Hb (T-state)] by regulation of pH or ionic strength. Crystallography has played a particularly important role in establishing the mechanisms of O_2 binding to Hb and Mb. However, other spectroscopic studies have proved useful in studying the dynamics of O_2 uptake and release (Figure 4.6).

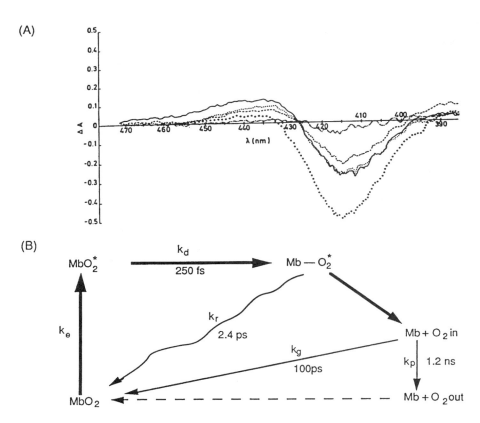

Figure 4.6 (A) Picosecond-difference spectra resulting from intermediate stages in the photodissociation reaction of $Mb \cdot O_2$. Data were taken at a series of time intervals after an initiation pulse at 307 nm: •••• $+1$ ps; —— $+4$ ps; ····· $+9$ ps; ----- $+50$ ps; -·-·-·- $+200$ ps. (B) A schematic model for the photodissociation process in $Mb \cdot O_2$ where (*) represents an excited heme state, $Mb–O_2^*$ is a transient state where O_2 has dissociated from the heme, and $Mb + O_2$ (in or out) represent stable O_2-free heme. Subscripts "in" or "out" imply that O_2 is close to or far from the heme active site, respectively. Subscripts on rate constants (k) indicate the following; e (excitation); d (dissociation); s (stable); r (relaxation); p (protein); g (ground state). The thicker arrows represent faster processes. [Adapted from J. L. Martin et al., *Biochem. Biophys. Res. Commun., 107*, 803–810 (1982).]

4.1.2   Hemerythrin and Hemocyanin

Although Hbs are found throughout the animal kingdom (and have also been isolated from leguminous plants, crustaceans, and worms, the nonheme O_2 carriers hemerythrin (Hr) and hemocyanin (Hc) have been found only in invertebrates (annelids, mollusks, arthropods). In contrast to heme O_2 carriers, only recently has crystallographic data become available for nonheme carrier proteins; thus, much of the available structural and reactivity data have been derived from a combination of chemical and physical studies. Difficulty in crystallizing these proteins derives in part from their tendency to form aggregates. For example, hemerythrin isolated from the blood of invertebrates is typically found as an octamer of eight 13,000 Da subunits [although a special monomeric hemerythrin (myohemerythrin) has been isolated from the muscles of invertebrates], and hemocyanin exists as a large aggregate of 50–70 kDa subunits yielding complexes with masses of the order of several million. Hr and Hc possess binuclear iron or copper centers, respectively (see Figure 4.1), and are further described according to the oxidation and ligation states of the metal centers. In particular, the terms *met* and *semi-met* refer to (oxidized, oxidized) and (reduced, oxidized) oxidation states, respectively.

$$(Fe^{2+} \cdot Fe^{2+}) \quad (Fe^{2+} \cdot Fe^{3+}) \quad (Fe^{3+} \cdot Fe^{3+}) \quad \text{Hemerythrin}$$
$$(Cu^{1+} \cdot Cu^{1+}) \quad (Cu^{1+} \cdot Cu^{2+}) \quad (Cu^{2+} \cdot Cu^{2+}) \quad \text{Hemocyanin}$$
$$\text{deoxy} \qquad\quad \text{semi-met} \qquad\quad \text{met}$$

Oxygen binding is essentially noncooperative for Hr,[5] although cooperative binding of O_2 is common for the multimeric Hc's. In both cases, O_2 binding is accompanied by redox chemistry. Unlike Mb and Hb, where O_2 binds as the neutral species,[6] the uptake of O_2 by either Hr or Hc is accompanied by two-electron oxidation of the reduced binuclear center to produce bound peroxide. Direct evidence for this was obtained from Raman data since the vibrational stretching frequency of O_2 is sensitive to its oxidation level (Table 4.3) and the bond order can be determined by referencing the frequency of this vibration to defined standards. For example, by monitoring the O–O vibration[7] that is coupled to the $O_2 \rightarrow Fe(III)$ charge-transfer transition at 500 nm, a peak at 844

Table 4.3   Stretching Frequencies for a Variety of Binuclear Oxygen Species

Species	Bond Order[a]	O–O Stretch Frequency
O_2^+	2.5	$1865 \ cm^{-1}$
O_2	2	$1560 \ cm^{-1}$
O_2^-	1.5	$1110 \ cm^{-1}$
O_2^{2-}	1	$850 \ cm^{-1}$

[a] The *bond order* refers to the number of formal bonds connecting the two atoms. Bond order = (number of bonding electrons − number of antibonding electrons)/2.

cm^{-1} was detected in the resonance Raman spectrum of oxy-Hr (Figure 4.7), indicating a peroxide-like oxidation level. Unsymmetrically labeled $^{16}O^{18}O$ was used to identify the coordination geometry from the various structural possibilities shown here. In structures I, II, and V, a single symmetric O–O stretch would be obtained. However, for III and IV, a doublet would be expected because the unsymmetrical ligand may bind to Fe(III) in two distinct modes (^{16}O-bound or ^{18}O-bound). A doublet was, in fact, observed, and structure IV was ultimately selected on the basis of polarized single-crystal absorption spectroscopy and later confirmed by high-resolution X-ray crystallography. Interested readers who require a more comprehensive review of spectroscopic studies on binuclear iron centers are directed to the excellent article by J-S. Loehr in the Further Reading section.

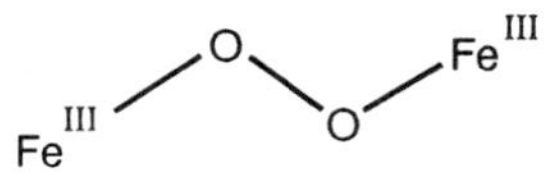

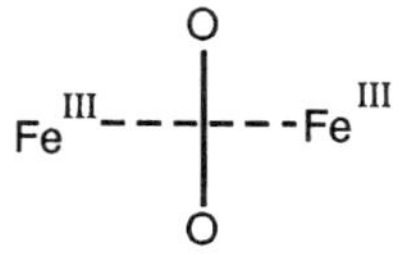

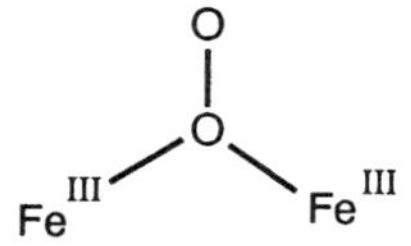

The chemistry of O_2 binding by Hr is summarized in Figure 4.8. On the basis of NMR and magnetic circular dichroism experiments, deoxy-Hr contains two high-spin ferrous ions that are weakly antiferromagnetically coupled through a hydroxyl ion. The bridging hydroxyl serves as the proton donor for peroxide after O_2 binding, resulting in formation of a μ-oxo bridge in oxy- and met-Hr. Molecular oxygen binds to the pentacoordinate Fe_B^{2+} at the vacant coordination site. Electrons are then transferred from the ferrous ions to generate the binuclear ferric site with bound peroxide, which can accept the proton from Fe_A–OH–Fe_B to facilitate μ-oxo bridge formation. In this regard, and in contrast to Hb, there is no proton release and no Bohr effect for O_2 binding to Hr. The bridging hydroxyl in the reduced enzyme can be exchanged by isotopically labeled water, and so the 486 cm^{-1} frequency of the F–O–Fe μ-oxo stretch is reduced by 9 cm^{-1} if an ^{18}O-labeled water is used in the exchange reaction.

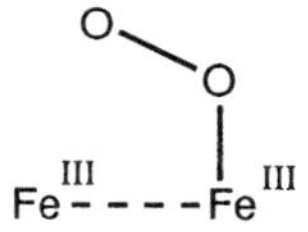

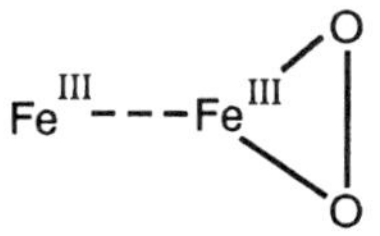

Although Hr and Hc apparently exhibit many similarities in the structural features of the O_2-binding domain, there are, in fact, some significant differences in their chemistries. Resonance Raman spectra again show that O_2 binds as peroxide; however, spectroscopic analysis suggests the symmetric side-on bridging mode shown in Figure 4.1, which has also recently been confirmed by crys-

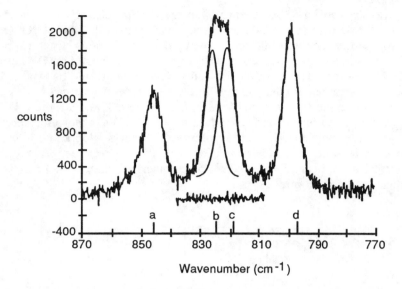

Figure 4.7 Resonance Raman spectrum in the ν_{0-0} region of oxyhemerythrin equilibrated with a mixture of the isotopic species $^{16}O_2$, $^{16}O^{18}O$, $^{18}O_2$. The smooth curves used for fitting the central peak ($^{16}O^{18}O$) represent the deconvolution of the 822 cm^{-1} feature into two discrete components (from Fe–^{16}O–^{18}O and Fe–^{18}O–^{16}O, respectively). The difference between observed and fitted curves is shown below the spectrum in the 822-cm^{-1} region. The vertical lines a–d show the calculated peak positions for models III and IV of Fe–$^{16}O_2$ (836 cm^{-1}), Fe–^{16}O–^{18}O (825 cm^{-1}), Fe–^{18}O–^{16}O (819 cm^{-1}), and Fe–$^{18}O_2$ (798 cm^{-1}), respectively. [Reprinted with permission from I. M. Klotz and D. M. Kurtz, *Acc. Chem. Res., 17,* 16–22; copyright (1984) American Chemical Society.]

tallography. In hemocyanin, the peroxide stretching frequency is 100 cm^{-1} lower than that obtained in Hr since now peroxide is bound by two metal ions, which weakens the O–O bond. Although the deoxy enzyme is colorless, oxidation of the Cu^{1+} centers to Cu^{2+} after O_2 uptake is obvious from the bright blue color (hence the name *hemocyanin*) of the oxy-enzyme that gives rise to the "blue blood" common to arthropods (spiders, crabs, lobsters, etc.) and mollusks (clams, octopus, squid). The two Cu^{2+} centers are also strongly antiferromagnetically coupled (see Section 2.3) to yield a diamagnetic complex.

Reversible oxygen binding derives from two features of the binding pocket. First, the two Cu^+ ions are each bound by three His residues. The geometry is pseudo-tetrahedral with the His forming a trigonal base and the Cu at the apex. This favors the reduced state since Cu^{2+} typically adopts a square planar geometry. Second, the cavity is lined with hydrophobic residues that disfavor the buildup of charged intermediates in the pocket. The mechanism of cooperativity is also built into the stereochemical design of the binding pocket. After O_2 uptake, the ionic radius of copper shrinks from 0.95 å for Cu^{1+} to 0.72 Å for Cu^{2+}. Furthermore, the distance between the copper ions decreases by about 0.2–0.3 Å. Overall, the distance between opposite helices may change by about 0.6–0.7

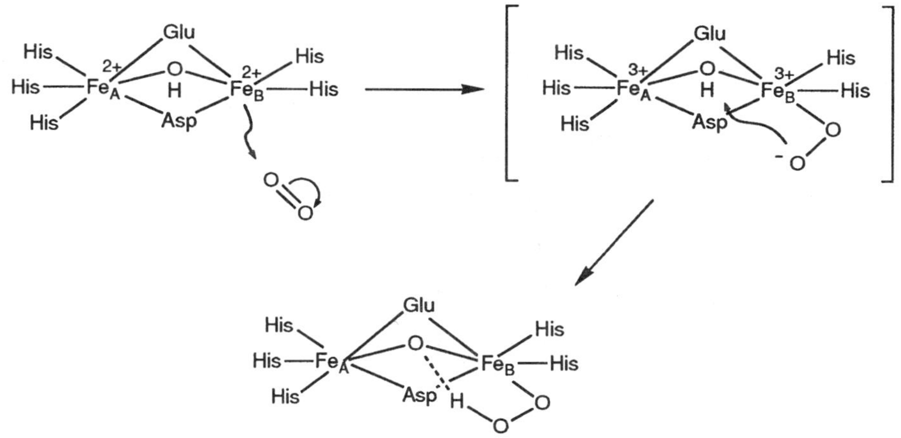

Figure 4.8 Pathway for oxygen binding to hemerythrin. In the deoxy form, the HS ferrous ions are coupled through the hydroxyl ion. After O_2 binding, the hydroxyl is deprotonated by the peroxo intermediate, and an oxygen bridge is formed between the two ferric centers. The hydroperoxide forms a hydrogen bond to the bridging oxo ligand. When O_2 binds to Hr, there is no net release of protons and no change in pH (i.e., no Bohr effect).

Å, which (as for Hb) can trigger the change from the T to an R quaternary structure.

The interested reader should refer to Perutz's book in the Further Reading section for an excellent summary of O_2-binding proteins and a general overview of mechanisms of cooperative interaction in proteins and enzymes.

Summary of Section 4.1

1. Transition metals are used in a variety of functional roles, reflecting their diverse chemical properties.
2. The nonheme O_2-binding proteins Hr and Hc (binuclear iron and copper proteins, respectively) differ from Hb and Mb insofar as the former carry out redox chemistry. Both bind O_2 as a peroxide complex: Hr uses an end-on mode, whereas Hc adopts a side-on bridging arrangement.
3. Hb is an allosteric protein, displaying cooperative binding of four O_2 molecules. Binding is accompanied by H^+ release (Bohr effect) and is also dependent on concentrations of charged counterions (especially DPG and Cl^-).
4. The allosteric mechanism in Hb ultimately depends on structural changes at the iron porphyrin site as a result of changes in spin state following O_2 binding.

4.2 Hydrolase Enzymes

Enzymes are classified into one of six general classes: (1) oxidoreductase, oxidation/reduction; (2) transferase, transfer of a group from one compound to another; (3) hydrolase, hydrolysis; (4) lyase, nonhydrolytic addition or removal of groups; (5) isomerase, conversion of a substance into an isomeric form; (6) ligase, synthesis of a large molecule from two smaller ones. In this section we review the chemistry of a selection of hydrolase enzymes. Enzymes that catalyze redox chemistry are introduced in the next chapter.

In the introduction to this chapter, we noted that both Zn^{2+} and Mg^{2+} are commonly used as Lewis acids in the enzymatic catalysis of hydrolysis reactions. Of these two metal ions, zinc is the stronger Lewis acid (reflected in the relative pK_a's for bound H_2O: $Zn–OH_2 \rightarrow Zn–OH^- + H^+$ $pK_a = 8.8$; $Mg–OH_2 \rightarrow Mg–OH^- + H^+$ $pK_a = 11.4$). As a general rule, Zn^{2+} is employed where the substrates bear a carbonyl (C=O) functional group (esters, amides, CO_2), and Mg^{2+} is found as a cofactor for enzymes that catalyze the hydrolysis (or formation) of phosphate esters. By way of introduction, we shall briefly discuss two structurally well-defined systems (carboxypeptidase A and alkaline phosphatase). In both cases, Zn^{2+} is the catalytic metal ion. These examples also illustrate the general function of such metal cofactors: (1) to stabilize an intermediate (e.g., carboxypeptidase A); (2) to stabilize a leaving group (e.g., alkaline phosphatase); (3) to bind simultaneously two reactive substrates and facilitate reaction through a proximity (template) effect (e.g., alkaline phosphatase).

$$M^{n+} + S \rightarrow M^{n+}–Y \rightarrow M^{n+} + P \tag{1}$$

$$M^{n+} + SX \rightarrow M^{n+}–XS \rightarrow M^{n+} X + S \tag{2}$$

$$M^{n+} + S + X \rightarrow M^{n+} \begin{array}{c} /S \\ \backslash X \end{array} \rightarrow M^{n+} + SX \tag{3}$$

4.2.1 Carboxypeptidase A

Bovine carboxypeptidase A (CPA) is a zinc exopeptidase of molecular mass 34.5 kDa. The enzyme catalyzes the hydrolysis of C-terminal amino acids from polypeptide chains and shows a preference for substrates with aromatic side chains. The active site contains a hydrophobic pocket that promotes binding of

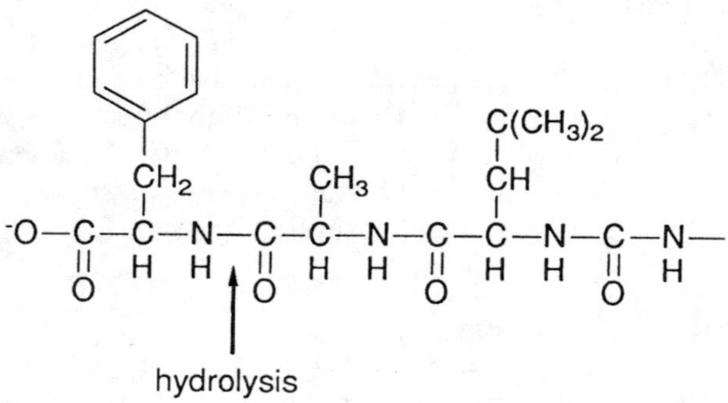

hydrolysis

such side chains. CPA was one of the first zinc enzymes (in fact, one of the first metalloenzymes) to be discovered and has been extensively investigated by kinetic, structural, and spectroscopic methods. As a result, it is one of the best-understood hydrolytic enzymes, providing a useful reference for related zinc enzymes. Zinc ion is bound in a tetrahedral geometry (Section 9.2 of Chapter 9 describes how the coordination chemistry of the zinc site can be investigated by use of Co^{2+} as a probe ion). The reaction pathway was first believed to involve a covalent acyl enzyme intermediate following nucleophilic attack by a protein side chain on the amide. However, there is now strong evidence in support of an alternative pathway involving direct attack by water, promoted by Zn^{2+} and assisted by a neighboring glutamate (Figure 4.9). Recent structural work has lead to a proposed mechanism that accommodates the available chemical and structural data. The mechanism favors the promoted-water pathway where nucleophilic water is deprotonated by Glu 270. A combination of the zinc ion and neighboring positively charged residues lower the pK_a of the bound water ($pK_a \sim 7$). Note that there is no direct coordination of the carbonyl oxygen to Zn^{2+} prior to addition of water. The role of the Zn^{2+} is therefore to stabilize negatively charged intermediates formed during hydrolysis rather than to polarize a bound carbonyl, which would have the adverse effect of increasing the pK_a of bound water. Instead, the carbonyl interacts with a neighboring Arg 127, which also stabilizes the negatively charged intermediates.

4.2.2 Alkaline Phosphatase

Escherichia coli alkaline phosphatase is a dimer of 94-kDa subunits that hydrolyzes a variety of phosphate esters. The enzyme has optimal activity around pH 8, hence the name. The basic reaction scheme is as follows and is seen to proceed by way of a covalently phosphorylated enzyme intermediate ($E-PO_3^{2-}$).

$$E + ROPO_3^{2-} \underset{k_{-1}}{\overset{k_1}{\rightleftharpoons}} E \cdot ROPO_3^{2-} \xrightarrow[ROH]{k_2} E - PO_3^{2-} \xrightarrow[H_2O]{k_3} E + HPO_4^{2-}$$

Burst kinetics is observed (Figure 4.10) since k_3 is rate-determining, and so ROH is rapidly produced in the early pre-steady state of the reaction.

Each subunit contains two Zn^{2+} ions (separated by a distance of 4 Å) and one Mg^{2+} that is located 5–7 Å from the binuclear zinc site. The two Zn^{2+} ions form the catalytic site. The role of Mg^{2+} is uncertain (it may perhaps be structural) but appears unrelated to catalysis. A high-resolution structure of the enzyme–phosphate complex is available (Figure 4.11), and, from this and related spectroscopic and kinetic data, the reaction pathway has been mapped out in some detail (Figure 4.11). The two Zn^{2+} ions are bound by His and Asp. One Zn^{2+} is positioned close to the serine hydroxyl. This decreases the pK_a of the hydroxyl and facilitates nucleophilic attack at the phosphate ester. The other Zn^{2+} stabilizes the developing negative charge on the leaving alkoxide group.

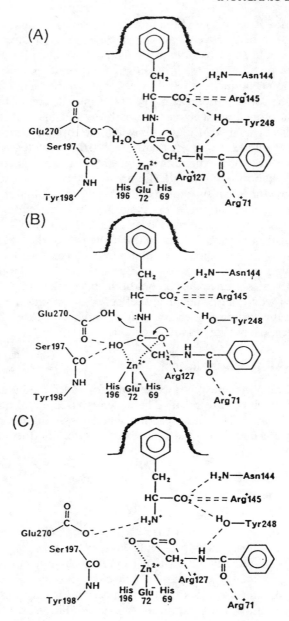

Figure 4.9 Proposed mechanism for carboxypeptidase A-catalyzed hydrolysis of an amide bond. (A) The amide carbonyl is hydrogen-bonded to Arg, and attack by a water molecule is assisted by glutamate and Zn^{2+}. Note that Zn^{2+} does not interact directly with the carbonyl oxygen. (B) The tetrahedral intermediate is stabilized, and then collapses with proton abstraction from glutamic acid. (C) Electrostatic interactions between the product carboxylate and Glu 270 may assist with product release. [Adapted from D. W. Christianson and W. N. Lipscomb, *Acc. Chem. Res. 22*, 62–69 (1989).]

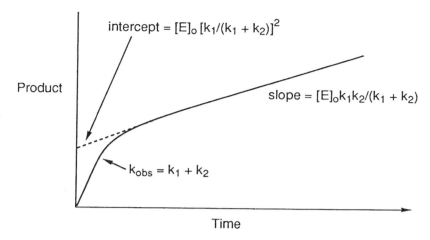

Figure 4.10 Schematic illustration of burst kinetics, showing a characteristic product versus time plot. The initial rise may occur very rapidly (within milliseconds) and may not be observed unless rapid kinetic methods (such as stopped-flow) are used.

The five-coordinate trigonal bipyramidal intermediate is bound and stabilized by both Zn^{2+} and a neighboring positively charged side chain (an arginine guanidinium group), and the resulting serine phosphate ester remains bound to Zn^{2+}. The pK_a of the aquated Zn(II) ion is 8.8, and so the coordination of H_2O to Zn(II) can produce an HO^- nucleophile to displace the phosphoryl group from the phosphoserine intermediate to yield a binuclear Zn^{2+} site bridged by the product metaphosphate. The pH dependence of the hydrolase activity provides evidence for the existence of such a zinc–hydroxide species. Because product phosphate results from two nucleophilic displacements (by Ser and HO^-), the reaction proceeds with the net retention of configuration at phosphorus. This fact will become important in the context of a comparison with the reaction catalyzed by purple acid phosphatase discussed in the next section.

4.2.3 (Purple) Acid Phosphatase

Acid phosphatases, so-termed because their pH optimum normally lies in the range 4.9–6.0, hydrolyze orthophosphate monoesters and are widespread in nature.

$$R–CH_2–O–PO_3^{2-} + H_2O \rightarrow R–CH_2–OH + HPO_4^{2-}$$

Two members of the mammalian acid phosphatase family have been extensively studied, uteroferrin and bovine spleen acid phosphatase. Each possesses a coupled binuclear iron center (shown later) that exhibits an intense purple color, and, thus, their name *purple acid phosphatases*. The different pH optima for the zinc-binding alkaline phophatase and the iron-binding acid phosphatase arise

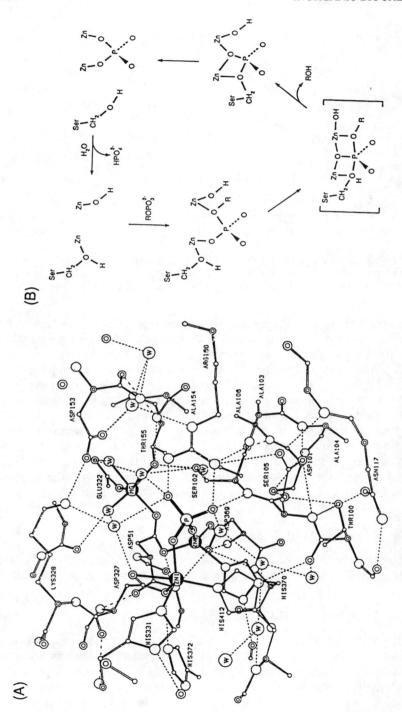

Figure 4.11 (A) Active site of *E. coli* alkaline phosphatase, and (B) a proposed mechanism incorporating the dinuclear zinc site. [Adapted from E. E. Kim and H. W. Wyckoff, *J. Mol. Biol.*, *218*, 449–464 (1991) and J. B. Vincent et al., *Trends Biochem. Sci.*, *17*, 105–110 (1992).]

in part from the distinct pK_a's of the metal-bound H_2O [Fe^{3+} $pK_a = 2.2^8$; Zn^{2+} $pK_a = 8.8$]. Both reaction pathways proceed by way of metal–OH^- species and possess a similar catalytic apparatus that includes a binuclear metal site to bind the substrate prior to a final metal-activated hydrolysis of the phosphate ester to yield a metal-bound phosphate (see Figures 4.11 and 4.15). The enzyme is active in the reduced form [Fe^{3+}–Fe^{3+} (oxidized, inactive); Fe^{3+}–Fe^{2+} (reduced, active)].

$$Fe^{3+}-Fe^{2+} \qquad Zn^{2+}-Zn^{2+}$$
$$| \qquad\qquad |$$
$$^-OH \qquad\qquad ^-OH$$

$$\text{acid phosphatase} \quad \text{alkaline phosphatase}$$

Certain plant phosphatases also contain binuclear Fe^{3+}–Zn^{2+} centers. However, on the basis of iron content and color, these are also designated purple acid phosphatases. The plant enzymes are insensitive to mild reductants, and there is little change in the electronic absorption spectrum of oxidized and reduced enzymes. This indicates that the optical characteristics derive from the common ferric iron center, which must also be insensitive to mild reducing conditions. We shall see shortly that the purple or pink color derives from a tyrosinate-to-Fe(III) charge transfer.

Many mammalian acid phosphatases have oligosaccharides bound to the protein surface and, therefore, fall into the class of glycoproteins. The oligosaccharide chains help make it possible to locate the enzyme in the lysosomes of spleen, bone, or tumor cells. In mammalian cells, phosphatases are abundant in digestive organelles (i.e., lysosomes). An alternative role as an iron carrier has been suggested for the acid phosphatase isolated from porcine uterine fluid. This protein, called uteroferrin, shows little enzymatic activity at the pH of the fluid and may only serve to transport iron from the sow to the fetal pig. Uteroferrin has a molecular mass of 36 kDa.

Binuclear iron centers are common to a variety of proteins and enzymes (Table 4.4) and have been extensively studied by a variety of spectroscopic meth-

Table 4.4 Enzymes and Proteins Containing Binuclear Iron Centers

Enzyme	Reaction
Hemerythrin	$Hr + O_2 \rightleftharpoons Hr \cdot (O_2)$
Acid phosphatase	$RCH_2OPO_3^{2-} + H_2O \rightarrow RCH_2OH + HPO_4^{2-}$
Methane monooxygenase[a]	$CH_4 + O_2 + NAD(P)H + H^+ \rightarrow CH_3OH + NAD(P)^+ + H_2O$
Ribonucleotide reductase[b]	$NDP + R(SH)_2 \rightarrow dNDP + RS_2 + H_2O$

[a] Methane monooxygenase, unlike the other enzymes listed, does not contain an oxo-bridge. The presence of an oxo-bridge has not yet been firmly established for acid phophatase enzymes.

[b] The substrate NDP is a general ribonucleotide diphosphate. Refer to Figure 1.20 for a typical structure.

ods in an effort to elucidate the coordination environment of each metal ion. We have already considered a role for this center in the O_2 carrier hemerythrin. Figure 4.12 summarizes the crystallographically defined or proposed active sites for two of the systems listed in Table 4.4. As expected for the mixed valent (Fe_A^{3+}/Fe_B^{2+}) active reduced form of the enzyme, Mossbauer spectra of ^{57}Fe-enriched uteroferrin show noticeable differences in isomer shifts (Fe_A δ = 0.52 mm/s, ΔE_Q = 1.83 mm/s; Fe_B δ = 1.27 mm/s, ΔE_Q = 2.66 mm/s] and are supported by optical spectroscopy. There is little difference between the absorption spectra of the Fe–Fe and mixed-metal (Fe–M) derivatives,[9] and the integrated intensity of the visible bands in the native enzyme is insensitive to whether the enzyme is oxidized (Fe^{3+}–Fe^{3+}) or reduced (Fe^{3+}–Fe^{2+}). This suggests that only one of the iron centers has a visible chromophore. Resonance-enhanced tyrosyl vibrations have been detected in Raman studies. These bands are retained in the reduced pink form of the enzyme, indicating that Tyr coordination is maintained in both the oxidized and reduced states (Figure 4.13). Since Fe(II)–Tyr complexes do not show LMCT in the visible region, Tyr must be coordinated to the iron that remains oxidized. The purple and pink colors in the oxidized and reduced enzyme, respectively, may then be assigned to a Tyr-to-Fe(III) charge transfer.

In the oxidized state, the enzyme is EPR-silent and almost diamagnetic. The two HS ferric ions (d^5, S = 5/2) are antiferromagnetically coupled. The reduced enzyme yields a rhombic EPR spectrum (Figure 4.14) with an integrated spin intensity corresponding to one unpaired electron. Again there is antiferromagnetic coupling between the ferric (S = 5/2) and ferrous (S = 2) ions. The identity

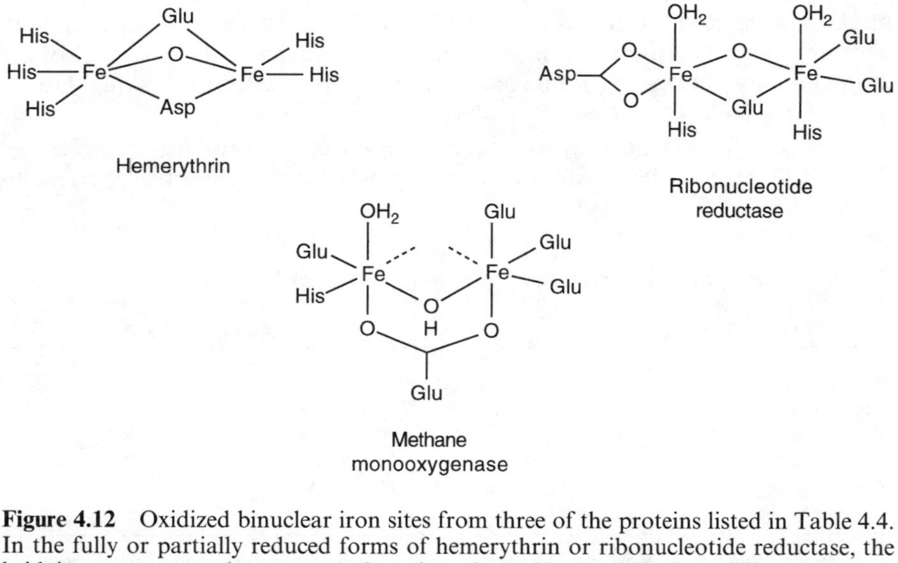

Hemerythrin

Ribonucleotide reductase

Methane monooxygenase

Figure 4.12   Oxidized binuclear iron sites from three of the proteins listed in Table 4.4. In the fully or partially reduced forms of hemerythrin or ribonucleotide reductase, the bridging oxygen may be protonated to give a bound hydroxide. For methane monooxygenase the free two sites at the rear are bridged by exogenous ligand.

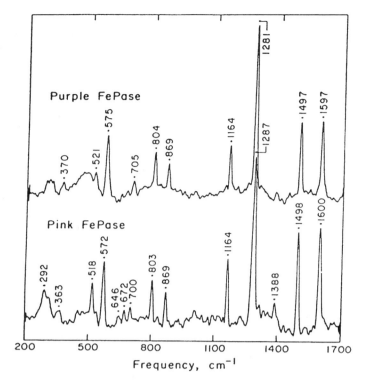

Figure 4.13 Resonance Raman spectra of oxidized (top) and reduced (bottom) purple acid phosphatase obtained using 514.5 nm excitation in the tyrosinate–Fe(III) LMCT band to emphasize tyrosyl vibrations. [Reprinted with permission from B. A. Averill et al., *J. Am. Chem. Soc., 109,* 3760–3767; copyright (1987) American Chemical Society.]

of the bridging ligand may be investigated by measuring the isotropic exchange coupling constant (J) (Table 4.5). These coupling parameters are defined by the following equation:

$$H = H_o - 2JS_1 \cdot S_2 + \text{other terms}$$

and may be measured by magnetic susceptibility experiments[10] or by monitoring the temperature dependence of either EPR spectral intensities or the chemical shifts of paramagnetically shifted resonances. The extent of antiferromagnetic coupling between spins S_1 and S_2 is described by the Hamiltonian $H = -2JS_1 \cdot S_2$. The values estimated for uteroferrin ($J_{ox} \geqslant 80$ cm^{-1}, $J_{red} \sim 10$ cm-1) are consistent with a μ-oxo bridge and a μ-hydroxo or μ-phenoxo bridge in the oxidized and reduced enzyme, respectively (Table 4.5). It should be noted, however, that, for the examples noted in Table 4.4, a μ-oxo bridge has been firmly established for oxidized hemerythrin and ribonucleotide reductase alone. The final verdict is awaited for the acid phosphatase, whereas the bridging ligand

(A)

(B)

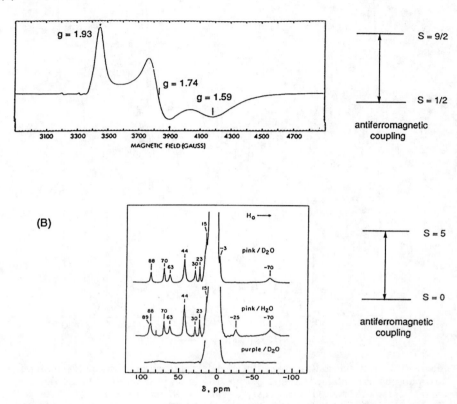

Figure 4.14 (A) EPR spectrum of reduced (pink) uteroferrin pH 4.9. The data are typical of a rhombic $S = \frac{1}{2}$ system that arises from antiferromagnetic coupling of the $S_1 = \frac{5}{2}$ and $S_2 = 2$ spins. [From Antanaitis and Aisen, *Adv. Inorg. Biochem., 5,* 130 (1983).] (B) ^{1}H NMR spectra of reduced (pink) and oxidized (purple) uteroferrin. By comparing spectra obtained in H_2O and D_2O, exchangeable protons neighboring the active site can be detected. In the oxidized form the enzyme is diamagnetic because of antiferromagnetic coupling of the two ferric centers, and so the paramagnetically shifted resonances are absent. [From Lauffer et al., *J. Biol. Chem., 258,* 14212–14218 (1983).]

Table 4.5 Exchange Coupling Constants for Coupled Binuclear Ferric Centers

Bridge	$-J(\text{cm}^{-1})$
μ-oxy	>80
OH and OR	~5–20
S-bridge	>60

for methane monooxygenase is a hydroxyl anion for both the reduced and oxidized enzyme.

The molecule p-nitrophenylphosphate is a convenient substrate for the reaction since the p-nitrophenolate product ion is colored and the reaction can be

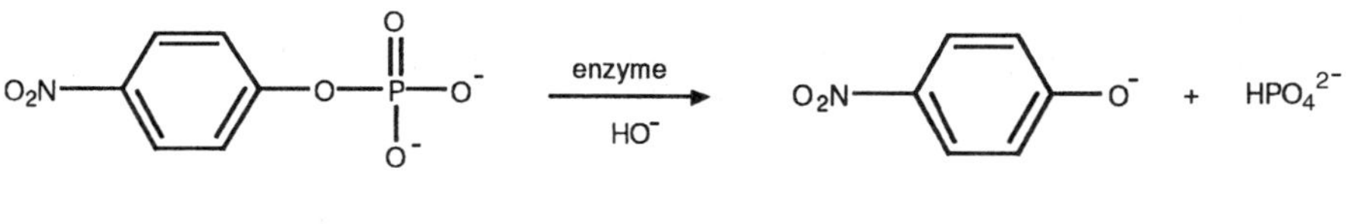

Colorless   Yellow

followed spectrophotometrically. The enzyme may be activated by reducing agents (e.g., 2-mercaptoethanol, ascorbate, glutathione, dithiothreitol) and ferrous ion (to inhibit loss of Fe^{2+} from the enzyme). Only one of the iron ions (the more labile of the two) is reduced, and so the mixed metal derivatives Fe^{3+}–Zn^{2+} and Fe^{3+}–Co^{2+} are active without reduction.

Several mechanisms have been proposed for this reaction, including metal-catalyzed release of metaphosphate (Figure 4.15A), direct attack of a metal-coordinated hydroxide (Figure 4.15B), and attack by a protein residue to give a phosphoenzyme intermediate that is subsequently hydrolyzed. Mechanistic studies using chiral phosphate, obtained by labeling P with ^{16}O, ^{17}O, ^{18}O, and S, indicate a net inversion of configuration at phosphorus and support either of the two schemes illustrated in Figure 4.15. In this regard purple acid phosphatase enzymes differ from other broad substrate phosphatases, such as alkaline and prostatic (from the prostate gland) phosphatases, which all proceed via a covalent intermediate as evidenced by the retention of configuration at phosphorus.

Summary of Section 4.2

1. Acid phosphatases contain a ubiquitous binuclear center that is also found in ribonucleotide reductase, hemerythrin, and methane monooxygenase.

(A)   (B)

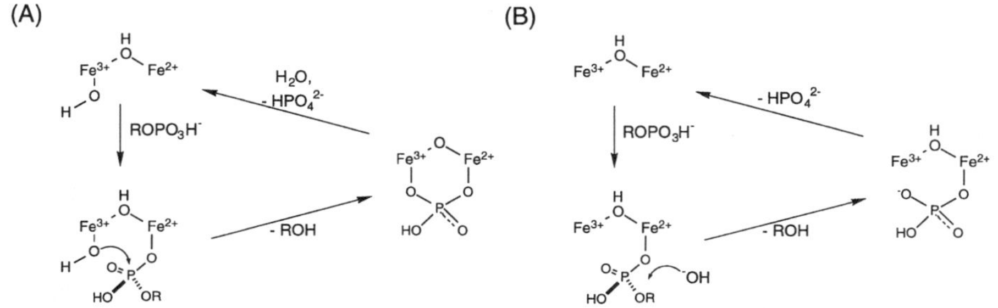

Figure 4.15 Two possible mechanisms for phosphate ester hydrolysis that result in a net inversion of configuration.

2. The two ions are antiferromagnetically coupled.
3. One of the iron centers is labile in the reduced state. This is a possible substrate binding site.
4. The ferric site binds tyrosine and a tyrosinate–Fe^{3+} ligand–metal charge-transfer band is the source of the purple (or pink) color of the enzyme in the oxidized (reduced) state.
5. Reaction occurs with inversion of configuration at phosphorus, consistent with direct transfer to water.

4.3 Hydro-Lyase Enzymes–Aconitase

Hydro-lyases fall into the general class of enzyme called lyases, but they catalyze the addition and elimination of hydro (or water) functionality. Aconitase is the second enzyme of the citric acid cycle (Figure 4.16), catalyzing the interconversion of citrate and isocitrate via *cis*-aconitate. The discovery that aconitase was an iron–sulfur (Fe–S) protein was quite unexpected inasmuch as all previously characterized Fe–S proteins functioned as electron-transfer agents. The studies that unraveled the structure of the active center and the mechanistic biochemistry are a beautiful demonstration of the combined approach of chemical and biophysical analysis to solve an important problem in biology.

The reaction that aconitase catalyzes is summarized in Figure 4.17. Characterization of the structural chemistry of the Fe–S prosthetic center (summarized as follows) was essential for the successful elucidation of the enzymatic reaction pathway, and, with the benefit of hindsight, clarifies much of the prior

Inactive   Active

spectroscopic and physicochemical data that had been collected on this enzyme. For example, the EPR spectrum of aconitase obtained from cellular extracts shows an isotropic signal with $g \sim 2$ that is very similar to signals previously reported for $[Fe_3S_4]^+$ clusters, while inactive aconitase shows a distinct spectrum that can revert to the original if the enzyme is incubated in the presence of both ferrous ion and reductant. The Mossbauer spectrum of inactive aconitase shows a single quadrupole doublet ($\Delta E_Q = 0.71$ mm/s, $\delta = 0.27$ mm/s) arising from a set of antiferromagnetically coupled ferric ions with a total spin of 1/2. After reduction, two quadrupole doublets are observed with a relative ratio of 2:1.

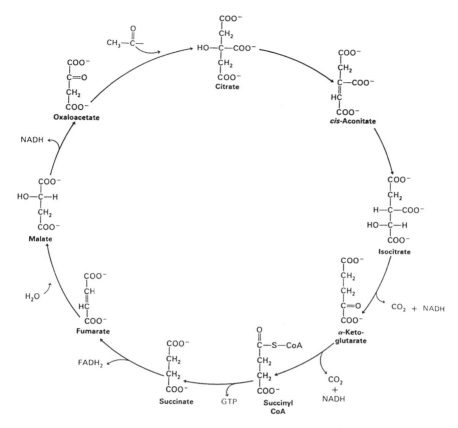

Figure 4.16 Kreb's cycle (also called the citric acid or tricarboxylic acid cycle). In mammals, this takes place in the mitochondrion and is a vital metabolic pathway for the conversion of fats and sugars to NADH- and FADH$_2$-reducing equivalents, which subsequently produce ATP through the mitochondrial electron-transport chain (see Chapters 5 and 10).

The low concentration signal has been assigned to Fe^{3+} (δ = 0.30 mm/s) whereas the high concentration signal with the larger isomer shift (δ = 0.46 mm/s) suggested a charge of +2.5 (i.e., one electron is shared by two Fe^{3+}). These data were unlike anything previously observed for Fe–S clusters. The 2:1 ratio of signals in the reduced protein indicated a total of three irons. Around this time, a novel 3Fe cluster was identified in crystallographic data from a ferredoxin (an electron-transport protein) isolated from the bacterium *Azotobacter vinelandii*. By careful quantitations of Fe and labile S^{2-}, the cluster was established as an [Fe$_3$S$_4$] center.[11] This represented a new class of Fe–S clusters in nature. Later it was demonstrated that ferredoxin II from the anaerobic bacterium *Desulfovibrio gigas* also contained such a center. The Mossbauer spectra of each of these reduced proteins are compared in Figure 4.18.

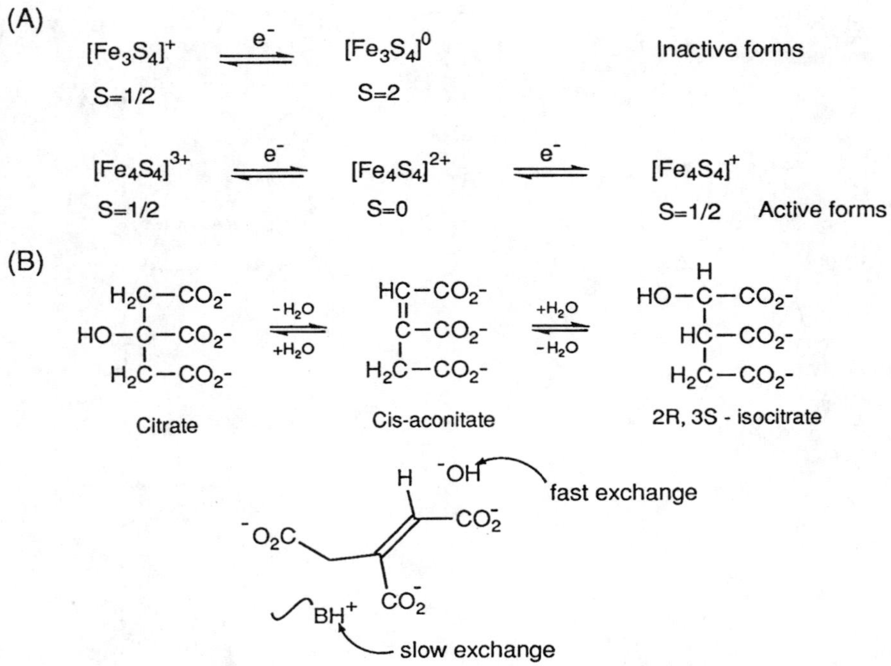

(A)

$$[Fe_3S_4]^+ \; \underset{}{\overset{e^-}{\rightleftharpoons}} \; [Fe_3S_4]^0$$

$S=1/2$ $S=2$ Inactive forms

$$[Fe_4S_4]^{3+} \; \underset{}{\overset{e^-}{\rightleftharpoons}} \; [Fe_4S_4]^{2+} \; \underset{}{\overset{e^-}{\rightleftharpoons}} \; [Fe_4S_4]^+$$

$S=1/2$ $S=0$ $S=1/2$ Active forms

(B)

Figure 4.17 Several early observations on enzyme activity included the following. (A) The enzyme is readily inactivated, although this can be reversed by addition of ferrous ion and reductant. (B) There is *trans*-addition of H_2O across the double bond in *cis*-aconitate to produce citrate or 2R,3S-isocitrate. A single base is involved in proton removal from substrate and protonation of aconitate. The proton exchange rate of the catalytic base with solvent is much slower than the enzyme turnover rate of ~15 s^{-1}, whereas the hydroxyl group undergoes rapid exchange with solvent during the reaction.

Once established that the inactive form of the enzyme possesses an $[Fe_3S_4]$ center, the requirement for additional iron and reductant to activate the enzyme becomes clear. The reduced $[Fe_3S_4]$ cluster binds Fe^{2+} to form an $[Fe_4S_4]$ center.[12] This fact can be used to good advantage in investigations of the chemistry of the labile iron site by Mossbauer spectroscopy. Enrichment of this site with $^{57}Fe^{2+}$ by addition to the iron-deficient aconitase (containing $[^{56}Fe_3S_4]$) yields a doublet in the Mossbauer spectrum with parameters ($\Delta E_Q = 0.80$ mm/s, $\delta = 0.45$ mm/s) that are very characteristic of $[Fe_4S_4]^{2+}$ clusters. This ^{57}Fe-enriched cluster can also be used to probe the substrate-bound complex (described later). The mechanism of iron incorporation has also been investigated. The rate of uptake of Fe^{2+} (determined by the rate of recovery of enzyme activity) was found to be much slower than the instantaneous reduction of the $[Fe_3S_4]^+$ center, and so the uptake of ferrous ion is dependent on a rate-limiting conformational change of the enzyme, following reduction. This conformational change, which

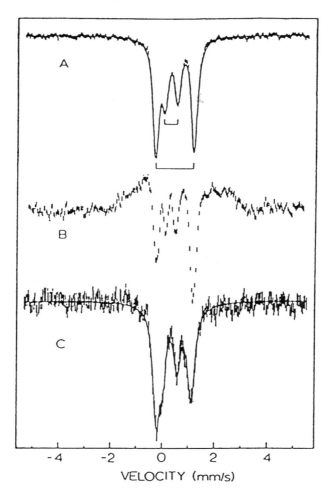

Figure 4.18   Mossbauer spectra of the reduced [Fe$_3$S$_4$] centers in (a) *Desulfovibrio gigas* ferredoxin II, (b) *Azotobacter vinelandii* ferredoxin, (c) beef heart mitochondrial aconitase. (Adapted from T. Kent et al., in *Electron Transport and Oxygen Utilization*, Elsevier, 1982)

precedes the insertion of iron ion, was also monitored directly by use of tryptophan fluorescence.[13]

The mechanism shown in Figure 4.19 provides a succinct summary of the many experiments that were targeted to elucidate the reaction chemistry of this interesting enzyme. In particular, detailed study of the enzyme–substrate complex was made possible by specific labeling of the exchangeable ferrous ion (Fe$_x$) [^{57}Fe$_x$^{56}Fe$_3$S$_4$] (described earlier). Addition of citrate to both of these active enzymes demonstrates substrate binding to the labile ferrous center (Fe$_x^{2+}$; Figure

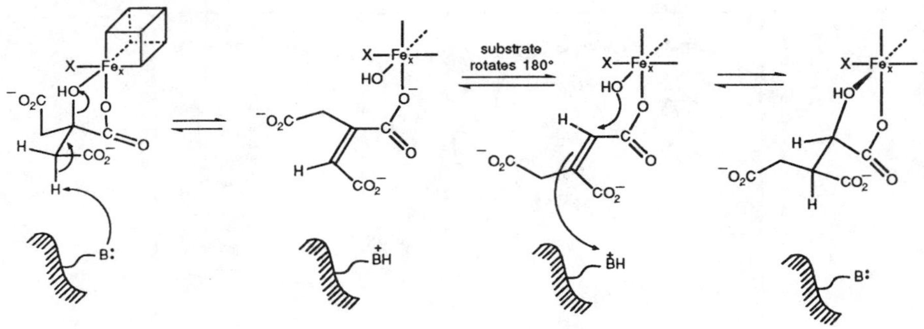

Figure 4.19 Proposed catalytic mechanism for aconitase. Citrate binds through the carboxylate and hydroxyl shown. After the deprotonation step, release of hydroxide is promoted by binding to the cluster. The substrate rotates 180° and the bound hydroxyl attacks the *cis*-aconitate intermediate to yield isocitrate as shown. The Fe–S cluster has acted as a Lewis acid, promoting the dehydration and rehydration of substrate and intermediate.

4.20). Moreover, comparison of the Mossbauer parameters [$\Delta E_Q = 1.26$ mm/s, $\delta = 0.84$ mm/s] with literature values (see Table 2.4) clearly shows that the ferrous ion is high-spin 6-coordinate. In principle, there are a variety of potential ligand atoms that might bind to ferrous ion (namely, three carboxylates and hydroxyl on the substrate, and water). The coordination chemistry was clarified by a combination of EPR and, most importantly, ENDOR experiments on aconitase with ^{17}O-labeled substrates and analogs. In the ENDOR experiment, NMR transitions of nuclei that are magnetically coupled to the electron spin are detected by the intensity change of an EPR transition that is simultaneously irradiated (Figure 4.20). Table 4.6 lists some results for key experiments in this study. Those ^{17}O-labeled ligands that bind directly to Fe_x in the cluster give rise to hyperfine coupling that can be measured.[14] These results demonstrate that both the hydroxyl and central carboxylate coordinate directly to the cluster, with water providing the sixth ligand.

Summary of Section 4.3

1. Aconitase is a key enzyme in the Kreb's cycle, catalyzing the isomerization of citrate and isocitrate.
2. An $[Fe_4S_4]^{2+}$ cluster forms the active site. Citrate binds to one of the iron centers. This iron is labile and is readily lost by mild oxidation to form an $[Fe_3S_4]^+$ cluster. The enzyme is activated by incubation with Fe^{2+} and reductants.

Notes

1. Hbs have also been found in the roots of leguminous plants and certain invertebrates (crustaceans, insects, and worms). Plant Hbs frequently contain many (>10) hemes.

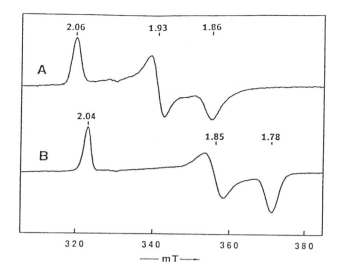

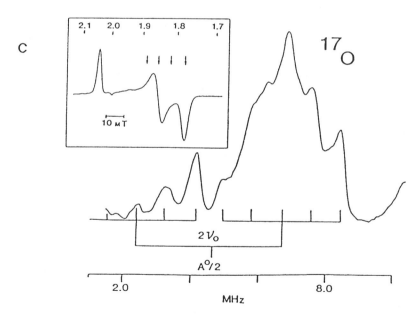

Figure 4.20 EPR spectra of aconitase. (A) no substrate added and (B) 1 mM citrate. [Reprinted with permission from M. H. Emptage, in *Metal Clusters in Proteins* (L. Que, ed.), ACS Symposium Series No. 372, pp. 343–371; copyright (1988) American Chemical Society.] (C) An ENDOR experiment. The inset shows the EPR spectrum of reduced aconitase in the presence of citrate. The EPR resonance is saturated at the position indicated and the intensity of the signal is perturbed by inducing relaxation at appropriate nuclear frequencies. Irradiation at $g = 1.88$ gave the ^{17}O ENDOR spectrum shown. [Adapted from J. Telser et al., *J. Biol. Chem., 261,* 4840–4846 (1986).]

Table 4.6 EPR and ENDOR Experiments with ^{17}O-Labeled Substrates and Water to Deduce the Ligand Atoms on Fe_x

---

Results from EPR and ENDOR experiments with ^{17}O-labeled substrate, analogs and water

Compound Added	^{17}O Labeling (●)	EPR Line width (mT)	EPR Broadening (mT)	Hyperfine Coupling Constant, A (MHz) Water	Hydroxyl
cis - Aconitate (88% citrate)	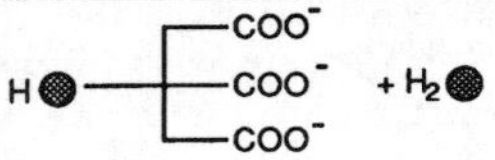 H●—C(—COO⁻)(—COO⁻)(—COO⁻) + H₂●	2.01	0.50	8.4	0
Nitroisocitrate	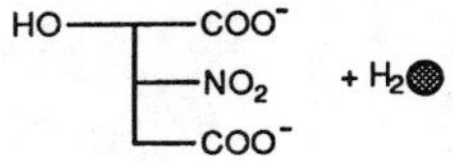 HO—(—COO⁻)(—NO₂)(—COO⁻) + H₂●	2.10	0.41	8.4	0
Nitroisocitrate	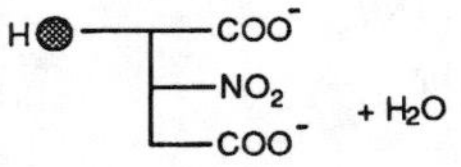 H●—(—COO⁻)(—NO₂)(—COO⁻) + H₂O	2.10	0.27	0	9.2
Nitroisocitrate	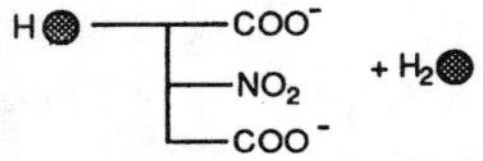 H●—(—COO⁻)(—NO₂)(—COO⁻) + H₂●	2.10	0.77	8.4	9.2

Results from ENDOR experiments with ^{17}O-labeled carboxylates of substrates and analogs

Compound Added	^{17}O Labeling (●)	Hyperfine Coupling Constant
Citrate	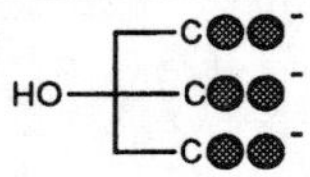 HO—C(—C●●⁻)(—C●●⁻)(—C●●⁻)	15
Citrate	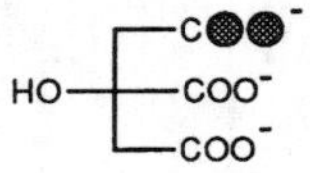 HO—C(—C●●⁻)(—COO⁻)(—COO⁻)	0
Citrate	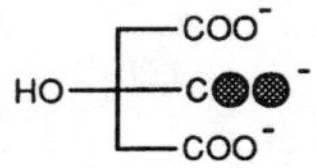 HO—C(—COO⁻)(—C●●⁻)(—COO⁻)	15
Isocitrate	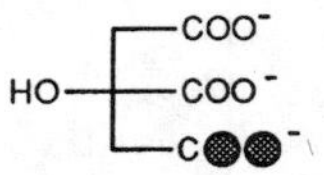 HO—C(—COO⁻)(—COO⁻)(—C●●⁻)	0
Nitroisocitrate	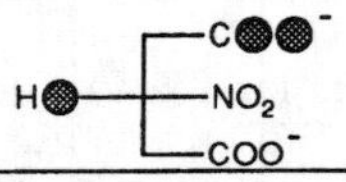 H●—C(—C●●⁻)(—NO₂)(—COO⁻)	9, 13

Source: Adapted from M. Emptage, in *Metal Clusters in Proteins*, ACS Symposium Series 372, (L. Que, ed.), American Chemical Society, 1988, pp. 343–371.

2. Cooperative binding implies that the binding of successive O_2 ligands becomes progressively stronger. The shapes of the curves in Figure 4.4 can be fit by a general equation of the form $f = KP^n/(1 + KP^n)$, where f is the fraction of Hb with bound O_2, P is the partial pressure of O_2, and K is the equilibrium binding constant for O_2 binding to heme for no cooperativity. For no cooperativity, $n = 1$, which gives the hyperbolic curve observed for Mb. For Hb, $n = 2.6$, indicating partial cooperativity. For total cooperativity, $n = 4$, and then Hb and $Hb(O_2)_4$ would be the only species found in solution.

3. For Mb, there is some controversy concerning the oxidation levels of the heme and oxygen species in the oxy-form. Optical data suggest O_2 binding to a low-spin ferrous heme. However, both Mossbauer and resonance Raman data indicate superoxide binding to a low-spin ferric center. The details of these issues need not concern us here, although the reader should be aware of the controversial points. As with most areas of inorganic biochemistry, the field is dynamic.

4. Christian Bohr was the father of Neils Bohr, who was awarded the Nobel Prize for physics in 1922.

5. Only one example is known of cooperative O_2 binding by an Hr isolated from a brachiopod.

6. For both Hb and Mb, there is actually some evidence for significant superoxide character. For example, the O–O stretch frequency is ~ 1107 cm^{-1}, which is close to the value for superoxide ion.

7. The O–O stretch vibration contains only minor contributions from other motions.

8. The enzyme-bound ferric iron is ligated by carboxylate and is unlikely to have a pK_a that is quite as low as the value quoted here for $Fe(H_2O)_6^{3+}$.

9. A variety of mixed-metal (Fe–M) species may be formed that provide useful spectroscopic probes of structure and reactivity. Exposure to reductants (ascorbate, dithionite, β-mercapto-ethanol) and iron chelators for short time intervals removes the labile ferrous ion (HS d^6), which can be replaced with Zn^{2+}, Cu^{2+}, Cd^{2+}, Co^{2+}, Hg^{2+}, and so on.

10. The magnetic susceptibility (χ_m), a measure of the inherent magnetism of a compound, is related to the magnetic moment (μ), where $\mu = e^{-J/kT}$. A plot of χ_m versus $1/T$ can be used to determine J.

11. Iron and sulfide ion are most easily detected by colorimetric tests, where a reagent (e.g., bath-ophenanthroline or ferrozine) that forms a colored complex with the ion of interest is added to the solution and the absorbance at a defined wavelength is compared with a calibration chart prepared from data collected against solutions of known concentration. For example, see Siegel, L. M. *Anal. Biochem.*, *11*, 126–132, (1965) for S^{2-} and Fish, W. W. *Methods in Enzymol.*, *158*, 357–364 (1988) for $Fe^{3+/2+}$.

12. This type of metal exchange $[Fe_3S_4] + M \rightleftharpoons [MFe_3S_4]$ for reduced 3Fe clusters is now well understood and many metal derivatives have been characterized ($M = Zn^{2+}, M^{2+}, Cd^{2+} \ldots$). The $[Fe_4S_4]$ center in clusters that can readily lose iron are sensitive to oxidative damage. Two-electron oxidation by ferricyanide yields the oxidized $[Fe_3S_4]$ center and Fe^{3+}.

13. Refer to Ramsay and Singer, *Biochem.*, *J. 221*, 489–497 (1984) for further details.

14. The ^{17}O-labeled compounds in Table 4.6 were synthesized by chemical or enzymatic methods. The hydroxyl was labeled by taking advantage of the fact that aconitase rapidly exchanges both citrate and isocitrate hydroxyl with solvent ($H_2^{17}O$). The nitro analog of citrate and isocitrate have a lower pK_a for the important central proton. Furthermore, the hydroxyl group on the nitro analogs do not exchange with solvent since the alternative equilibrium (as shown here) is not favored, and so the coordination of H_2O and substrate hydroxyl to the cluster may be distinguished.

Further Reading

Oxygen Carriers

Ellerton, H. D., N. F. Ellerton, and H. A. Robinson. Hemocyanin—A current perspective. *Prog. Biophys. Molec. Biol. 41*, 143–248 (1983).

Klotz, I. M., and D. M. Kurtz. Binuclear oxygen carriers: Hemerythrin, *Acc. Chem. Res., 17*, 16–22 (1984).

Kurtz, D. M. Oxo- and hydroxo-bridged diiron complexes: A chemical perspective on a biological Unit. *Chem. Rev. 90*, 585–606 (1990).

Perutz, M. *Mechanisms of Cooperativity and Allosteric Regulation in Proteins*, Cambridge, 1990.

Hydrolase Enzymes

Antanaitis, B. C., and P. Aisen. Uteroferrin and purple acid phosphatases, *Adv. Inorg. Biochem., 5*, 111–136 (1983).

Cheblowski, J. I., and J. E. Coleman. Zinc and its role in enzymes, in *Metal Ions in Biological Systems*, Vol. 6 (H. Sigel, ed.), Dekker, 1976.

Christianson, D. W., and W. N. Lipscomb. Carboxypeptidase A, *Acc. Chem. Res., 22*, 62–69 (1989).

Kim, E. E., and H. W. Wyckoff. Reaction mechanism of alkaline phosphatase based on crystal structures, *J. Mol. Biol., 218*, 449–464 (1991).

Loehr, J.-S. Binuclear iron proteins, in *Iron Carriers and Iron Proteins*, (Loehr, T. M., ed.), VCH, New York, 1989, pp. 373–468.

Mueller, E. G. et al. Purple acid phosphatase: A diiron enzyme that catalyzes a direct phosphogroup transfer to water. *J. Am. Chem. Soc. 115*, 2974–5 (1993).

Vincent, J. B., G. L. O-Lilley, and B. A. Averill. Proteins containing oxo-bridged dinuclear iron centers: A bioinorganic perspective, *Chem. Rev., 90*, 1447–1462 (1990).

Aconitase

Emptage, M. H. Aconitase: Evolution of the active site picture, in *Metal Clusters in Proteins*, ACS Symposium Series 372 (L. Que, ed.), American Chemical Society, 1988, pp. 343–371.

Haile, D. J. et al. Cellular regulation of the IRE-BP: Disassembly of the cubane iron–sulfur cluster results in high-affinity RNA binding. *Proc. Natl. Acad. Sci. USA 89*, 11734–39 (1992).

Kennedy, M. C. et al. Purification and characterization of cytosolic aconitase from beef liver and its relationship to the iron-responsive element binding protein. *Proc. Natl. Acad. Sci. USA 89*, 11730–34 (1992).

Lauble, H. et al. Crystal structures of aconitase with bound isocitrate and nitroisocitrate. *Biochemistry 31*, 2735–2748 (1992).

Zheng, L. et al. Mutational analysis of active site residues in pig heart aconitase. *J. Biol. Chem. 267*, 7895–7903 (1992).

Problems

1. Both CO and O_2 bind to reduced heme centers. Draw Lewis structures for CO and O_2 and explain why these ligands bind to the heme in a linear and bent arrangement, respectively.

2. (a) The table below lists on and off rate constants for ligand binding to the isolated subunits of human hemoglobin at a solution pH of 7.0 and 20 °C. Discuss any trends you observe, providing molecular insight on the reasons behind the trend and its significance.

Ligand	β Subunit		α Subunit	
	Association ($\times 10^{-6}$ M^{-1} sec^{-1})	Dissociation (sec^{-1})	Association ($\times 10^{-6}$ M^{-1} sec^{-1})	Dissociation (sec^{-1})
O_2	60	16	50	28
CO	4.5 (4.5)	0.008	4.5	0.013
NO	30	2.2×10^{-5}	30	4.6×10^{-5}
Alkyl isocyanides (R-CH$_2$NC)				
Methyl	0.79	7.0	0.39	3.9
Ethyl	0.30	0.80	0.14	0.17
n-Propyl	0.083	0.54	0.040	0.14
n-Butyl	0.34	1.2	0.059	0.31
n-Pentyl	0.45	1.0	0.083	0.41
n-Hexyl	1.00	1.5	0.12	0.32
tert-Butyl	0.0088	0.23	0.0012	0.06

(b) Evaluate the binding free energy for the oxygen, carbon monoxide, and nitric oxide adducts of the Hb β-subunit.

[Olson, J. S. *Methods in Enzymol.* 76, 631–651 (1981)]

3. Several redox active proteins contain a vacant metal binding site in an Fe$_3$S$_4$ cluster that can be populated by a variety of metals

$$[Fe_3S_4]^0 + M^{2+} \overset{k_1}{\underset{k_2}{\rightleftharpoons}} [Fe_3MS_4]^{2+}$$

Binding constants follow the order $Cd^{2+} > Zn^{2+} > Co^{2+} > Fe^{2+} > Mn^{2+}$.

(a) Comment on the relative binding parameters with reference to the hardness and softness of the metal ions and ligands, and the Irving-Williams series.

(b) In the case of aconitase, an observed equilibration rate constant k_{eq} can be evaluated. This is given by $k_{eq} = k_1[M^{2+}] + k_2$. Derive this equation from first principles. In the pseudo-first-order approach described in Chapter 1, we obtained an equivalent expression for $k_{obs} = k_1[M^{2+}]$. What assumption was made in the derivation of that simpler expression?

(c) In the presence of SO_4^{2-}, the expression for k_{eq} in the case of ferrous ion changes to $k_{eq} = k_1[Fe^{2+}] + k_2 + k_3[FeSO_4]$. Explain.

(d) It is found that $k_1 = 2.3$ M^{-1} s^{-1} and $k_3 = 276$ M^{-1} s^{-1} Comment on this fact. Given that three histidine residues lie in close proximity to the metal binding pocket, suggest a plausible reason for the difference in rate constants.

[Faridoon, K. Y. et al., *Inorg. Chem.* 33, 2209 (1994)]

4. The diiron (II) state of the hydroxylase subunit of soluble methane monooxygenase reacts with O_2 to form an intermediate with a broad visible absorption at $\sim$ 600–650 nm with an extinction coefficient $\varepsilon_{625} = 1500$ M^{-1} cm^{-1}. Irradiation of this intermediate at 647 nm gives a resonance Raman spectrum with a peak at 905 cm^{-1}. The signal moves to lower energy with ^{18}O-labeled dioxygen.

(a) Suggest a chemical formulation for the diiron–oxygen complex.

(b) Is the absorbance centered around 625 nm due to a d–d or charge-transfer transition? Explain your reasoning.

[Liu, K. E. et al., *J. Am Chem. Soc. 117*, 4997 (1995)]

5. Steady-state kinetic data obtained for the reaction of carboxypeptidase A are consistent with the reaction scheme shown, where K_s is the substrate

$$E + S \overset{K_S}{\rightleftharpoons} E.S_1 \underset{k_{-2}}{\overset{k_2}{\rightleftharpoons}} E.S_2 \overset{k_3}{\rightarrow} E + P$$

binding constant, with rapid formation of an initial intermediate ES_1, followed by formation of a second distinct intermediate ES_2 with an observed rate constant k_{obs}. A schematic representation of the fluorescence trace obtained from pre-steady-state turnover of a dansyl-labeled substrate is shown.

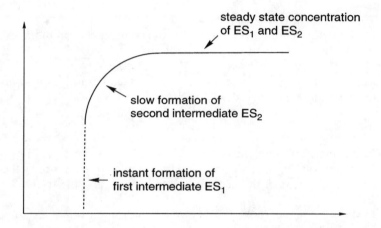

(a) Derive equations for k_{cat}, K_m and k_{obs} in terms of K_s, k_2, k_{-2}, k_3, and $[S]$.

(b) For a substrate with $k_{cat} = 1.18$ s^{-1} and $K_M = 13.5$ μM, a plot of $1/(k_{obs} - k')$ versus $1/[S]$ yields a straight line with y intercept 0.0306 s, and a slope of 3.13 μM s^{-1}, where k' is the limiting value of k_{obs} at low $[S]$ and has a value of 4.8 s^{-1}. Calculate the values of K_s, k_2, k_{-2}, and k_3.

(c) Provide an explanation for the pH dependence of log (k_{cat}/K_M), log k_{cat}, and $-$log K_M, shown below. Estimate the pK_as and identity of the functional groups that are being titrated.

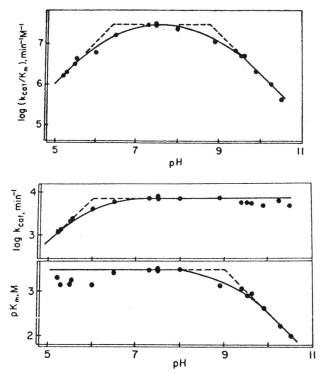

(d) Carboxypeptidase A can be activated by a variety of metal cofactors and is capable of hydrolyzing both amide and ester substrates. Comment on the kinetic parameters summarized in the table below for turnover of specific ester and amide substrates.

metal	Amide		Ester	
	k_{cat} (min^{-1})	K_M (mM^{-1})	k_{cat} (min^{-1})	K_M (mM^{-1})
Co^{2+}	6000	1500	39000	3300
Zn^{2+}	1200	1000	30000	3000
Mn^{2+}	230	2800	36000	660
Cd^{2+}	41	1300	34000	120

[Galdes, A. et al., *Biochemistry 22*, 1888 (1983); Auld, D. S. and B. L. Vallee, *Biochemistry 9*, 4352 (1970); and Chapter 2 of *Zinc Enzymes,* Spiro T. G., ed., Wiley (1983)]

6. The structure of urease has recently been determined. This landmark protein was first crystallized in 1926 by James B. Sumner and provided a proof that enzymes were, in fact, protein molecules. It was later to become the first characterized nickel enzyme. Only very recently, however, was the crystal structure finally solved. The active site contains two nickel ions. One is bound by water, whereas the other is coordinately unsaturated.

  (a) We have already reviewed some enzymes that possess a similar array of metals in the active site. Name them.

  (b) Propose a role for each nickel site and draw a mechanism that is consistent with your proposal.

  (c) The two nickel centers are bridged. A lysine residue lies close to both nickels. Is this likely to bridge the nickel centers? It has been demonstrated that CO_2 is required to promote nickel binding to metal-free urease. Explain this observation.

[Evelyn, J. et al., *Science 268,* 998–1004 (1995)]

5

Metalloproteins and Metalloenzymes: (II) Redox Chemistry

5.1 Introduction

Electron-transfer reactions have been a continuing source of fascination to chemists and have proved central to the development of modern coordination chemistry. It should therefore come as no surprise that redox proteins and enzymes are among the best characterized and most widely studied by inorganic biochemists. They display interesting electronic and magnetic phenomena and demonstrate a varied redox chemistry that raises many important and fundamental questions concerning reaction mechanisms. In this chapter we shall study a diverse range of redox proteins and try to understand the fundamental principles underlying their reactivity.

By way of introduction, we consider two essentially different types of biological pathways that use redox proteins or enzymes. First, there are respiratory pathways (illustrated by scheme A) that are involved in the production of chemical energy within the cell. A series of electron-transfer steps are coupled to chemical reactions through membrane-bound, multisubunit, enzyme complexes that contain several prosthetic centers. A reactive substrate molecule, such as NADH, acts as an initial electron source, and these electrons are ultimately trapped by a terminal electron acceptor. This energetically favorable process is coupled to, and provides the driving force for, a series of chemical redox reactions. These reactions are not in themselves the net products of the respiratory electron-transfer chain. Rather, these redox reactions are devices used to pump

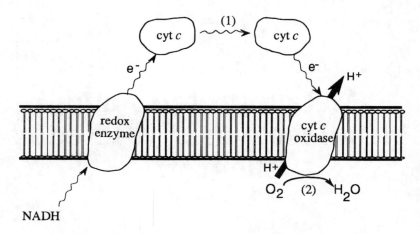

Scheme A

ions (especially H^+) and metabolites across the bacterial or mitochondrial membrane. These transmembrane concentration gradients, in turn, drive the synthesis of ATP, and so electron-transfer pathways are ultimately transformed into a useful form of chemical energy! Second, there are anabolic and catabolic pathways (scheme B) where we define anabolic reactions as those that build large molecules from smaller substrates, while catabolic reactions break down bio-

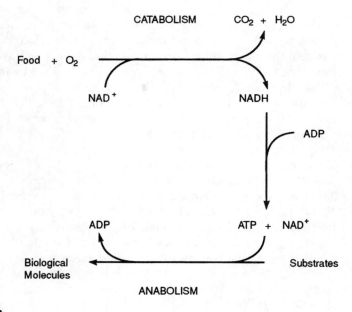

Scheme B

logical macromolecules into smaller fragments that may then be further used in cellular metabolism.

For either of these two pathways we can identify two types of metalloredox protein or enzyme, as follows.

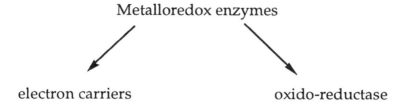

Metalloredox enzymes

electron carriers   oxido-reductase

First, there are electron-transfer proteins that serve only to shuttle electrons between a donor and acceptor [e.g., cytochrome *c* in scheme A, step (1)]. Second, one can identify oxido-reductase enzymes that catalyze redox reactions and combine an electron acceptor–donor function with a net chemical transformation [e.g., cytochrome *c* oxidase, step (2)].

In this chapter we first become familiar with the common redox centers found in a variety of metalloredox proteins and enzymes, and then consider the functional role of proteins containing these centers in respiratory and other metabolic pathways. Topics related to cellular detoxification mechanisms are described in Chapters 8 and 10.

5.2 Prosthetic Centers, Cofactors, and Coenzymes in Metalloredox Proteins and Enzymes

It is clear that nature employs a large number of oxido-reductase enzymes, a reflection of the necessarily complex biological environment in which they function. The polypeptide chains that make up the protein regulate both the physicochemical properties of the redox centers and the interactions of the protein with other biological molecules. For example, membrane-spanning helices might position the protein in a membrane. Protein–protein contacts define interactions in multisubunit enzymes and the binding of natural redox partners. Protein side chains may enhance the reactivity of a metalloprosthetic group. Channels that facilitate the coupling of redox chemistry to proton pumping must be defined. Surface attachments of lipids or sugar molecules further define the eventual location of a protein in a cell. In spite of the obvious complexities of these molecules, there are a few simplifying features; for example, many possess common prosthetic centers with related functional properties. For this reason it is worth reviewing the chemistry of the common redox prosthetic centers, cofactors and coenzymes, before considering the detailed biochemistry of metalloredox enzymes. First we give a few definitions. A prosthetic center is a collection of exogenous metal ions and/or organic ligands that are firmly held by the protein or enzyme and typically remain bound during isolation and purification procedures. Cofactors and coenzymes differ inasmuch as they tend to be less tightly bound, are normally absent from freshly isolated protein, and must be added to promote activity. The common redox centers are introduced in the following sections. Their chemistry, the proteins in which they are found, and the biochemical roles of these redox proteins will be elaborated on later.

5.2.1 Major Redox Centers

We have already introduced the two main categories of redox centers: (1) electron carrier sites, and (2) sites that catalyze oxidation-reduction chemistry on substrate molecules. We begin our review of the most common redox prosthetic centers and coenzymes by noting some general characteristics that underlie their functional role. The chemistry of a redox site in a metalloprotein depends on which of these two classes it belongs to.

Metal redox centers in *electron-carrier* proteins are typically found in coordination sites that undergo minimal structural change as a result of oxidation or reduction. This minimizes the activation barrier associated with electron exchange (discussed in Section 5.5) and optimizes electron-transfer rates. For example, heme centers in redox proteins, such as cytochrome c, are six-coordinate, resulting in a low-spin $(d_{xy})^2(d_{yz})^2(d_{xz})^2$ configuration in the reduced state. These orbitals do not point directly at the bound ligands (compare d_{z^2} and $d_{x^2-y^2}$), and so there is a minimal requirement for structural rearrangement, after electron loss, to form ferric heme. Similarly, Fe–S centers undergo minimal structural changes. Although the iron centers are now high spin, the ligand set forms a tetrahedral geometry that again does not point directly at the metal d-orbitals. In the case of the "blue copper" proteins, the ligand set is firmly held by the protein backbone in a distorted tetrahedral geometry that is intermediate between the preferred ligand geometries for Cu^{2+} and Cu^+. Electron-carrier proteins are therefore characterized by redox centers that are structurally designed to facilitate rapid and efficient electron transfer.

We can contrast this with the situation for enzymes that catalyze *oxido-reductase* chemistry on bound substrate molecules. In these cases, the redox cofactor will possess, in at least one oxidation state, a vacant coordination site for substrate binding. The electronic structure of the cofactor also promotes facile binding and release of substrate and product, respectively. Frequently, this coordination site is occupied by a weakly bound water molecule that can be readily displaced by substrate. Examples include the square pyramidal or tetragonal copper sites in a number of oxidases or reductases, the high-spin, five-coordinate heme center in cytochrome P-450, or the special labile iron in Fe_4S_4 clusters in aconitase (see Section 4.3). In all these examples, there exists a vacant coordination site on a labile metal center.

Many of the common redox centers can be classified according to their reduction potentials. Since the relative reduction potentials of a series of redox proteins serves to define the direction of electron flow, this type of categorization can provide useful insight on why specific redox proteins are used in certain metabolic pathways. Table 5.1 lists reduction potentials for a variety of redox centers. These are ordered from most positive to most negative potentials in the following series. Although there are several exceptions, the general trend in reduction potential, noted as follows,

$$Cu > heme > Fe–S \sim flavin$$
$$E°(V) \quad + \quad 0 \quad -$$

helps to explain the tendency of Cu and heme to serve as oxidizing centers, and Fe–S clusters and flavin to serve as reducing centers in redox enzymes. For example, for copper, positive potentials indicate that the reduced state of the cofactor is most stable, and so Cu^{2+} will take electrons from (or oxidize) a substrate molecule. The reverse holds true for the flavin redox center. Furthermore, the preceding scheme rationalizes the relative ordering of Cu, heme, or Fe–S/flavin electron-carrier proteins in particular metabolic cycles. Remember that the relative $E°$'s of biological redox centers, for the most part, simply define the direction of electron flow.

Iron–Sulfur Clusters

Cluster-containing proteins are normally categorized according to the number of iron ions in the prosthetic center: (1Fe) rubredoxin (Rd); (2Fe) ferredoxin (Fd); (4Fe) ferredoxin (Fd). The class of $[Fe_4S_4]$ ferredoxins may be further subdivided into low- and high-potential ferredoxins.[1] The wide variety of structural forms of these iron–sulfur sites and their characteristic EPR signatures are illustrated in Figure 5.1. The clusters are bound to the enzyme by coordination to cysteine and are arranged as a cluster with bridging sulfides.

Table 5.1   Reduction Potentials for Selected Redox Proteins and Enzymes

Protein	Redox Cofactor	$E°$ (mV vs. NHE)
Plastocyanin	$Cu^{2+/1+}$	370
High-potential iron protein	$[Fe_4S_4]^{3+/2+}$	360
Cytochrome c	Heme ($Fe^{3+/2+}$)	260
Azurin	$Cu^{2+/1+}$	250
Stellacyanin	$Cu^{2+/1+}$	180
Cytochrome b	Heme ($Fe^{3+/2+}$)	60
Myoglobin	Heme ($Fe^{3+/2+}$)	50
Rubredoxin	$Fe^{3+/2+}$	−60
Cytochrome c peroxidase	Heme ($Fe^{3+/2+}$)	−194
Cytochrome P-450	Heme ($Fe^{3+/2+}$)	−300
Ferredoxin	$[Fe_4S_4]^{2+/1+}$	−400
Flavodoxin	Flavin	−420
Ferredoxin II	$[Fe_2S_2]^{2+/1+}$	−430

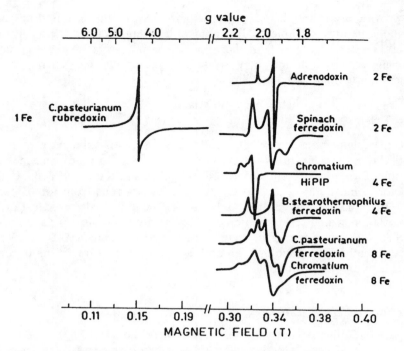

Figure 5.1 Iron–sulfur centers and their characteristic EPR spectra.

Two additional Fe–S centers that have been characterized are the $[Fe_3S_4]$ core found in aconitase (see Section 4.3) and Ni-Fe hydrogenase, among others, and the Rieske center identified in mammalian mitochondria.[2] Clusters most commonly serve as simple electron carriers and have been structurally designed for this purpose. However, we have already seen one exception (the $[Fe_4S_4]$ center in active aconitase) where the coordination environment of one iron site has evolved to participate in the chemical transformation of a substrate molecule.

Flavins

The prosthetic center of these redox proteins may uniquely serve as either a one- or two-electron carrier. Flavodoxins and ferredoxins can often be used interchangeably in biological redox cycles, and so an organism grown in an iron-deficient medium will produce additional flavodoxin to make up for the deficiency of ferredoxin. The various oxidation states of the flavin chromophore (Figure 5.2) are most readily characterized by electronic absorption spectroscopy, although these spectral characteristics do show some dependency on the flavin derivative that is bound to the protein, that is, riboflavin, flavin mononucleotide (FMN), or flavin adenine dinucleotide (FAD). The fully reduced forms of the flavin are denoted $FMNH_2$ and $FADH_2$, respectively. The biosyn-

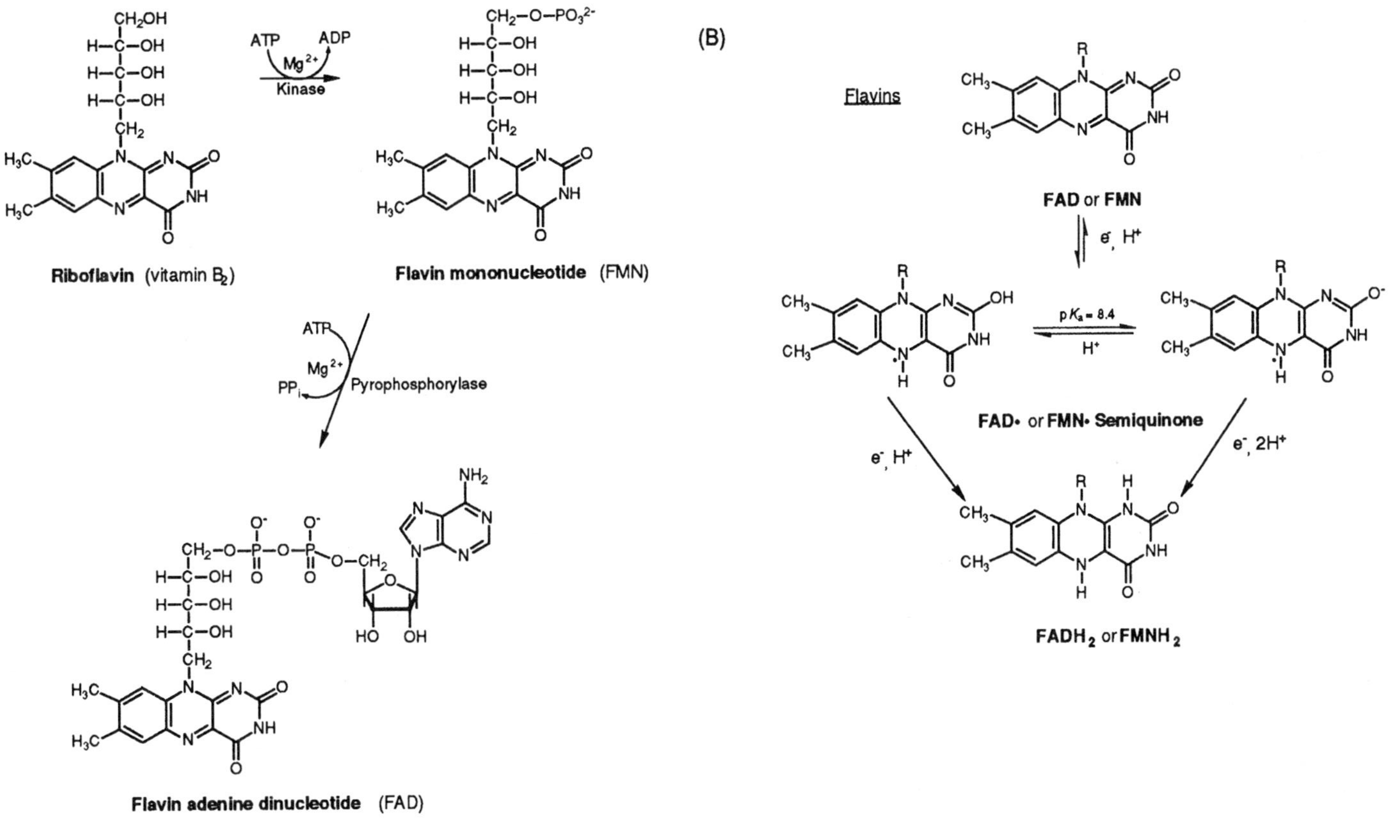

Figure 5.2 (A) Biosynthetic pathways linking flavin derivatives. (B) Oxidation states of the flavin cofactor showing the one- and two-electron reduced ring.

thetic relationships linking the most common flavin derivatives are shown in Figure 5.2. The nature of the flavin ring allows for facile coupling of atom-transfer and electron-transfer redox chemistry.

$$FAD + 2H^+ + 2e^- \rightarrow FADH_2 \qquad \text{electron transfer}$$
$$FAD + H^+ + \text{``}H^-\text{''} \rightarrow FADH_2 \qquad \text{atom transfer}$$

Hemes

Hemes perhaps represent the archetypal electron-transfer unit. The heme center is so pervasive in discussions of electron-transfer proteins or enzymes that those iron-containing proteins that do not contain heme iron are generally referred to as *non-heme* iron proteins. Figure 5.3 illustrates the most common types of heme centers according to their classification as *a*-, *b*-, *c*-, or *d*-type cytochromes. In common with other redox centers, hemes are used as both electron carriers and catalytic sites for oxidation-reduction chemistry. Cytochrome *c* is probably the most thoroughly studied example of any class of redox protein. As explained in the introduction, the hexacoordinate low-spin heme shows minimal structural change following electron transfer and the low activation barrier promotes facile electron transfer. In contrast, the catalytic hemes in cytochrome *P*-450 and hexaheme nitrite reductase are high-spin pentacoordinate with vacant coordination sites for substrate. The high-spin configuration (resulting from the pentacoordinate geometry) promotes the rapid exchange of substrate and product at heme iron since there is a smaller contribution from ligand field stabilization to the activation energy (review Sections 1.6, 1.7, and 1.11 in Chapter 1).

Copper Ion

Copper binding sites are categorized as type I, type II, and type III according to their coordination environment and geometry. Type I refers to the distorted tetrahedral center depicted in Figure 5.4. Type II centers are characterized by a square planar or tetragonal geometry around copper, which is ligated by *N*- or *O*-ligands. Type III copper centers contain a pair of antiferromagnetically-coupled bridged copper ions. The diversity of the coordination environments is reflected in their biochemical roles (Table 5.2).

Type I, Blue Copper Proteins. The name derives from the unusually intense blue color of proteins containing type I copper centers. This color derives from a strong Cys–Cu^{2+} LMCT band around 620 nm in the electronic absorption spectrum. Type I centers are not associated with any direct role in catalysis; rather they serve as electron-transfer sites in mobile electron-carrier proteins, such as azurin and plastocyanin. In more complex oxido-reductase enzymes that contain multiple functional centers, type I copper serves to deliver or take up electrons from the catalytic site. It is clear from Table 5.3 that nature has found ways of tuning the reduction potentials for copper centers over a large range. The structural mechanisms that achieve this are not yet clearly defined.

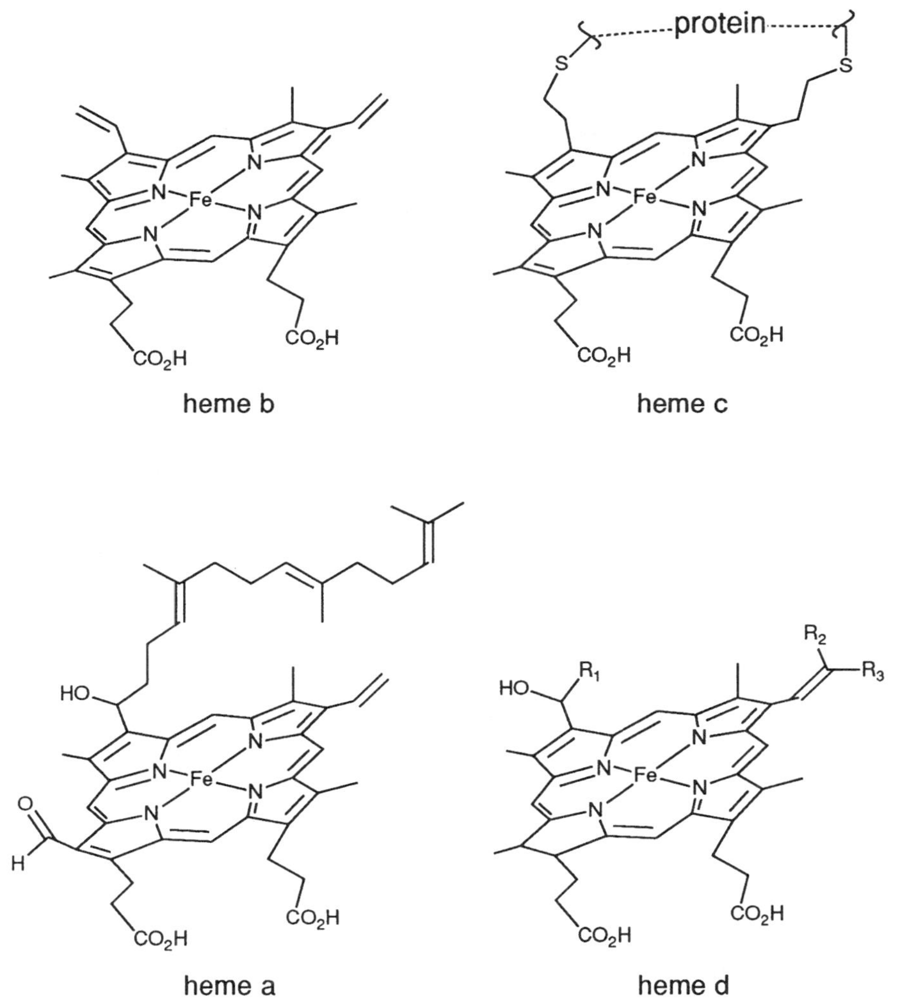

Figure 5.3 Classification of cytochromes according to heme type. The metal-free ligands are derivatives of the porphyrin ring system. Note that c-type hemes are covalently bonded to the protein through thioether linkages to cysteine residues. Heme d is a dihydro-reduced porphyrin ring. Heme b (protoporphyrin IX) is commonly found in myoglobins and hemoglobins.

Type II. Relative to type I sites, the reduction potentials in type II sites are generally more positive and there is a change to a N- and O-ligand environment. The weak bands in electronic absorption spectra and larger hyperfine coupling constants in EPR spectra are typical of tetragonal coordination (Table 5.4). Although type I centers are used only for electron transfer, type II copper has vacant coordination sites and can be involved in catalytic oxidation of substrate

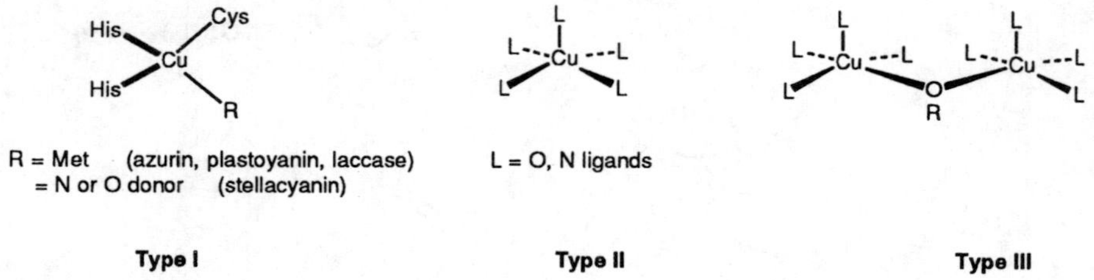

R = Met   (azurin, plastoyanin, laccase)   L = O, N ligands
  = N or O donor   (stellacyanin)

Type I   **Type II**   **Type III**

Figure 5.4   Types I, II, and III copper centers in metalloproteins.

Table 5.2   Biochemical Roles for Representative Copper Proteins

Protein[a]	Copper Content	M_r	Function[a]
Azurin (Az)	1 × type I	14,000	Az(ox) + e^- ⇌ Az(red)
Laccase	1 × type I, 1 × type II, 2 × type III	64,000	$2RH_2 + O_2 \rightarrow 2R + 2H_2O$ $4RH + O_2 \rightarrow 4R + 2H_2O$
Tyrosinase	2 × type III	46,000	phenol + O_2 + $2H^+$ + $2e^-$ ⇌ o-diphenol + H_2O 2-o-diphenol + O_2 ⇌ 2-o-quinone + $2H_2O$
Hemocyanin (Hc)	2 × type III	75,000	Hc + O_2 ⇌ Hc·(O_2)

[a] Azurin is an electron-carrier protein in eukaryotic respiratory pathways. Laccase catalyzes oxidative chemistry on a variety of substrates (especially p-diphenols). These reactions are particularly common in the formation of a protective polymeric lacquer after injury to trees. Tyrosinase catalyzes the hydroxylation of monophenols and/or the two-electron oxidation of diphenols to quinones. Hemocyanin is an oxygen-binding protein in mollusks and arthropods.

Table 5.3   Reduction Potentials Measured at pH 7 (vs. NHE) for Type I, II, and III Copper Centers in a Variety of Metalloredox Proteins[a]

Protein	$E°$ (mV)	Protein	$E°$ (mV)
Type I		**Type II**	
Azurin	230–280	Laccase (*P. versicolor*)	782
Plastocyanin (*P. vulgaris*)	360	Laccase (*R. vernicifera*)	434
Stellacyanin (*R. vernicifera*)[b]	191	Galactase oxidase	410
Laccase (*P. versicolor*)	785		
Laccase (*R. vernicifera*)	415	**Type III**[c]	
		Laccase (*P. versicolor*)	570
		Laccase (*R. vernicifera*)	390

[a] $E°$ for Cu^{2+}(aq) is $+153$ mV.
[b] Stellacyanin lacks a methionine ligand, and a N- or O-donor ligand may substitute.
[c] Data for type III sites is open to error.

Table 5.4 Optical and EPR Spectral Characteristics for Types I and II Copper Sites

Protein	Prominent Absorption Bands [nm, (mM^{-1} cm^{-1})]	EPR Parameters[a]
Type 1		
Azurin	625, (3.5)	Axial $g_\perp$ = 2.052, $g_\parallel$ = 2.29, $A_\parallel$ = 6
Plastocyanin	597, (4.9)	Axial $g_\perp$ = 2.053, $g_\parallel$ = 2.26, $A_\parallel$ = 5
Laccase	607, (9.7)	Rhombic g_x = 2.030, g_y = 2.055, g_z = 2.300, $A_\parallel$ = 4.3
Ascorbate oxidase	614, (5.2)	Rhombic g_x = 2.036, g_y = 2.058, g_z = 2.227, $A_\parallel$ = 45.8
Type II		
Laccase	788, (0.9)	Axial $g_\perp$ = 2.053, $g_\parallel$ = 2.237, $A_\parallel$ = 20.6
Ascorbate oxidase	760, (3.6)	Axial $g_\perp$ = 2.053, $g_\parallel$ = 2.242, $A_\parallel$ = 19.9

[a] Parallel hyperfine coupling constants ($A_\parallel$) are noted in units of 10^3 cm^{-1}. The perpendicular component is too small to measure.

molecules. Type II may act in isolation (e.g., dopamine-β-hydroxylase) or in concert with a pair of type III coppers (e.g., ascorbate oxidase, Figure 5.5).

Type III. Several proteins contain binuclear copper sites that are nondetectable by EPR. These proteins can accept as many electrons as there are copper ions present, and so the diamagnetism cannot be attributed to a Cu(I) oxidation level. Rather, the pair of cupric ions is strongly antiferromagnetically coupled (refer to Section 2.3.3), and so the spins cancel. Type III centers are frequently found in proteins that also possess type I and/or type II copper. Ascorbate oxidase (Figure 5.5) illustrates how a type III pair may be closely associated with a type II copper to form a more complex functional unit. Type III centers

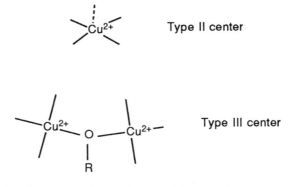

Figure 5.5 Active sites in ascorbate oxidase and laccase. A type II copper lies in close proximity to a type III dimer. [See A. Messerschmidt et al., *J. Mol. Biol.,* 206, 513–529 (1989).]

constitute the O_2-reducing site in many oxidase enzymes. In this way the thermodynamically unfavorable formation of superoxide $(O_2^{-\cdot})$ (Table 4.2) may be avoided since reduced type III centers have two electron equivalents for direct reduction of O_2 to O_2^{2-}.

Molybdenum

Molybdenum is an essential cofactor in a number of redox enzymes, commonly termed *molybdoenzymes* (Table 5.5). With the exception of the molybdenum center in nitrogenase, which takes the form of a Fe_7Mo cluster, the molybdenum cofactors (Mocos) have been identified as molybdenum–pterin complexes (Figure 5.6). Both L_3MoO_2 and L_3MoOS (L = ligand) sites are known, depending on the enzyme, and three oxidation states $(6+, 5+, 4+)$ are available. The $Mo=O$ and $Mo=S$ functionality may change to $Mo–OH$ or $Mo–SH$ according to solution pH and the oxidation state of Mo. In each case, the Mo center couples electron-transfer to atom-transfer chemistry, and so there is typically a latent coordination site (a labile Mo-bound ligand X that can be readily displaced by substrate).

5.2.2   Other Redox Cofactors and Coenzymes

In addition to the five classes of redox centers noted earlier, several other redox cofactors and coenzymes may be essential for enzymatic activity. The cofactors to be discussed first are of particular interest since the biosynthetic route to many of these molecules derives from a common pathway to heme (Figure 5.7).

Table 5.5   Representative Molybdoenzymes

Enzyme	Reaction
Nitrogenase[a]	$N_2 + 8H^+ + 8e^- \rightarrow 2NH_3 + H_2$
Nitrate reductase	$NO_3^- + 2H^+ + 2e^- \rightarrow NO_2^- + H_2O$
Sulfite oxidase	$SO_3^{2-} + H_2O \rightarrow SO_4^{2-} + 2H^+ + 2e^-$
DMSO reductase	$(CH_3)_2SO + 2H^+ + 2e^- \rightarrow (CH_3)_2S + H_2O$
Trimethylamine N-oxide reductase	$(CH_3)_3NO + 2H^+ + 2e^- \rightarrow (CH_3)_3N + H_2O$

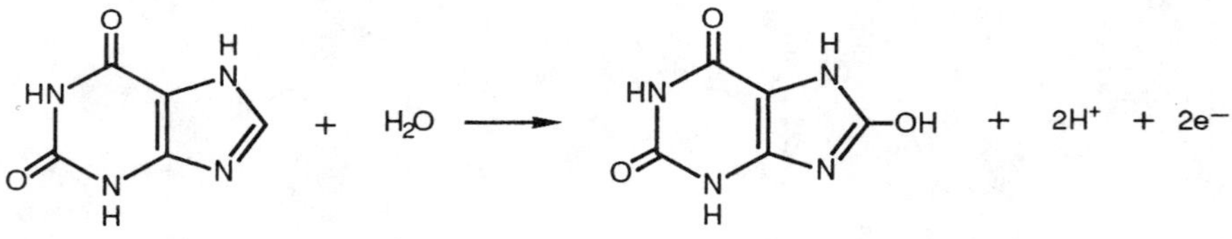

Xanthine   Uric acid

Xanthine oxidase

[a] Nitrogenase contains a Fe_7Mo-cluster cofactor (Figure 5.21). All other enzymes contain Mo-pterin (Figure 5.6).

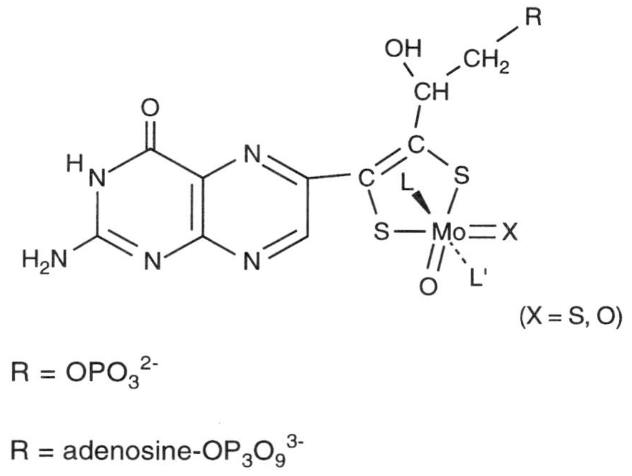

$$R = OPO_3^{2-}$$

$$R = \text{adenosine-OP}_3O_9^{3-}$$

Figure 5.6 Pterin cofactors in molybdoenzymes. The identity of the ligands L and L′ and the precise stereochemical details of the coordination at Mo remain unclear.

Cobalamins

The term *cobalamin* is generally used to describe a cobalt derivative of the corrin ring system that carries a benzimidazole ligand on one face and one of a number of possible ligands (R) on the other (Figure 5.8). Of these the two alkyl derivatives (methyl and adenosyl) are the most important and are formed by oxidative addition of the activated alkyl to Co(I) corrin, yielding the Co(III) derivative. Cyanocobalamins (vitamin B_{12}) are enzymatically reduced from Co(III) to Co(I) species and alkylated with adenosine triphosphate (ATP) or *S*-adenosylmethionine (SAM). Bond cleavage sites are shown earlier. Adenosylcobalamins find use as coenzymes for rearrangement reactions (Figure 5.9, Table 5.6), whereas methylcobalamin is utilized as a methylating agent (a source of "CH_3^-"). In each case, the reaction mechanism appears to proceed by homolytic cleavage of the Co(III)–C bond to give an alkyl radical and Co(II). The B_{12} cofactor allows free-radical chemistry to proceed more readily, and at lower solution potentials, than might otherwise be possible since it is mediated by the cobalt center.

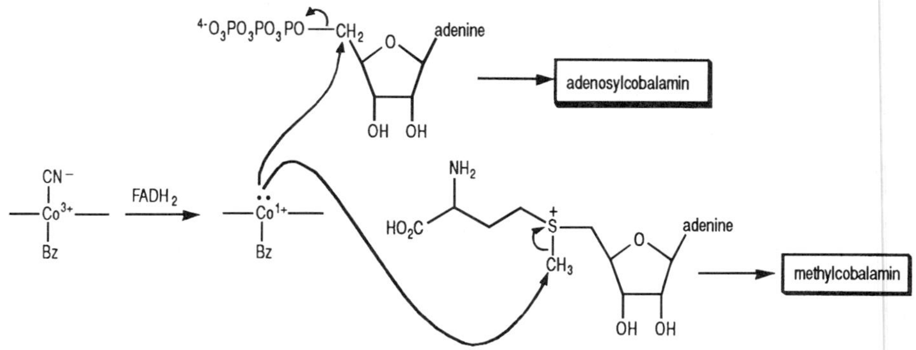

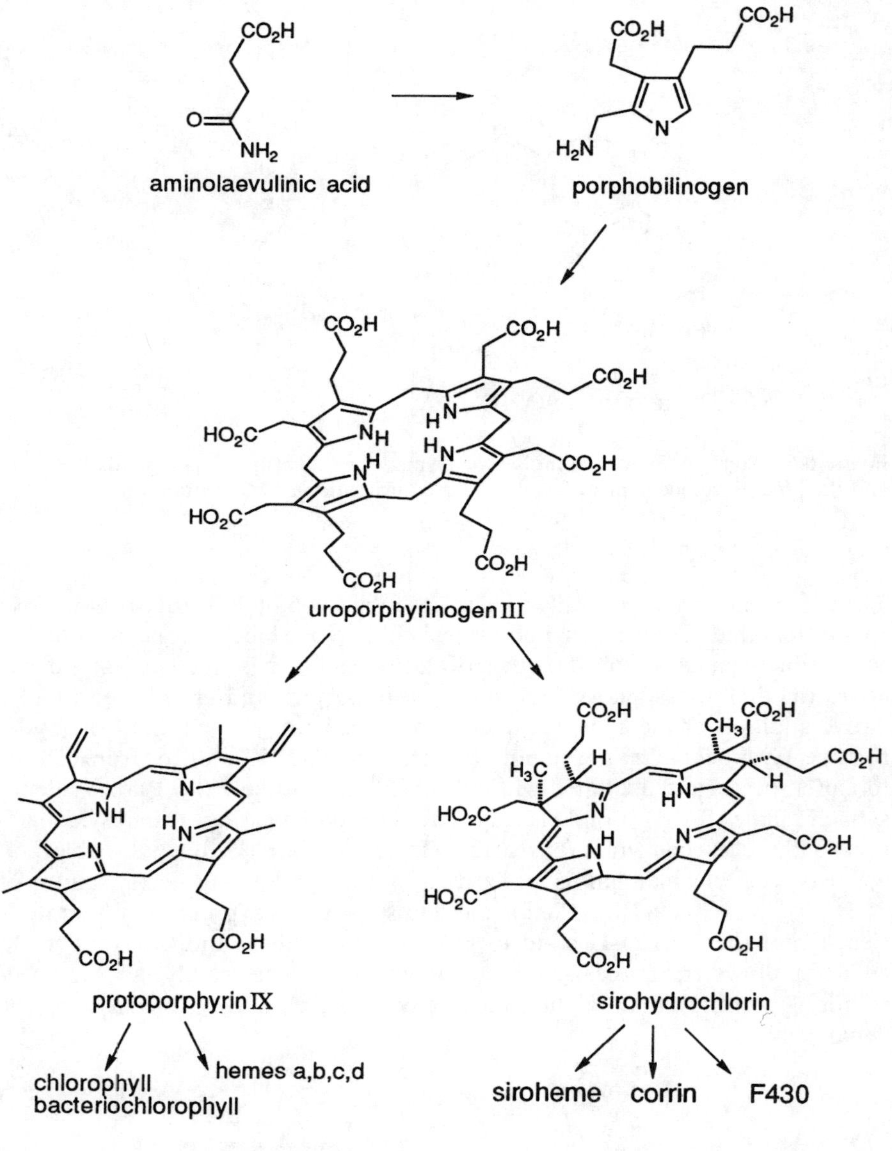

Figure 5.7   Biosynthesis of redox cofactors from common heme precursors.

Nickel Coenzymes and Cofactors

Factor F_{430} is a tetrapyrrolic ligand that binds nickel and is involved in the final step of methane formation in methanogenic bacteria. The coenzyme is bound by methyl-S-coenzyme-M-reductase, which catalyzes the reaction shown in Figure 5.10. A nickel center has also been implicated in two other important classes of enzymes, called hydrogenases and carbon monoxide dehydrogenases. These

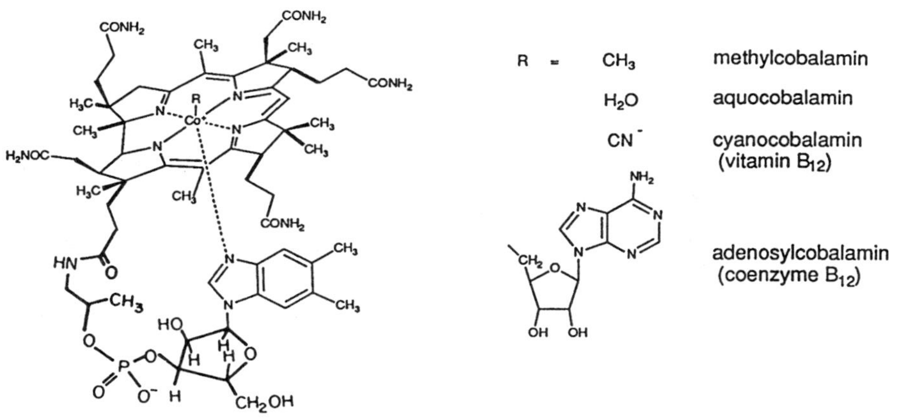

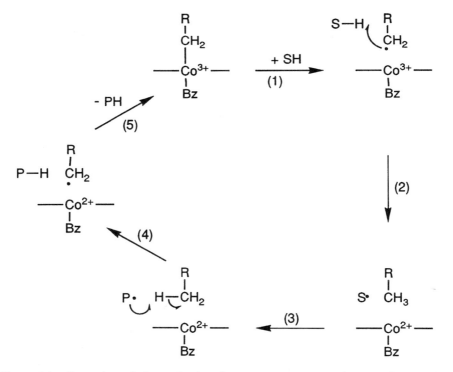

Figure 5.8 Cobalamin derivatives. A tabulation of electronic absorption data for these corrinoids can be found in B. Babior, ed., *Cobalamins*, Wiley, 1975.

Figure 5.9 General catalytic mechanism for rearrangement reactions. Bz is the axial benzimidazole ligand, and R is part of the adenosyl group. (1) Homolytic cleavage of the cobalt–carbon bond. (2) Hydrogen abstraction from the substrate (SH). (3) Rearrangement to product radical P·. (4) P· recovers the hydrogen. (5) The cobalt–carbon bond is reformed.

Table 5.6 Examples of B_{12}-Catalyzed Reactions

Enzyme	Reaction

Reduction

Ribonucleotide reductase

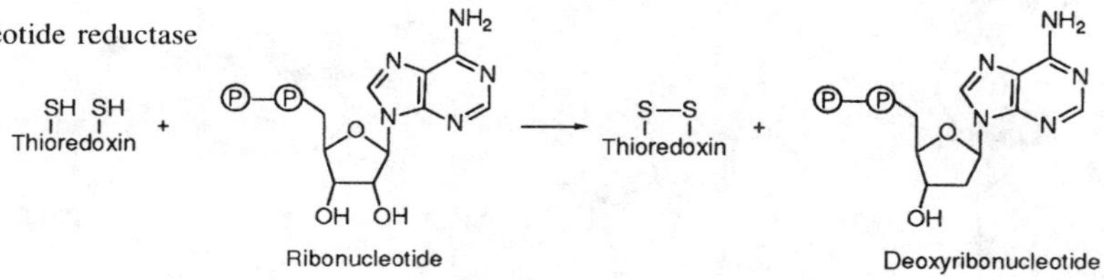

Rearrangement

Glutamate mutase

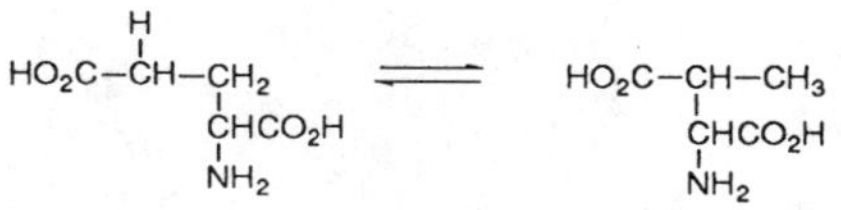

Diol dehydratase

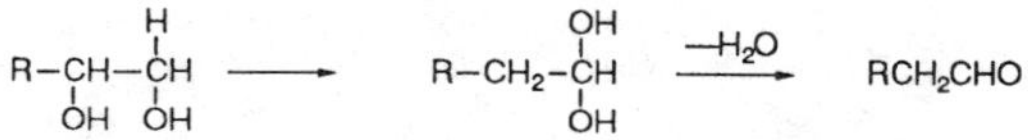

Methyl transfer

Methionine synthetase

enzymes catalyze oxido-reductase reactions of hydrogen and carbon monoxide ($H_2 \rightleftharpoons 2H^+ + 2e^-$, and $CO + H_2O \rightleftharpoons CO_2 + H_2$, respectively).

Reduced Porphyrin Derivatives

The fully oxidized porphyrin ring represents the maximal state of unsaturation in such a ring structure. Figure 5.11 shows some reduced forms of the porphyrin skeleton that are common in nature and are characterized by discrete optical and redox properties that tailor these and related molecules for specific functions (light harvesting, photoinduced electron transfer, SO_3^{2-} and NO_2^- nitrite reduction). Magnesium derivatives of chlorins[3] (chlorophylls) are present in green algae, cyanobacteria, and higher plants. Bacteriochlorophylls are located in green and purple bacteria. Both classes of molecules show strong absorption of visible light, and so play a key role in the harvesting of light energy during photosynthesis.

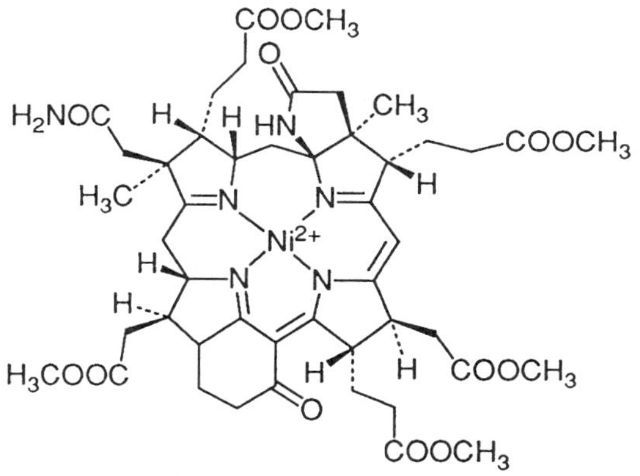

Figure 5.10 Factor F_{430} and its role in methane formation. Methyl-S-coenzyme-M $(CH_3SCH_2CH_2SO_3^-)$ is reduced to form CH_4 and coenzyme-M $(HSCH_2CH_2SO_3^-)$. F_{430} is a catalytic cofactor. The mechanism has not been established.

Organic Redox Cofactors

Figure 5.12 illustrates several of the more common organic redox cofactors (no metals!). Some of these, especially quinone and flavin derivatives, may participate in redox reactions by either electron-transfer or atom-transfer chemistry.

Summary of Sections 5.1–5.2

1. Biological electron-transfer reactions encompass both oxido-reductase enzymes that carry out chemical transformations on a substrate molecule and electron carriers that shuttle electrons between redox centers.

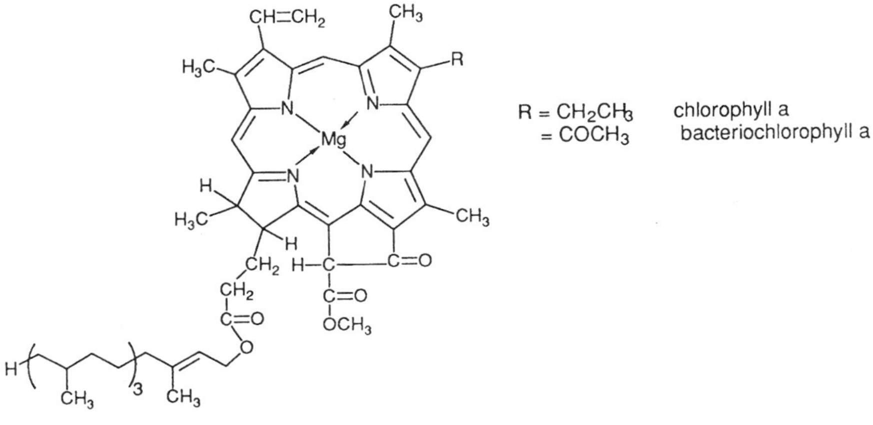

Figure 5.11 Reduced porphyrins associated with photosynthesis as ancillary pigments in the light-harvesting complex or redox centers in the reaction center.

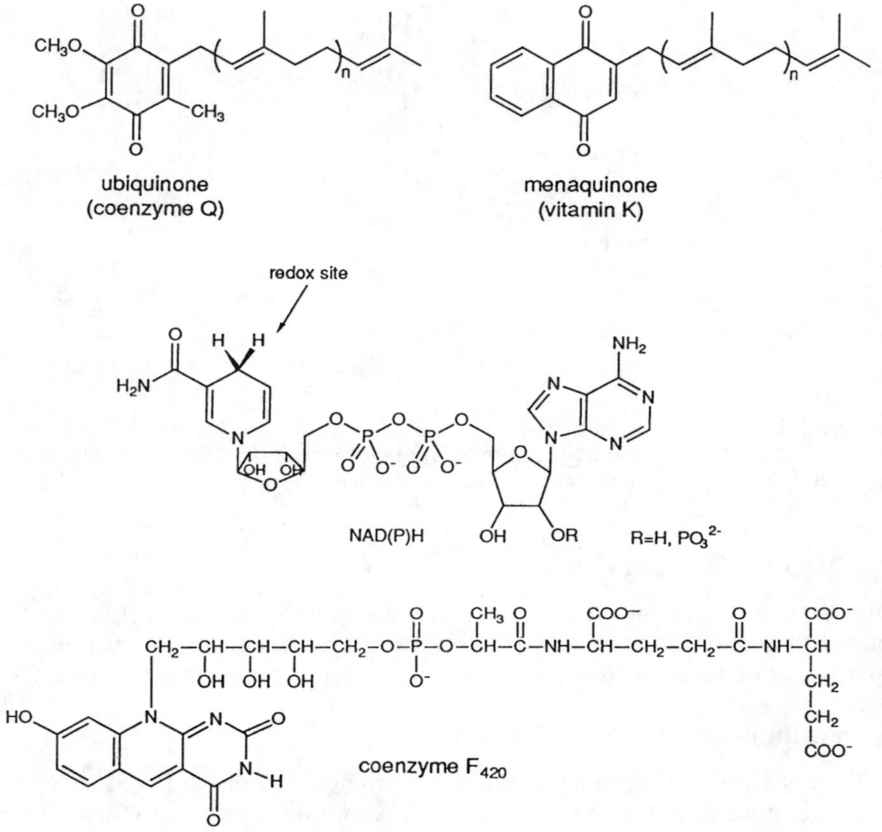

Figure 5.12 Other redox cofactors and coenzymes. Coenzyme F_{420} is isolated from methanogenic bacteria (see Figure 5.17).

2. Nature employs only a very few redox prosthetic centers (Fe–S clusters, hemes, flavins, copper, molybdenum–pterin) and several redox cofactors and coenzymes (cobalamins, F_{430}, quinones) to carry out a large range of chemical processes.
3. The protein matrix regulates the physicochemical properties of each center to match the required chemistry.

Review Questions

• Compare and contrast the chemistry of dinuclear copper centers (in ascorbate oxidase, cytochrome c oxidase, and hemocyanin), ferredoxin-like dinuclear iron, and other types of dinuclear iron (in MMO, acid phosphatase, hemerythrin, and ribonucleotide reductase).
[Solomon, E. I, et al., *Chem. Rev. 92,* 521 (1992); Feig, A. L. and S. J. Lippard, *Chem. Rev. 94,* 759 (1994)]

- Cyanide binds tightly to oxidized heme but weakly to reduced heme, whereas it binds weakly to oxidized siroheme but tightly to reduced siroheme. Explain. [Lui, S. M. et al., *Biochem. J. 304*, 441–447 (1994)]

- From Table 5.5 note that molybdoenzymes serve as oxidases or reductases by transferring oxygen to or from the substrate, respectively. Comment on this fact, with particular emphasis on the thermodynamics of each type of reaction. [Holm, R. H. *Chem. Rev. 87*, 1401 (1987)]

5.3 Electron-Transfer Pathways in Respiratory Metabolism

Having introduced the basic components of biological redox systems, we shall now turn our attention to the manner in which an array of such protein-bound centers can be fashioned into a functional electron-transport chain that performs useful work. We saw, in the introductory Section 5.1, that biological electron-transfer reactions may be broadly classified into one of two groups. The first of these is involved in the production of chemical energy (i.e., ATP) via an external energy input (light, food). This category includes photosynthetic and respiratory pathways, which generate transmembrane proton gradients that drive ATP synthesis. The second group consists of oxido-reductase enzymes that catalyze the chemical transformation of organic substrates in anabolic and catabolic pathways. In this section of the chapter, our attention will focus on an overview of important metabolic pathways. This will best illustrate the interdependence of various redox cycles and provide some "meaning" to the reactions of individual proteins and enzymes.

5.3.1 Mitochondrial Respiration

Ultimately the mitochondrion serves to produce chemical energy in the form of ATP by way of the powerful reducing agents NADH and $FADH_2$. These are the products of the citric acid cycle (Kreb's cycle in Figure 5.13). The mitochondrion also contains redox enzymes and ATP synthetase required for oxidative phosphorylation (so called since it links oxidation of NADH to the phosphorylation of ADP).[4] Many of the proteins to be considered are located in the inner mitochondrial membrane. In aerobic respiration, reduced coenzyme NADH is the primary electron donor and molecular O_2 is the terminal electron acceptor.

$$NADH + \tfrac{1}{2} O_2 + H^+ \rightleftharpoons H_2O + NAD^+ \qquad \Delta E = +1.14 \text{ eV}$$

The energy produced in this reaction is used to drive protons across the inner membrane by way of a series of electron-transfer reactions that couple redox changes at protein prosthetic centers to vectorial proton transport. Figure 5.14 summarizes the electron-transfer sequence and the energetics of the process. The electron-transport chain consists of four respiratory complexes. Complex IV is

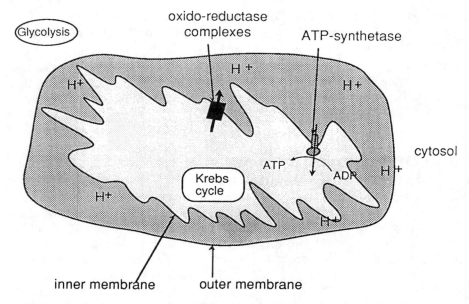

Figure 5.13 Structural features and chemistry of the mitochondrion.

better known as cytochrome c oxidase, the chemistry of which will be discussed in Chapter 10. These complexes contain a variety of Fe–S centers, hemes, flavins, and other coenzymes.

5.3.2 Respiratory Pathways in Nitrate- and Sulfate-Reducing Bacteria

Formation of ATP in the anaerobic sulfate-reducing bacterium *Desulfovibrio vulgaris* is again driven by a proton motive force, although in this case proton translocation is *not* coupled directly to electron transfer (Figure 5.15). Molecular hydrogen, formed from the transformation of lactate to pyruvate, diffuses through the cytoplasmic membrane and is oxidized by a periplasmic hydrogenase ($H_2 \rightarrow 2H^+ + 2e^-$). The electrons are transferred back across the membrane via cytochrome c_3 and are subsequently used in the cytoplasm for the reduction of sulfate to sulfide. By analogy with mitochondrial respiration in aerobic species, SO_4^{2-} may be considered the terminal electron acceptor. Note, however, that reduction of SO_4^{2-} is not coupled directly to proton translocation. When SO_4^{2-} is used as the terminal electron acceptor in this respiratory chain with the excretion of product S^{2-}, the process is termed *dissimilatory sulfate reduction.* If product S^{2-} is used in the synthesis of amino acids and other S-containing metabolites, the term *assimilatory reduction* is used.

The terms *dissimilatory* and *assimilatory* may also be used in nitrogen metabolism. For example, dissimilatory nitrate reductases catalyze the reduction

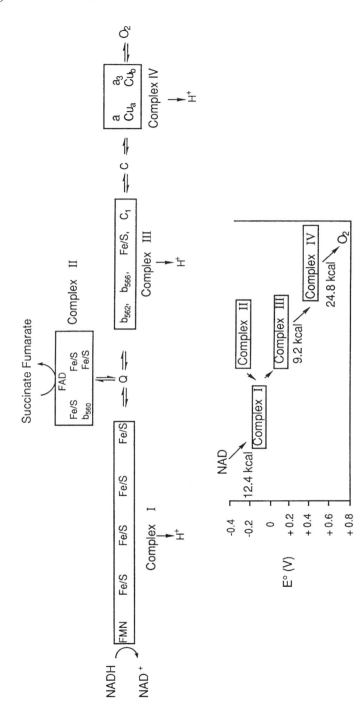

Figure 5.14 Mitochondrial electron transport is coupled to a transmembrane proton pump. The energetics of the reaction are shown. Energy, stored in the form of NADH, is used to generate the proton gradient. Oxygen is the terminal electron acceptor. Standard abbreviations are used for flavin cofactors (FMN, FAD), iron–sulfur centers (Fe/S), quinones (Q), and hemes (a, b, c).

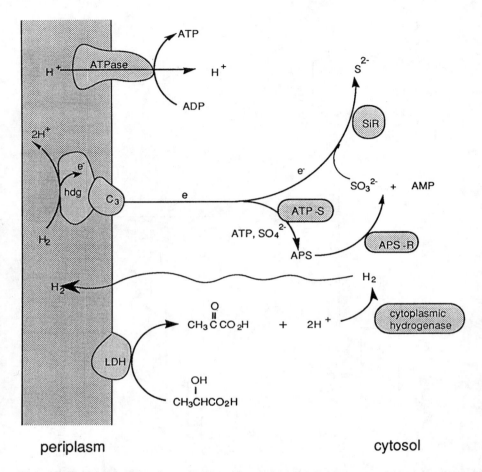

Figure 5.15 Metabolic pathway in the sulfate-reducing bacterium *Desulfovibrio vulgaris* (Hildenborough). Lactate is converted to pyruvate by lactate dehydrogenase (LDH). The protons released are reduced to $H_2(g)$ by an intracellular hydrogenase, which then diffuses into the periplasmic space. $H_2(g)$ is oxidized by a periplasmic hydrogenase (hdg) and the resulting transmembrane proton gradient drives ATP synthesis via ATP synthetase. Electrons released by oxidation of H_2 are fed back into the cell through the tetraheme cytochrome c_3 and used in the conversion of SO_4^{2-} to S^{2-}. Conversion of SO_4^{2-} to SO_3^{2-} proceeds via an adenosine phosphosulfate (APS-R) intermediate and is catalyzed by ATP sulfurylase (ATP-S). Subsequent reduction of SO_3^{2-} to S^{2-} is catalyzed by either an assimilatory or dissimilatory sulfite reductase (SiR).

of nitrate (usually to nitrite, which is further reduced to N_2O or N_2) and is coupled to ATP synthesis by a chemiosmotic mechanism.[5] In assimilatory nitrate reduction, product ammonia is used in the synthesis of amino acids and many other cellular materials. Nonrespiratory nitrogen metabolism is extremely important and will be one of the topics of discussion in Section 5.4. These pathways are summarized in Figure 5.16.

In this section our attention focuses on the dissimilatory reduction of nitrate to volatile products (N_2O or N_2) by way of NO_2^-. If molecular nitrogen is the major product, the process is called *denitrification* and is common to a variety of bacteria found in soils and sediments that use oxidized nitrogen species as a terminal electron acceptor (compare sulfate reducing bacteria earlier). These pathways have economic and ecological importance, since bacterial denitrification removes nitrogen sources in fertilizers from soil before it can be taken up by plants, while some of the intermediates produced are hazardous. NO_2^- reacts with secondary amines to form carcinogenic nitrosamines, while the release of N_2O could contribute to global warming and the destruction of the ozone layer.

$$(CH_3)_2NH^+ + NO_2^- + H^+ \rightarrow (CH_3)_2NNO + H_2O$$

The overall reaction scheme is summarized as follows:

$$NO_3^- \rightarrow NO_2^- \rightarrow (NO) \rightarrow N_2O \rightarrow N_2$$

The reactions are enzymatically catalyzed by nitrate reductase, nitrite reductase, and nitrous oxide reductase, respectively. The question of whether nitric oxide

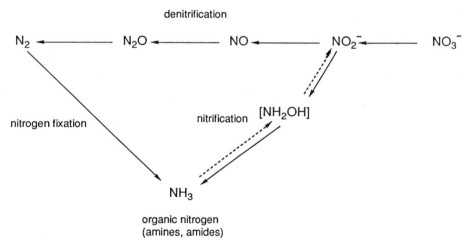

Figure 5.16 Biological nitrogen cycle. The reduction of NO_3^-, NO_2^-, and N_2 to NH_3, which can then be used in the formation of organic nitrogen compounds, is termed *assimilatory* reduction. *Dissimilatory* reduction is the corresponding series of enzymatic reactions in respiratory cycles. All nitrogen contained in biomolecules is derived from either atmospheric N_2 or nitrate salts. *Nitrogen fixation* ($N_2 \rightarrow NH_3$) is carried out by bacteria and legumes. *Denitrification,* so termed because it leads to nitrogen loss from soil, is the process of taking nitrogen salts (NO_3^-, NO_2^-) to $N_2(g)$. The reverse process, where reduced nitrogen (NH_3) is converted to oxyanions by microorganisms to complete the biological nitrogen cycle, is termed *nitrification*.

NO is a free intermediate remains a subject of debate and may be dependent on the particular organism in question. Again, this overall scheme is coupled to ATP formation by a chemiosmotic mechanism.

The reduction of NO_3^- to NO_2^- is catalyzed by a molybdoenzyme, nitrate reductase, while subsequent reduction of NO_2^- to N_2O or N_2 is catalyzed by nitrite reductase, which may contain either copper ions or heme cd as active cofactors.

5.3.3   Methanogenesis ($CO_2 \rightarrow CH_4$)

Methanosarcina barkeri is the textbook example of a methanogenic bacterium. Carbon dioxide is reduced to methane by the series of metabolic redox reactions (illustrated in Figure 5.17) that are used to synthesize ATP by the generation of transmembrane proton gradients.[6] The final step requires factor F430, a nickel tetrapyrrole derivative that is biosynthetically related to heme and corrin, which (like vitamin B12) may function as a methylating agent (Figure 5.10). The last step in this reduction scheme is coupled to transmembrane proton transport and the resulting proton motive force is used to drive ATP synthesis. Once again we see that an electron-transfer reaction may be linked to vectorial proton transport across a membrane.[7]

5.3.4   Photosynthesis

The principal goal of photosynthesis is the generation of the powerful chemical reductant NAD(P)H and a transmembrane proton gradient.[8] These then drive further cellular chemistry and ATP synthesis, respectively. Oxygen is evolved by the photosynthetic apparatus of green plants and cyanobacteria.

$$2H_2O + 2NAD(P)^+ \xrightarrow{\text{light}} O_2 + 2NAD(P)H + 2H^+$$

However, the goal of photosynthesis is not the production of O_2! This is a by-product to which nature has adapted its evolutionary pattern. In the preceding reaction scheme, water merely serves as a hydrogen donor for $NAD(P)^+$. In contrast, anaerobic phototrophic bacteria use alternative hydrogen donors to replace water, and so O_2 is not evolved.

$$2H_2S + 2NAD^+ \xrightarrow{\text{light}} 2S + 2NADH + 2H^+ \qquad \text{(purple sulfur bacteria)}$$

$$2H_2 + 2NAD^+ \xrightarrow{\text{light}} 2NADH + 2H^+ \qquad \text{(purple nonsulfur bacteria)}$$

Since the electrochemical potential for H_2O (1.229 V) is very different from the values for H_2S and H_2 (0.12 V and 0 V, respectively),[9] the chemical apparatus required to carry out mechanisms for these reactions must be modified to accommodate the greater demands of the water-splitting reaction (Figure 5.18).[10] All photosynthetic systems possess a light apparatus termed *photosystem I* (PS

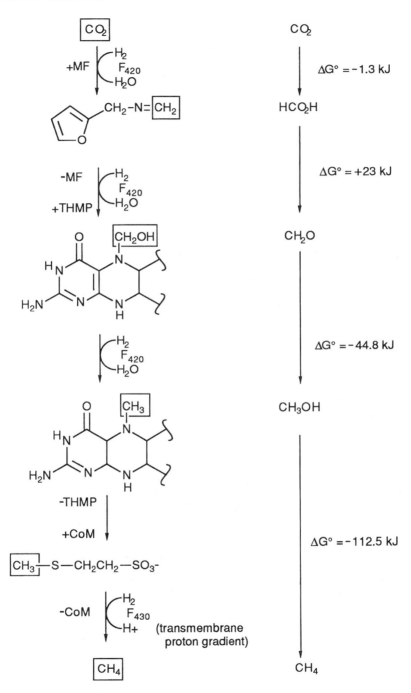

Figure 5.17 Reduction of CO_2 to CH_4. Abbreviations used are MF–methanofuran, THMP–tetrahydromethanopterin, CoM–coenzyme M. The structures of the F_{430} and F_{420} cofactors and coenzymes are summarized in Figures 5.10 and 5.12.

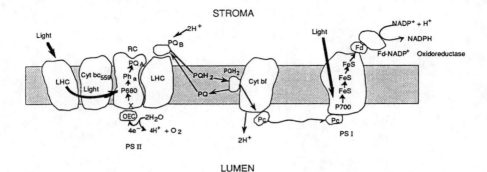

Figure 5.18 The apparatus for green plant photosynthesis (similar to that for cyano-bacteria) is compared with that for purple nonsulfur bacteria. The energetics of each step are displayed below the pathways. Note the more positive potentials for green plant photosynthesis, which generates a more powerful oxidant to carry out the water-splitting reaction ($H_2O \rightarrow O_2$). (A) In green plants, light is absorbed by the light-harvesting complexes (LHCs) and used to stimulate electron transfer in the reaction center (RC) complex from a special chlorophyll complex (P680) in PSII. The electron travels along an electron-transport chain involving pheophytin (Ph_a), plastoqinone (PQ), plastocyanin (Pc), and cytochrome *bf*, before delivery to PSI. Here the electron reduces the electron hole that is generated by electron loss following photoexcitation from the reaction center in PSII (P700). Subsequently the electron is carried to iron–sulfur centers (Fe–S), ferredoxin (Fd), and NADP+ oxido-reductase, generating a proton gradient and NADPH en route. The initial electron donor (Z) to the chlorophyll cation radical generated in PSI is a tyrosine residue. [Adapted from R. E. Blankenship and R. C. Prince, *Trends Biochem. Sci., 10,* 382 (1985).]

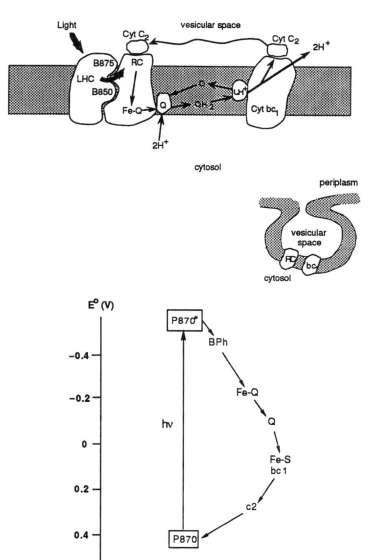

Figure 5.18 (B) The bacterial photosynthetic apparatus is located in a special vesicle associated with the cell membrane. The LHC contains bacteriochlorophyll and other pigment molecules that channel energy to the reaction center RC (see Figure 5.19). Again, photoinitiated electron loss results in an electron-transfer pathway that generates a proton gradient.

I) that spans a potential range from approximately $+400$ mV to -600 mV. However, oxygen-evolving pathways possess an additional light apparatus (PS II) spanning a potential range from about $+850$ mV to -50 mV that is coupled to a catalytic site (a Mn cluster) with a potential that is positive enough to oxidize H_2O.

All photosynthetic apparatus are essentially similar in design. The important light-capturing pigment molecules, proteins, and enzymes are membrane-bound, with a few additional mobile electron carriers. Light is harvested by pigment molecules (carotenes in bacteria; porphyrin-derived chromophores in O_2-evolving systems) that absorb visible light and transfer this energy via exciton coupling (Figure 5.19) to a reaction center complex containing chlorophylls or bacteriochlorophylls.[11] Figure 5.19(c) illustrates how this collection of diverse

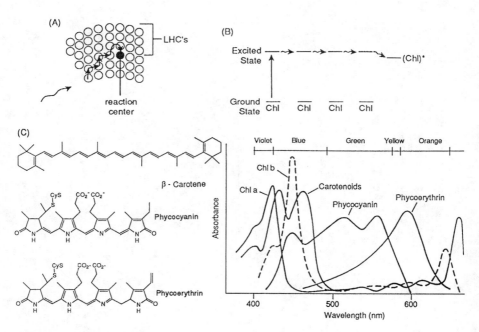

Figure 5.19   General features of a photosynthetic apparatus. (A) Light is captured by light-harvesting complexes (LHC). These are transmembrane polypeptides that contain a variety of chromophores (chlorophylls, carotenes, or other pigments, according to the photosynthetic system) that together absorb light over the entire range of the visible spectrum. Light absorption may be tuned by small changes in chromophore structure [e.g., note the difference in absorption spectrum for chlorophylls *a* and *b* in (C)]. (B) Energy is transferred to the reaction center by a mechanism known as exciton coupling where the energy stored in the excited state of one chromophore is used to promote the excitation of another. Eventually this energy is trapped by a special chromophoric unit (typically a special chlorophyll or bacteriochlorophyll complex) in the reaction center, from which a rapid charge separation occurs (see Figure 5.20). (C) Ancillary pigments and their absorption spectra. See Figure 5.11 for chlorophyll structures.

pigment molecules spans a broad absorption envelope that encompasses the entire range of the visible spectrum, maximizing energy capture from sunlight. The stored energy is transferred to a special chlorophyll or bacteriochlorophyll molecule. Figure 5.20 shows how an electron is subsequently promoted to an energetically excited state and transferred to an Fe–quinone complex. The electron is then transferred via a series of quinones to the cytochrome *bf* complex shown in Figure 5.18. At this point, electron transfer is coupled to proton release into the lumen. The electron hole generated by electron loss from the special pair is filled by electrons derived by oxidation of H_2O to O_2 at a site containing a manganese cluster.[12]

Electrons are subsequently carried from the cyt *bf* complex via a blue-copper plastocyanin to reduce the electron hole formed by photoexcitation of the reaction center in photosystem II (PS II). As for PS I, light-harvesting pigment molecules absorb energy from sunlight. This energy is then transferred to the PS II reaction center (P700) where photoinduced electron loss results in an electron-transport chain that culminates in the reduction of the iron–sulfur protein ferredoxin, which ultimately acts as the electron source for the enzyme $NADP^+$-reductase. Clearly, at several points in this scheme, there is net transfer of H^+ into the lumen. The controlled release of this transmembrane proton gradient drives the synthesis of ATP via ATP synthetase.[13] The net effect, then, is to generate a highly reducing molecule NAD(P)H and a chemical energy store (ATP) that can then drive cellular metabolism.

The redox steps associated with the photosynthetic reaction centers are among the most rapid biological electron transfers known (10^{12} s^{-1}). The structure of the reaction center from the purple nonsulfur bacterium *Rhodopseudomonas viridis* has been determined (Figure 5.20). This shows how the relative geometries of the light-harvesting, bacteriochlorophyll dimer and electron-carrier pigments regulate the extent of orbital overlap among these chromophores that is important in the control of electron-transfer rates.

Summary of Section 5.3

1. Respiratory pathways couple electron-transport chains to transmembrane proton flow. These proton gradients drive ATP synthesis.
2. Sources of reducing equivalents include NADH (from metabolic degradation of nutrients) or light energy (photosynthesis).

5.4 Electron-Transfer Pathways in Nonrespiratory Metabolism

In the previous section, we reviewed some important respiratory pathways. Now our attention turns to redox reactions in catabolic and anabolic metabolism. We shall introduce this topic by considering a few reactions from the biological nitrogen cycle, summarized in Figure 5.16, and the carbon cycle.

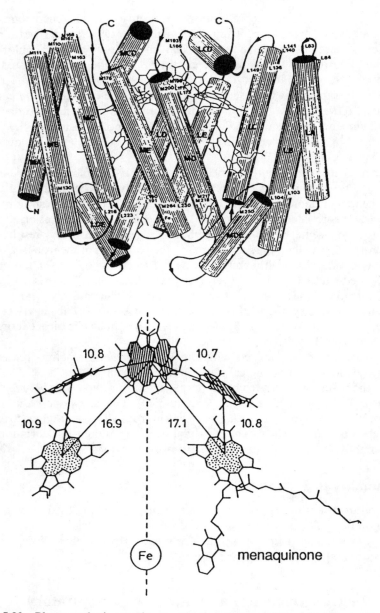

Figure 5.20 Photosynthetic reaction center complex from *Rhodopseudomonas viridis*. Energy from the light-harvesting complexes are directed toward the special bacterio-chlorophyll dimer from which an electron is lost in a photoinitiated electron transfer. The relative distances (in angstroms) between chromophores are indicated. [Adapted from M. E. M.-Beyerle et al., *Biochim. Biophys. Acta, 932,* 52–70 (1988), and H. Michel and J. Diesenhofer, *Biochemistry, 27,* 1–7 (1988).]

5.4.1   Nitrogenase

Microorganisms in soil can assimilate molecular N_2 by reduction to NH_3 using the enzyme nitrogenase. The overall reaction is summarized in equation (5.1), where Pi = inorganic phosphate.

$$N_2 + 8H^+ + 8e^- + 16\,MgATP \rightarrow 2NH_3$$
$$+ H_2 + 16\,MgADP + 16Pi \quad (5.1)$$

Note that H_2 evolution is an integral component of the nitrogenase catalytic apparatus and continues even in the absence of N_2 if MgATP and a reductant are present. Carbon monoxide, a potent inhibitor of N_2 reduction, does not inhibit H_2 evolution, and so binding of each substrate (N_2 and H^+) at the catalytic site is distinct. The enzyme is relatively nonspecific and a variety of small unsaturated molecules are reduced, including $C_2H_2 \rightarrow C_2H_4$, $N_2O \rightarrow N_2$ and H_2O, $N_3^- \rightarrow NH_4^+$ and N_2, and $HCN \rightarrow CH_3NH_2$. All substrates are reduced in steps that require multiples of two electrons.

Figure 5.21 shows that nitrogenase is composed of an "iron-protein" (α_2 dimer) that couples ATP hydrolysis to electron transfer, and a "molybdenum–iron protein" ($\alpha_2\beta_2$ tetramer) that contains the dinitrogen binding site. The Fe–protein dimer contains one ferredoxin-like $[Fe_4S_4]^{2+/1+}$ cluster that bridges the two subunits, while the MoFe-protein binds two FeMo clusters (M-clusters) and a total of four unusual $[Fe_8S_8]$ centers (P-clusters). The MoFe centers are presumed to be the catalytically active sites. Figure 5.21 summarizes recent crystallographic data that has defined these cluster units and their disposition within the protein subunits. The FeMo-cofactor (M cluster) contains an Fe_4S_3 cluster bridged to a $MoFe_3S_3$ cluster by three nonprotein S^{2-} ligands. Six of the seven Fe sites have an unusual trigonal coordination geometry. The remaining Fe is tetrahedral, while the Mo site is octahedral with ligation from three sulfides in the cofactor, a histidine imidazole side chain, and two oxygen atoms from homocitrate. The P-clusters are essentially 8-Fe units that are made up of two Fe_4S_4 clusters bridged by two cysteine thiolates. The other ligands are cysteine or inorganic sulfide, as indicated. Since neither the M- or P-clusters are solvent exposed, the sequence of events that couple substrate binding, electron transfer, and substrate release must be carefully orchestrated.

A working model for nitrogenase function can be outlined as follows (reference can be made to Figure 5.21). Two molecules of MgATP bind to the Fe-protein, causing a change in both protein conformation and the redox potential of the cluster that facilitate binding and electron transfer to the FeMo-protein. Synchronously with the electron-transfer event, there arises the hydrolysis of MgATP to MgADP, which results in the dissociation of the Fe-protein, thereby inhibiting the back-electron-transfer reaction. This molecular switching ("gating") mechanism serves two functions. First, MgATP hydrolysis occurs only when the complex is formed, and so nonproductive electron loss is prevented. Second, the mechanism allows unidirectional flow of electrons to the MoFe-protein, which must store up to eight electron equivalents prior to reduction of

(A)

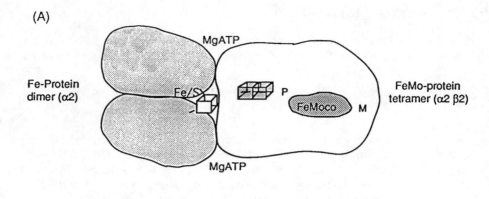

(B)

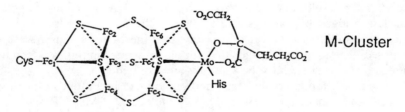

P-Cluster

M-Cluster

Figure 5.21   (A) Schematic illustration of nitrogenase showing the arrangement of subunits and prosthetic centers. The Fe-protein contains a regular [Fe$_4$S$_4$] cluster. Each α,β-pair of the FeMo-protein binds an [Fe$_8$S$_8$] P-cluster and an FeMo-cofactor (M-cluster). (B) Structures for the P- and M-clusters from recent crystallographic studies of the FeMo-protein.

substrate molecules. Subsequently, MgADP dissociates and the Fe-protein is reduced and binds another two molecules of MgATP. The process is repeated to deliver a second electron. This occurs eight times to deliver a total of eight reducing equivalents. A detailed molecular mechanism for N_2 reduction and the obligatory production of one equivalent of H_2, has yet to be established.[14] Dinitrogen reduction was once believed to proceed by an asymmetric binding mode via species such as $Mo = N-NH_2$, and $Mo = NH$ after cleavage of the N–N bond. However, the more recent crystallographic data suggest that the coordinatively saturated octahedral Mo site may not bind N_2 and that substrate molecules might bind inside the cluster. Also, since neither the P-clusters nor FeMo-site are solvent exposed, the entry of substrate, electron transfer, and product release must be regulated by a series of structural changes along the lines of the gating mechanism described earlier.

5.4.2   Nitrate Reductase

Inorganic nitrogen species can be converted to biologically useful forms by fixation of N_2 ($N_2 \rightarrow NH_3$) or by the assimilation of nitrate ($NO_3^- \rightarrow NH_3$). Nitrogen fixation was considered earlier. The first step in the assimilatory reduction of NO_3^- is the two-electron reduction of nitrate to nitrite catalyzed by nitrate reductase. Nitrate reductase is a multimeric complex containing FAD, heme b_{557}, and a Mo–pterin complex as prosthetic centers (Figure 5.22). The physiologic electron donor in eukaryotes is NADPH and the relative redox potentials for these sites (noted in Figure 5.22) suggest that the flow of electrons proceeds from NADPH $\rightarrow$ FAD $\rightarrow b_{557} \rightarrow$ Mo.

A likely reaction mechanism is suggested by the coordination of the Mo center shown here.

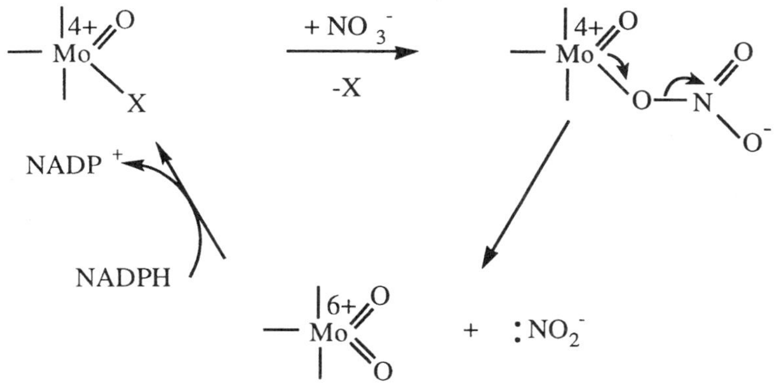

Nitrate can displace ligand X (H_2O ?) and bind to reduced Mo(IV). Two-electron reductive cleavage of a N–O bond produces nitrite. Subsequent six-electron reduction of nitrite to ammonia can be catalyzed by a variety of nitrite reductases, depending on the bacterial or plant source. A number of distinct

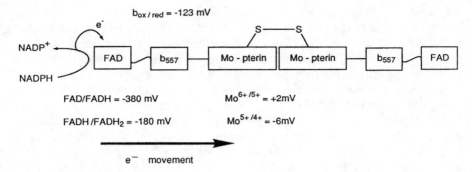

Figure 5.22 Assimilatory nitrate reductase from spinach is a multimeric enzyme that contains FAD, heme b_{557}, and Mo-pterin prosthetic centers in three distinct subunits. The natural direction for electron movement from an exogenous electron donor (NADH) is NADH → FAD → b_{557} → Mo-pterin.

prosthetic centers have been identified in these enzymes, including a coupled siroheme–[Fe$_4$S$_4$] cluster complex and an array of six hemes (one of which is HS five coordinate with an available substrate binding site).

5.4.3   Nitrite (Sulfite) Reductase

The coupled [Fe$_4$S$_4$]-siroheme prosthetic center shown in Figure 5.23 is used to catalyze the six-electron reduction of both nitrite and sulfite (to NH$_3$ and HS$^-$, respectively) and is found in enzymes that can be isolated from a variety of aerobic and anaerobic bacteria, and higher plant sources, such as spinach.

The siroheme is more reduced than the b- and c-type hemes found in myoglobin and cytochromes (illustrated in Figure 5.3), which are derived from the porphyrin ligand. Siroheme is the iron derivative of the isobacteriochlorin ligand, described earlier in Figure 5.7, and is characterized by reduction of two *cis* pyrrolic double bonds by addition of the elements of methane (CH$_4$). [Can you spot these in Figure 5.23?] As a result, the ring is more electron rich, and this influences both its reduction potential (for example $E_{s'heme} \sim -50$ to -300 mV and $E_{heme} \sim +50$ to $+300$ mV) and coordination chemistry. Because the ring is electron rich, the reduced iron center in ferrous siroheme is able to bind π-acceptor ligands, such as the substrates nitrite (NO$_2^-$) or sulfite (SO$_3^{2-}$). In turn, π-backbonding from the ferrous ion pushes electron density into the antibonding N–O or S–O bonds in the substrate, which weakens these bonds and facilitates their reductive cleavage (Figure 5.24). One might ask what purpose the cluster serves. A probable answer is that Nature has designed this coupled catalytic center so that it contains the very efficient electron/donor acceptor unit (the Fe$_4$S$_4$ cluster from Section 5.2.1), which can capture electrons from natural redox partners and deliver them to the siroheme. In turn, the latter has a vacant coordination site in the reduced state to which substrate can bind and undergo

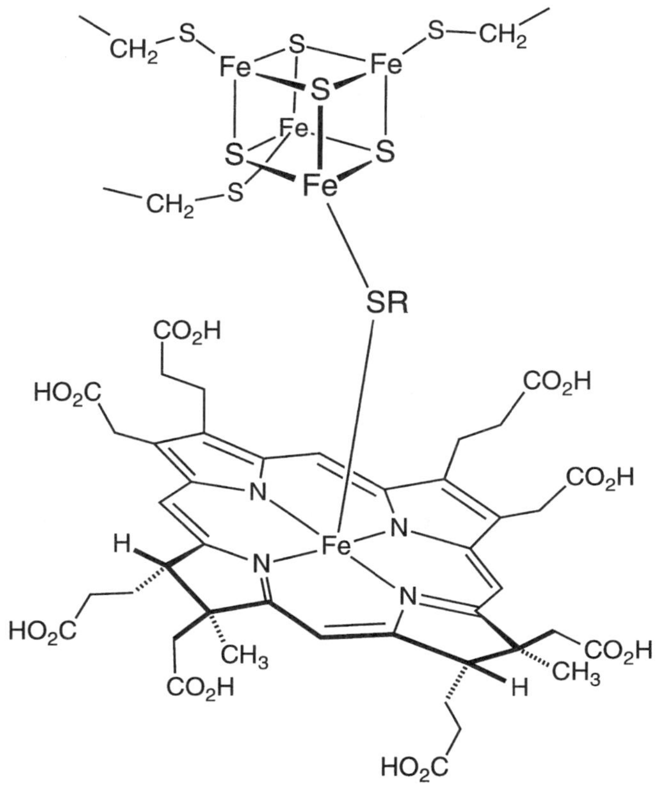

Figure 5.23 The coupled [Fe$_4$S$_4$]-siroheme prosthetic center found in both nitrite and sulfite reducing enzymes.

a series of reductive cleavage reactions of N–O or S–O bonds that results in formation of the products (NH$_3$ or HS$^-$).

Closer examination of the reactions catalyzed by this enzyme emphasizes the need for an uptake of protons to combine with both the oxygen centers that are released as water and the nitrogen or sulfur heteroatoms.

$$SO_3^{2-} + 6e^- + 7H^+ \rightarrow HS^- + 3H_2O$$
$$NO_2^- + 6e^- + 7H^+ \rightarrow NH_3 + 2H_2O$$

One might expect that the molecular mechanism employed by the enzyme would promote efficient transfer of H$^+$ equivalents to the active site. Moreover, because these proton transfer steps arise in the context of electron-transfer chemistry at the prosthetic center, it is reasonable to assume that the redox chemistry and proton transfer chemistry might be intimately coupled. This is indeed the case.

Figure 5.25 shows the variation of the siroheme reduction potential with pH,

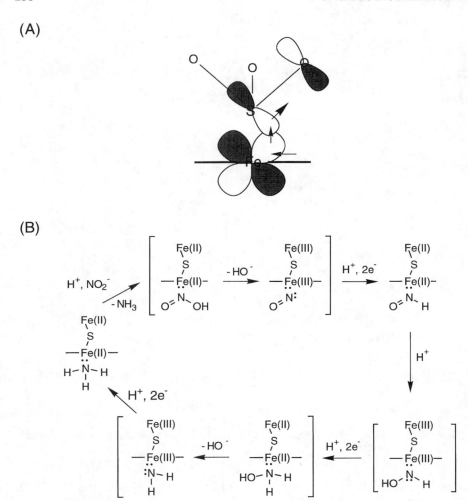

Figure 5.24 (A) Illustration of the synergic bonding between the iron-siroheme center and the π-acceptor nitrite (or sulfite) substrate. The arrows indicate how electron density flows from the siroheme through the iron to populate antibonding orbitals and weaken the S–O (or N–O) bonds of the substrate. (B) An overview of the reaction chemistry at the siroheme, illustrating the sequence of three two-electron reductive cleavages of N–O bonds. A similar scheme can be drawn for sulfite reduction. Note that only the bridging iron of the cluster is shown for clarity.

which demonstrates that each electron equivalent is, in fact, coupled to the uptake of one proton. This follows from the gradient of the plot and the modified Nernst equation when ionization accompanies electron transfer.

$$\text{ox} + n e^- + H^+ \rightleftharpoons \text{redH}^{(n-1)-}$$

$$E = E^0 - (RT/nF)\log([\text{redH}^{(n-1)-}]/[\text{ox}]) - (RT/nF)\,\text{pH}$$

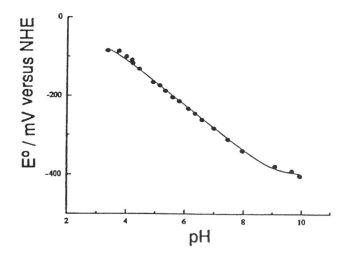

Figure 5.25 Dependence of $E°$ on pH for the siroheme center in *Desulfovibrio vulgaris* sulfite reductase. The slope ($-RT/nF$) yields a value for n (the e^-/H^+ ratio) of ~1.

5.4.4   Methane Monooxygenase

In contrast to methanogenic bacteria which produce methane gas (CH_4) as a by-product of their respiratory cycle (Figure 5.17), methanotrophic bacteria, such as *Methylosinus trichosporium* or *Methylococcus capsulatus,* consume methane. Methanotrophic bacteria play an essential role in the carbon cycle by retaining carbon in a usable form in the biosphere and also by limiting the release of methane to the atmosphere where it may act as a *greenhouse* gas.

$$CH_4 + NADH + H^+ + O_2 \rightarrow CH_3OH + NAD^+ + H_2O$$

The first step in the oxidation of methane is catalyzed either by a soluble or particulate (membrane-bound) form of methane monooxygenase (sMMO or pMMO, respectively). The former possesses a binuclear iron center at the catalytic site (Figure 5.26), whereas the latter uses a complex cluster of copper ions. The enzymology of the soluble form is better understood and this will be the focus of our attention. For that reason, the abbreviation MMO will henceforth denote the soluble monooxygenase.

MMO has attracted considerable attention because it is capable of oxidizing a wide range of hydrocarbon substrates in addition to methane. This affords a broad spectrum of possible practical applications, including functionalization of hydrocarbons, bioremediation of land contaminated by oil and petroleum spills, oxidative removal of trichloroethylene from drinking water, and conversion of natural gas to methanol to facilitate transportation.

The functional enzyme complex consists of three proteins that include the hydroxylase ($M_r \sim 251K$), a reductase ($M_r \sim 38.6K$), and a coupling protein

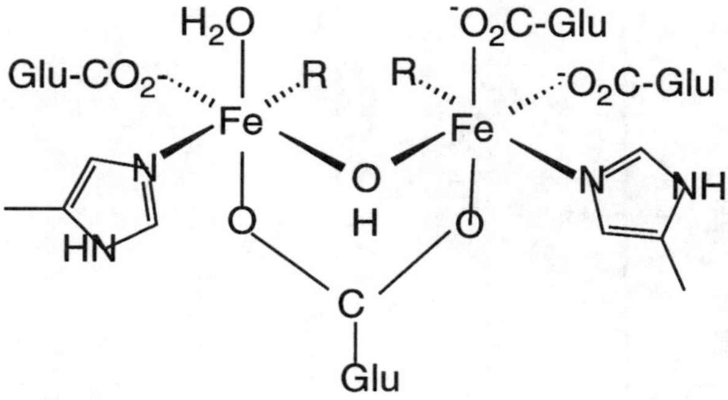

Figure 5.26 The binuclear iron center in the soluble form of methane monooxygenase. In the crystallographic study, the two sites labeled R were occupied by a bridging acetate ligand, which presumably derived from the crystallization buffer. These two sites, along with the axial water on one unique iron ion, are likely binding sites for dioxygen in the reduced hydroxylase, for peroxo and oxo intermediates during turnover, and the methoxide reaction product, as illustrated in Figure 5.27. An interesting feature of the enzyme is the design of the hydrophobic active site cavity in which the binuclear iron center lies, which promotes binding of methane and other hydrocarbon substrates with reasonable affinity.

($M_r \sim 15.5K$). The hydroxylase is an $\alpha_2\beta_2\gamma_2$ hexamer that contains two independent binuclear iron sites. The reductase transfers electrons from NADH to the iron sites in the hydroxylase, whereas the coupling protein serves to regulate overall activity. A mechanistic scheme consistent with the spectroscopic and mechanistic studies on MMO is illustrated in Figure 5.27. Many of the intermediates shown have been spectroscopically characterized (and form the basis for several problems in this and in other chapters). The ferryl oxo intermediate appears to be the critical species that catalyzes substrate oxidation. For methane and other alkyl substrates, oxidation arises from H atom abstraction to form a substrate radical, which then rapidly reacts with iron-bound OH to produce product. Product release appears to be the rate-limiting step.

Summary of Section 5.4

1. Anabolic reactions build large molecules from smaller substrates, whereas catabolic reactions break down biological macromolecules into smaller fragments that may then be further used in cellular metabolism. Many important anabolic pathways involve redox chemistry of small inorganic substrate molecules and anions (N_2, NO_3^-, SO_4^{2-}, CO_2).
2. Nitrogenase catalyzes the reaction $N_2 \rightarrow NH_3$, which is then incorporated into amino acids and other N-containing metabolites.
3. Nitrate reductase catalyzes the first reaction ($NO_3^- \rightarrow NO_2^-$) in assimilatory nitrate reduction ($NO_3^- \rightarrow NH_3$). Six-electron reduction of $NO_2^- \rightarrow NH_3$ is catalyzed by one enzyme, nitrite reductase. An analogous enzyme catalyzes sulfite reduction to H_2S.

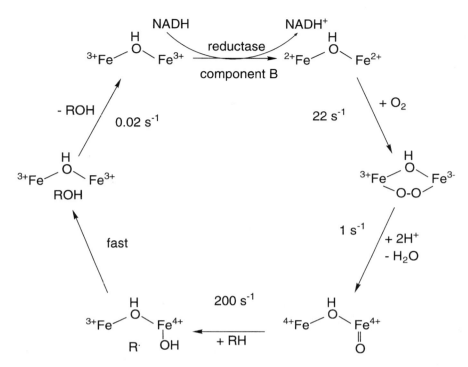

Figure 5.27 A possible mechanistic scheme for sMMO that is consistent with available evidence. The rates for substrate addition, conversion to the hydroxyl derivative, and product release are for a nitrobenzene substrate [refer to S-K. Lee et al. *J. Biol. Chem.* *268*, 21569–21577 (1993)].

4. Methane monooxygenase from methantrophic bacteria catalyzes oxidation of hydrocarbon substrates. The soluble form possesses a binuclear iron site.
5. Assimilatory reactions generate products that are employed in anabolic chemistry. Dissimilatory reactions produce products from respiratory pathways that are subsequently excreted.

Review Questions

- What is the formal valence state of N in NO_2^-, HNO, NH_2OH, and NH_3?
- Studies of the complex reduction chemistry of the sulfite/nitrite reductase illustrated in Figures 5.23 and 5.24 are simplified by the fact that some of the intermediates are stable molecules that can be used as substrates to probe later parts of the reaction pathway. The Michaelis constants for nitrite and hydroxylamine are 28 μM and 48 μM, respectively, whereas the binding affinity of ammonia is very slight. Discuss the relevance of the trend in these data for efficient enzyme turnover.

[Lui, S. M. et al., *J. Am. Chem. Soc. 115*, 10483–10486 (1993)]

5.5   Protein–Protein Electron Transfer

Efficient transport of electrons between biological redox centers is a central feature of any biological redox cycle. This includes both intermolecular and intramolecular ET (Figure 5.28). By this point it should be clear that the importance of protein–protein ET lies in the way in which these thermodynamically favorable reactions are coupled to (and drive) other cellular processes. Oxidation–reduction chemistry, membrane transport, activation of regulatory proteins, uptake and release of metals in storage and transport (e.g., ferritin/transferrin) are all controlled in one way or another through electron transfer by redox proteins. We therefore close this chapter by considering the question of how electron transfer per se is modulated by proteins. Figure 5.28 shows several examples of natural biological redox couples. Electron exchange (intra- or intermolecular) between redox centers clearly depends on a variety of factors, including the net driving force for the reaction ($\Delta G°$), the distance (d) between the centers, the conductivity of the protein matrix through which the electron must pass (β), the reorganization energy (λ) resulting from structural changes at the redox site, and general conformational changes of the protein backbone. Figure 5.29 defines the reorganization energy (λ) as the energy required to move the nuclei of all atoms from their equilibrium positions before electron transfer to the position they would occupy after electron transfer, while the electrons themselves remain fixed. The activation energy (ΔG^*) is the energy difference between the ground state of the reactants and the transition state. This includes components from electrostatic interactions between the reacting species, rearrangements of bond lengths and angles, and solvent rearrangement. The free energy for reaction ($\Delta G°$) is the energy difference between the ground states of the reactants and products. For intermolecular reactions, the contact region between the two proteins must be considered. The following equation represents the general situation for bimolecular electron transfer between a donor (D) and an acceptor (A) protein:

$$D + A \overset{K_p}{\rightleftharpoons} DA \overset{k_{et}}{\to} D^+A^- \overset{K_s}{\rightleftharpoons} D^+ + A^-$$

where K_p and K_s are formation constants for precursor and successor complexes (see Section 1.8) and k_{et} is the electron-transfer rate constant. Only the latter parameter is relevant for intramolecular electron transfer.

Inasmuch as the distances between redox centers in these systems are ≥ 10 Å, electron-transfer reactions may be thought of as outer sphere. Clearly, the Marcus relationship defined in Section 1.8 is not expressed in terms of the more fundamental parameters noted earlier. A semiclassical expression for the rate constant can be written in the form of equation (5.4):

$$k = \text{const} \cdot \exp\left[-\beta(d - d_0)\right] \cdot \exp(-\Delta G^*/kT) \qquad (5.4)$$

$$\Delta G^* = (\lambda + \Delta G°)^2/4\lambda \qquad (5.5)$$

where d_0 is the van der Waals radius (~ 3 Å, the minimal internuclear distance

(A)

(B)

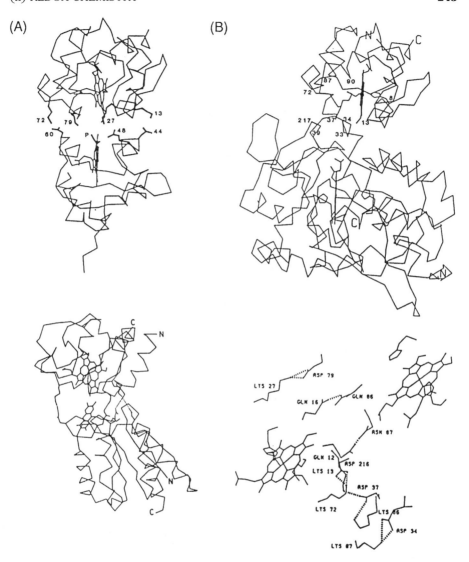

Figure 5.28 Protein–protein electron-transfer complexes formed between natural redox partners. (A) Models for the cytochrome c–cytochrome b_5 (upper) and cytochrome c_3–flavodoxin (lower) complexes. (B) Cytochrome c–cytochrome c peroxidase. Relevant factors controlling electron transfer between protein subunits include distance (d), driving force $\Delta G = -nF\Delta E$, and the efficiency of electronic coupling in the intervening medium (β). Some of these parameters (d and β) are dependent on the location of the interprotein contact region. The intermolecular hydrogen bond network that connects the two proteins is shown for cytochrome c–cytochrome c peroxidase. The "docking" of natural redox partners determines specificity in biological electron-transfer events. [Adapted from F. R. Salemme, *J. Mol. Biol., 102,* 563–588 (1976); R. P. Simondsen et al., *Biochemistry, 24,* 6366 (1982); and T. L. Poulos and J. Kraut, *J. Biol. Chem., 255,* 10322–10330 (1980).]

defined by repulsion of electron clouds) and β is an electronic coupling parameter that reflects the distance dependence of electron coupling. A large β implies a rapid fall-off in electron-transfer rate with distance, and so small β's are optimal for long-range electron transfer.

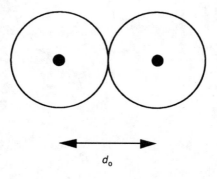

d_o

van der Waals contact

The activation free energy (ΔG^*) can be written as a function of the driving force (ΔG°) and reorganization energy (λ) [equation (5.5)]. The latter relationship can be readily derived by consideration of the potential energy profiles for reactants and products in Figure 5.29. By assuming that the potential energy profiles for reactant and product wells are parabolas (that is, molecular vibrations obey a simple harmonic oscillator model), it is possible to derive the simple relationship defined by equation (5.5), connecting λ, ΔG^*, and ΔG°.[15]

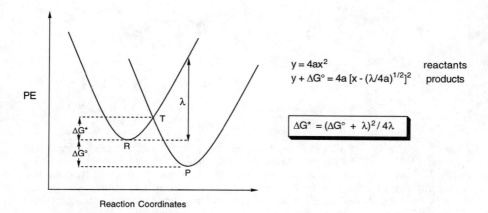

$$y = 4ax^2 \quad \text{reactants}$$
$$y + \Delta G^\circ = 4a\,[x - (\lambda/4a)^{1/2}]^2 \quad \text{products}$$

$$\boxed{\Delta G^* = (\Delta G^\circ + \lambda)^2 / 4\lambda}$$

Figure 5.29 Potential energy profiles for reactants (R) and products (P), showing the transition state (T), activation barrier (ΔG^*), driving force (ΔG°), and reorganization energy (λ) for conversion of reactants to products. The equations defining the parabolic curves for reactants and products are noted. These can be used to derive the Marcus relationship (boxed).

We close this chapter by summarizing three important concepts that often arise in discussions of electron-transfer processes.

Marcus Inverted Region

The variation of electron-transfer rate as a function of driving force is shown in Figure 5.30. This curve is anti-intuitive insofar as we normally expect the rate to keep increasing if we increase driving force. The inverted rate curve is a consequence of the change in ΔG^* and follows naturally from the equation $\Delta G^* = (\lambda + \Delta G^\circ)^2/4\lambda$.

Electron Tunneling

Consider Figure 5.29. At the transition point T, there is no change in the nuclear coordinates of either the reactant or product well. The only barrier to be overcome derives from the effectiveness of the electronic coupling between the two

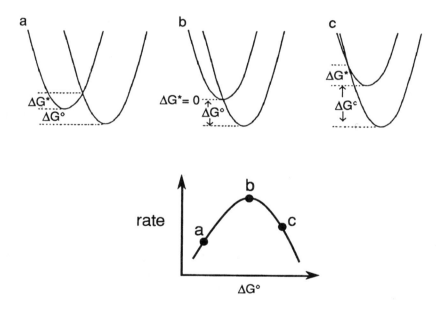

Figure 5.30 Marcus inverted region. A series of potential energy curves showing the effect of driving force (ΔG°) variation on the activation energy (ΔG^*). Since the rate of a reaction is dependent on ΔG^*, the theoretical rate profile shown can be easily obtained. At first ΔG^* decreases when ΔG° increases (i.e., becomes more negative). In (b) we see that $\Delta G^* = 0$, that is, electron transfer is activationless and the rate is maximal. (c) In the *inverted region*, a further increase in ΔG° results in an *increase* in ΔG^*, with a drop-off in rate. The existence of the inverted region has been experimentally verified. [See, for example, J. R. Miller et al., *J. Am. Chem. Soc.*, *106*, 3047 (1984) and G. L. Closs and J. R. Miller, *Science*, *240*, 440 (1988).]

redox sites. Since this involves electronic properties, movement across this barrier is referred to as *electron tunneling*. Thermal energy is required to attain the transition state T, and so electron tunneling is an *activated* process.

Nuclear Tunneling

Figure 5.31 shows that there is overlap between the lower vibrational levels in the reactant and product wells, and so it is possible to transfer from the reactant to product wells by this pathway. In effect, we are moving from the nuclear geometry of the reactants to the nuclear geometry of the products by way of molecular vibrations. The barrier to be overcome is related to the coupling of nuclear motions, and so the term *nuclear tunneling* is used. Since we are moving horizontally on the PE profile, no thermal activation is required, and there is no activation barrier. Both nuclear and electron tunneling can occur together. However, nuclear tunneling is activationless and becomes more evident at low temperature, where electron tunneling is minimal, since there is insufficient energy to overcome the activation barrier to reach the transition state T.

Summary of Section 5.5

1. Long-range electron transfer between protein redox centers is outer sphere and depends on driving force, distance, and the efficiency of electronic coupling in the intervening medium. Redox chemistry on substrates usually (but not necessarily) involves inner sphere chemistry.
2. A theoretical framework exists for the quantitation of biological electron-

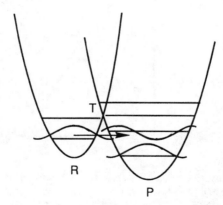

Figure 5.31 Nuclear tunneling. Overlap between the ground vibrational state in the reactant well (R) and upper vibrational levels in the product well (P) provides a mechanism for electron transfer from reactants to products without passing through an activated transition state (T). This coupling is generally very weak and is evident only at low temperatures where activated transfer is negligible. Nuclear tunneling is an activationless process.

transfer chemistry. Much of this centers on a simple relationship that relates activation free energy (ΔG^*) to driving force ($\Delta G°$) and reorganization energy (λ), namely, $\Delta G^* = (\lambda + \Delta G°)^2/4\lambda$.

Notes

1. Normally referred to as high-potential iron proteins (HiPIPs). Low- and high-potential Fd's are related by the following redox scheme.

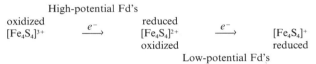

2. Named after J. S. Rieske, who first identified this distinct class of cluster centers, with two His replacing two Cys ligands.
3. Reduced porphyrin rings are sometimes referred to as chlorins. Doubly reduced (or tetrahydro) derivatives are called bacteriochlorins or isobacteriochlorins according to the disposition of the saturated pyrrole rings.
4. ATP might also be synthesized by normal fermentative metabolism. An example was noted in Figure 4.16 from the citric acid cycle (Kreb's cycle). ATP is generated by routes that do not depend on transmembrane proton gradients. The energy from breakdown of high-energy metabolites is used to phosphorylate ADP. Note the occurrence of aconitase (Section 4.3) in this cycle.
5. The term *chemiosmosis* generally indicates a process driven by a transmembrane concentration gradient (normally a proton gradient). In some cases, nitrate reduction that results in ATP formation is not coupled to proton gradients but is the result of a series of metabolic reactions (fermentative nitrate reduction).
6. Refer to *Bacterial Metabolism* by G. Gottschalk (Springer-Verlag), p. 257, for more details.
7. It can be easily demonstrated that methane formation and ATP generation are coupled via a transmembrane potential. Dicyclohexylcarbodiimide (DCCD) inhibits ATP synthetase, and so methane formation eventually stops, and the transmembrane potential is retained. By adding uncouplers (e.g., nigericin, Table 3.4) that dissipate the proton motive force, it is observed that methane formation continues but ATP synthesis is halted.
8. Prokaryotes use NAD^+ as oxidant while eukaryotes utilize both NAD^+ and $NADP^+$.
9. Although these literature values were taken at a nonphysiologic pH of 1, the general discussion is valid.
10. The Z-scheme for green plant photosynthesis and cyanobacteria is so called because of its zig-zag shape.
11. Chlorophylls for O_2-evolving systems and bacteriochlorophylls for non O_2-evolving systems.
12. The details of the structure and chemistry of the oxygen-evolving complex remain the subject of speculation. Refer to *Manganese Redox Enzymes* (V. L. Pecoraro, ed.) VCH for further details of the relevant manganese chemistry.
13. It is important to realize that, at the time of its initial inception by Peter Mitchell (Nobel laureate, 1978) in the 1960s, the idea of vectorially coupling transport to metabolic reactions was quite revolutionary. Mitchell coined the term *chemiosmosis* to describe this process that coupled chemical transformations to solute translocation. Prior to this idea, researchers had fervently searched for elusive (and ultimately nonexistent) enzymes that were required to support other proposed pathways connecting transport phenemena and metabolism.
14. Reductive addition to give a molybdenum dihydride center with subsequent oxidative elimination of H_2 is one of many possible pathways, that is,

$$\backslash \qquad\qquad\qquad \backslash \quad H^- \quad \backslash$$
$$-Mo + ne^- + 2H^+ \rightarrow -Mo \overset{/}{\underset{\backslash}{}} \rightarrow -Mo + H_2$$
$$/ \qquad\qquad\qquad / \quad H^- \quad /$$

15. The general equation for a parabola is $y - Y = 4a(x - X)^2$, where a is a constant and (X, Y) are the coordinates of the base of the parabola.

Further Reading

Redox Cofactors and Coenzymes

Babior, B. Ed. *Cobalamins*, Wiley, 1975.

Dolphin, D. B_{12}, Wiley, 1982, Vol. 1.

Lancaster, J. R. Ed. *The Bioinorganic Chemistry of Nickel*, VCH, 1988.

Solomon, E. I. Electronic structure contributions to function in bioinorganic chemistry. *Science* *259*, 1575–1581 (1993).

State of the Art Symposium: Bioinorganic Chemistry, *J. Chem. Educ., 62*, (issue 11) (1985).

Walsh, C. T., and W. H. Orme-Johnson. *Nickel Enzymes, Biochemistry, 26*, 4901–4906 (1987).

Wood, J. M., and D. G. Brown. Vitamin B_{12}-Enzymes, *Structure and Bonding, 11*, 47–105 (1972).

Oxido-Reductase Enzymes and Pathways

Cramer, W. A., and D. B. Knaff. *Energy Transduction in Biological Membranes*, Springer-Verlag (1990).

Diesenhofer, J., and H. Michel. The photosynthetic reaction center, *Chem. Scripta, 29*, 205–220 (1989).

Gottschalk, G. *Bacterial Metabolism*, Springer-Verlag, 1986.

Huber, R. Copper clusters in oxidases, *Ang. Chem. (Int. Ed.), 28*, 848–869 (1989).

Kim, J., and D. C. Rees. Crystallographic structure and functional implications of the nitrogenase molybdenum-iron protein from *A. vinelandii, Nature, 360*, 553–560 (1992).

Orme-Johnson, W. H. Molecular basis for nitrogen fixation, *Ann. Rev. Biophys. Biophys. Chem., 14*, 419–459 (1985).

Rosenzweig, A. C. and S. J. Lippard. Determining the structure of a hydroxylase enzyme that catalyzes the conversion of methane to methanol in methanotrophic bacteria. *Acc. Chem. Res. 27*, 229–36 (1994).

Youvan, D. C. and B. L. Marrs. Molecular mechanism of photosynthesis. *Sci. Amer. 256*, 42–48 (1987).

Electron-Transfer Proteins

Adman, E. T. Copper protein structures. *Adv. in Protein Chem. 42*, 145–198 (1991).

Fee, J. Copper proteins—Systems containing the blue-copper center, *Structure and Bonding, 23*, 1–60 (1975).

Long-Range Electron-Transfer in Biology, *Structure and Bonding, 75* (1991).

Marcus, R. A., and N. Sutin. Electron-transfer in chemistry and biology, *Biochem. Biophys. Acta, 811*, 265–322 (1985)

Moore, G. R., and G. W. Pettigrew. *Cytochrome c: Biological Aspects*, Springer-Verlag, 1987.

Moore, G. R., and G. W. Pettigrew. *Cytochrome c: Structural and Physicochemical Aspects*, Springer-Verlag, 1990.

San Pietro, A., ed. *Non-Haem Iron Proteins*, Antioch, 1965.

Williams, R. J. P. Electron-transfer in biology. *Molec. Phys. 68*, 1–23 (1989).

Worked Problem

Question 1: Compare and rationalize the reduction potentials for the following copper complexes and comment on the relevance of these observations for the understanding of the redox chemistry of copper proteins.

R	$E°$ (NHE)	X	$E°$ (NHE)
CH_3-	-0.90 V	O	-1.21 V
$(CH_3)_3C$-	-0.66 V	S	-0.83 V

Solution 1: The more negative the value of $E°$, the more stable is the oxidized Cu^{2+} complex. We recall from Table 1.3 that square pyramidal or tetragonal coordination is preferred by O and N-ligated Cu^{2+} ion, and that a tetrahedral geometry is favored by the reduced Cu^+ ion. For complexes **1**, the ligands cannot arrange themselves in a planar fashion when R = $(CH_3)_3C$- as a result of steric hindrance. Relative to the methyl derivative, the complex is forced to adopt a distorted tetrahedral geometry that is more favorable for the reduced ion, and so the $E°$ is less negative. For complexes **2**, the softer sulfur ligand atoms are preferred by Cu^+, which serve to stabilize this state relative to the case where X = O. As a result, the potential is less negative when X = S. Comparing complexes **1** and **2**, the ligand in **2** is more rigid and constrains the geometry to a square planar geometry which tends to favor Cu^{2+}, and so the potentials for **2** are more negative relative to **1**. Note that, in each case, the arguments are based on a comparison of one complex relative to another.

Because proteins are able to define both the ligand atoms and coordination geometries with a high degree of precision, they enjoy exquisite control of all aspects of the redox chemistry of the metal center. Blue-copper proteins stabilize reduced copper because the coordination sphere consists of softer sulfur and nitrogen centers that are arranged in a pseudotetrahedral fashion with no flexibility to distort to a square planar arrangement. As a result, the Cu^+ is strongly favored for type I copper centers, which is reflected by the positive $E°$ noted in Table 5.3. Surprisingly, however, the potentials for type II and III copper centers are generally even more positive, which is not consistent with the O and N ligand atoms or the square planar or square pyramidal geometries. Clearly Nature employs other methods to tune the potentials of these centers.

Problems

1. The reactions of a variety of electron-transfer proteins with small-molecule redox reagents have been studied. The table summarizes the second-order rate constants k_{12} for oxidation of the listed proteins by $[Ru(NH_3)_5py]^{3+}$. The reduction potentials (vs. NHE), radius, and overall charges of the reduced and oxidized proteins are also summarized.

	Redox Center	Overall Charge	ΔE (mV)	k_{12} (M^{-1} s^{-1})
Ru(NH$_3$)$_5$py	Ru$^{2+/3+}$	+2/+3	370	—
cytochrome c	Fe$^{2+/3+}$	+6.1/+7.1	260	1.86×10^4
stellacyanin	Cu$^{1+/2+}$	+2.4/−1.4	184	1.94×10^5
azurin	Cu$^{+/2+}$	−2.0/−1.0	328	2.00×10^3
HiPIP	[Fe$_4$S$_4$]$^{2+/3+}$	−3.5/−2.5	350	1.10×10^3
plastocyanin	Cu$^{+/2+}$	−10.0/−9.0	350	7.10×10^3

[Note: Ru(NH$_3$)$_5$py, $k_{22} \sim 3.38 \times 10^2$ M^{-1}s^{-1}].

   (a) Determine the protein self-exchange rates from simple Marcus theory.
   (b) Compare these results with the corrected values: cytochrome c (2.15×10^4 M^{-1}s^{-1}), stellacyanin (9.7×10^4 M^{-1}s^{-1}), azurin (2.7×10^3 M^{-1}s^{-1}), HiPIP (2.09×10^3 M^{-1}s^{-1}), plastocyanin (1.22×10^4 M^{-1}s^{-1}). Why are the values different? Explain the relative ordering.
   (c) Why do the total protein charges differ from the charge on the metal
[Cummins, D. C. and H. B. Gray, *J. Am. Chem. Soc. 99,* 5158 (1977)]
2. (a) Many cytochromes are found to possess multiple heme centers. Assign the features in the EPR spectrum shown below to high- and low-spin components. Predict the coordination states of each type of heme.

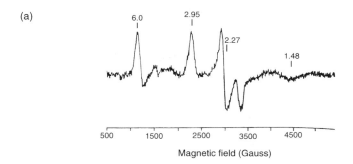

(a)

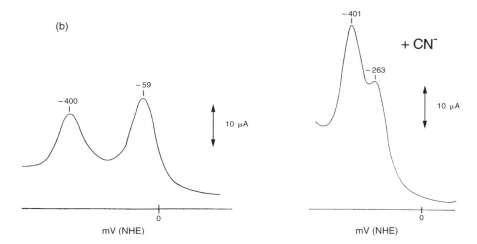

(b)

+ CN⁻

mV (NHE)   mV (NHE)

(b) The electrochemical traces above were obtained for this protein. Compare and succinctly discuss the relative values of the potentials. Are the changes observed after addition of CN⁻ consistent with the coordination assignments you made in (a)?

[Tan, J. and J. A. Cowan, *Biochemistry, 29,* 4886–4892 (1990)]

   3. Hexaheme cytochromes that display nitrite reductase activity ($NO_2^- \rightarrow NH_3$) have been isolated from a variety of bacterial sources, including *Wolinella succinogenes.* This enzyme contains five low-spin hemes (bis-histidine coordinated) and one high-spin heme (mono-histidine coordinated).

   (a) Where do you think the substrate binds and what is the role of the other heme centers? Is this cytochrome well designed for its catalytic role?

   (b) One of the low-spin hemes is found to be strongly coupled to the high-spin heme. Propose a detailed chemical mechanism for the reaction ($NO_2^- \rightarrow NH_3$). You may assume that electrons are provided from a natural redox partner.

[Blackmore, R. S. et al., *FEBS Lett. 219,* 244 (1983); Blackmore, R. S. et al.,

Biochem. J. 233, 553–557 (1986); Blackmore, R. S. et al., *Biochem. J. 271*, 259–264 (1990)]

4. (a) Flavins are unique redox cofactors. Not only can they engage in one- and two-electron transfer chemistry, they may also directly participate in chemical transformation. Consider the following reaction between adenosine phosphosulfate and reduced flavin adenine dinucleotide. Propose a detailed chemical mechanism for this reaction pathway. [Hint: The nitrogen indicated by * is a good place to start].

$$APS + FADH_2 \rightleftharpoons SO_3^{2-} + FAD + ADP + 2H^+$$

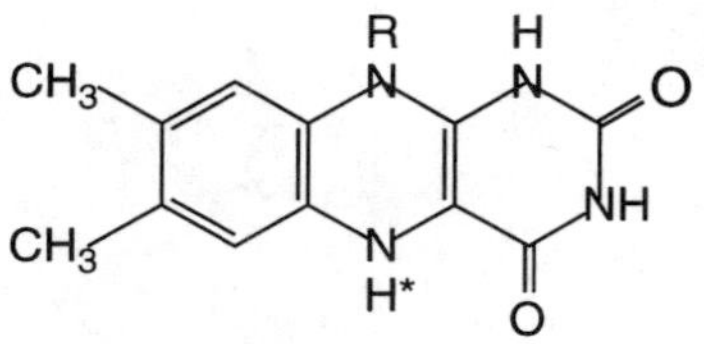

FADH$_2$
where R = adenosine dinucleotide

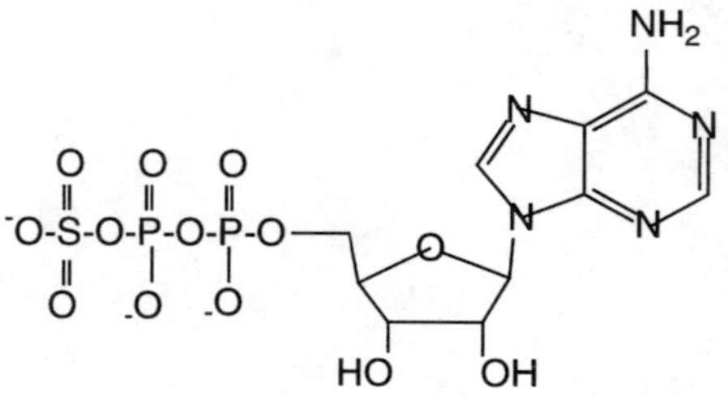

APS

5. Ribonucleotide reductase (RR) removes the 3′-OH from the ribose ring in ribonucleosides to form the deoxy ring that forms the component units of DNA. This enzyme possesses a binuclear iron center much like that found in hemerythrin. However, activation of native RR results in formation of a Tyr radical that is formed adjacent to the iron core. The native protein with the radical center can be reconstituted by reaction of apo-RR with ferrous ion and molecular oxygen,

$$2Fe^{2+} + Tyr122 + O_2 + 2H^+ + e^- \rightarrow Fe^{3+} - O^{2-} - Fe^{3+}$$
$$+ \ ^{\cdot}Tyr122 + H_2O$$

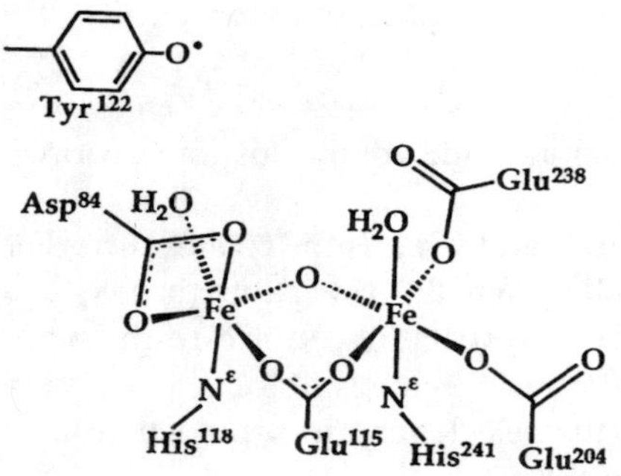

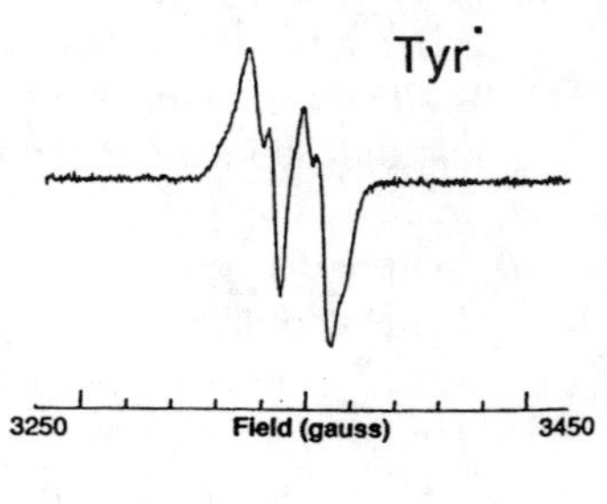

(a) Suggest two possible pathways for this reaction.

(b) One of these pathways should show an intermediate that reminds you of a high valent species that we covered earlier in the chapter. Which one?

(c) List the key differences that distinguish these two pathways.

(d) Which mechanism do you think is correct? Propose experiments to critically distinguish these two pathways.

[J. M. Bollinger et al., *Science 253*, 292–298 (1991), Figure is reprinted with permission; L. Que *ibid* 273–274; J. A. Cowan *Chemtracts (Inorg. Chem.) 4*, 318–322 (1992)]

6. (a) Xanthine oxidase is a molybdoenzyme that catalyzes the oxidation of xanthine to uric acid. Note the absence of molecular oxygen in this pathway. Experiments with isotopically labeled oxygen show that the molybdenum-bound oxygen rather than solvent water is transferred to the substrate in such reactions.

$$RH + H_2O \rightarrow ROH + 2e^- + 2H^+$$

This is an important reaction in nucleic acid biosynthesis. Propose a detailed chemical mechanism for this reaction, assuming that substrate binding and catalysis occur at the molybdenum center.

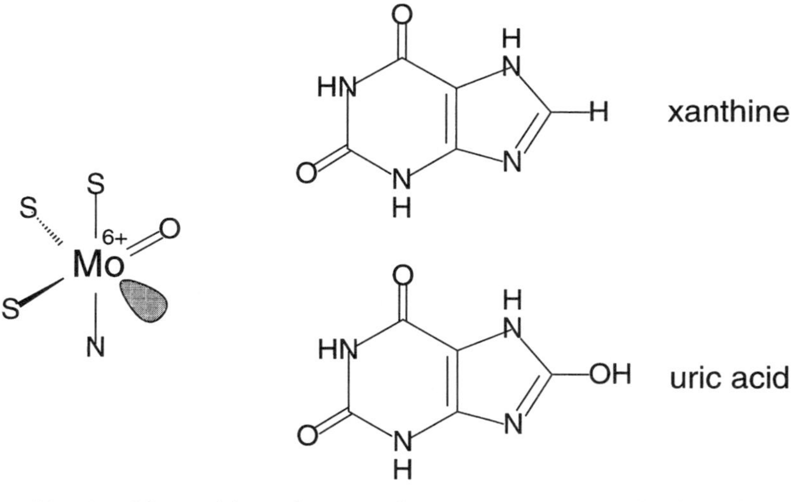

xanthine

uric acid

(b) Xanthine oxidase also contains Fe/S centers and flavin cofactors. Comment on the role of these redox sites for enzyme function.

(c) An EPR active Mo species is often observed during rapid-freeze-quench turnover experiments. What valence state might give rise to an $S = 1/2$ signal. Is this state relevant for catalysis?

[Hille, R. et al., *Biochemistry 32*, 3973–3980 (1993)]

7. Methane monooxygenase catalyzes the hydroxylation of many hydrocarbon substrates. The catalytic site contains a binuclear iron center reacts with O_2 and substrate as shown

$$Fe^{2+}-Fe^{2+} + O_2 \rightarrow \text{intermediate (Q)} + S \rightarrow \text{product}$$

The rate of the reaction with O_2 is independent of the concentration and identity of the substrate (S). Reaction with O_2 produces an intermediate Q which reacts with substrate to produce product. Other species are produced as side products from the reaction with O_2, but these are not catalytically relevant. The table summarizes Mossbauer parameters for intermediate Q and the side-product species.

(a) Suggest redox states for each iron. [Hint: The oxidation level of each iron in the binuclear center is identical for each pair.]

Intermediate	δ mm/s	ΔE_Q mm/s
Q	0.17	0.53
species 1	0.50	1.05
species 2	1.3	2.4–3.1

(b) If one of the iron centers in Q is ligated to an oxy (O^{2-}) ligand, propose a mechanism for formation of intermediate Q from Fe^{2+}–Fe^{2+} and O_2.

(c) What other methods might be used to support the structure you propose for intermediate Q?

(d) The binuclear iron center (Fe^{2+}–Fe^{2+}) shows an EPR signal at $g \sim 16$. This disappears during a (pseudo)-first-order reaction with O_2.

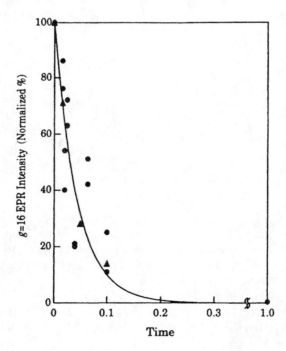

Estimate the apparent first-order rate constant. If k_{cat} for substrate turn-over is typically slower than this, what is the implication for subsequent reaction steps?

[Lee, S-K. et al., *J. Am. Chem. Soc. 115*, 6450 (1993); Lee, S-K. et al., *J. Biol. Chem. 268*, 21569 (1993), Figure is reprinted with permission]

8. To evaluate intramolecular electron-transfer rates in proteins, the one-electron reduced product from ruthenium (III)-labeled ferric myoglobin has been studied. Rapid introduction of one electron equivalent results in an approsimately statistical mixture of reduced Ru^{2+} and Fe^{2+}.

$$Ru^{3+}-Mb\text{–}Fe^{3+} \xrightarrow{e^-} \underset{x}{Ru^{2+}-Mb\text{–}Fe^{3+}} \underset{k_{-1}}{\overset{k_1}{\rightleftharpoons}} \underset{y}{Ru^{3+}-Mb\text{–}Fe^{2+}}$$

Subsequently, intramolecular electron-transfer takes place to form the thermodynamically favored ratio of x and y.

(a) Derive an expression for k_{obs} in terms of k_1 and k_{-1}.

(b) If $E°(Fe^{3+/2+}) = 65.4$ mV and $E°(Ru^{3+/2+}) = 85.8$ mV and $k_{obs} = 0.06$ s^{-1} at 298 K, calculate the values of k_1 and k_{-1}.

[Ellis, W. R. et al., *J. Am. Chem. Soc. 107*, 5002, 1985]

9. (a) In contrast to the binuclear iron centers of hemerythrin, purple acid phosphatase, and ribonucleotide reductase, the bridging oxygen atom in the met-state of MMO is always hydroxide and never an oxo-group. Suggest a reason for this. [Hint: Consider the pK_a of the bridging hydroxyl.]

(b) The bridging hydroxide in the met-form of MMO was characterized, in part, by the rather small coupling constant ($J \sim 7.5$–15 cm^{-1}) between the two antiferromagnetically coupled ferric centers. What magnitude of coupling constant would have been expected for an oxo bridge? What other methods might be used to evaluate the identity of the bridging ligand in the met and semi-met states?

[Thomann, H. et al., *J. Am Chem. Soc. 115*, 8881–8882 (1993); DeRose, V. J. et al., *J. Am. Chem. Soc. 115*, 6440–6441 (1993)]

6

Alkali and Alkaline Earth Metals

6.1 Overview of the Biological Chemistry of Group IA and IIA Metals

The alkali and alkaline earth ions (Na^+, K^+, Mg^{2+}, Ca^{2+}) are the most abundant metal ions in biological systems. As bulk electrolytes, their chemistry not only regulates energy metabolism and signaling mechanisms through transmembrane concentration gradients but also serves to stabilize the structures of proteins, cellular membranes, and skeletal mass. Often these metals are required in specific stoichiometries to promote enzyme activity and protein function. In Chapter 3 we also saw that transmembrane concentration gradients of Na^+ and K^+ play a particularly important role in a variety of transport mechanisms used by cells to accumulate and expel nutrients and ions.

With the possible exception of magnesium, the distribution of ions inside and outside of cells is not uniform. In fact, Table 6.1 shows that K^+ and Mg^{2+} are the major intracellular ions, whereas Na^+ and Ca^{2+} are found in high concentrations outside of the cell. The biological roles of these ions (Table 6.2) reflect these concentration differences and their distinct chemical properties.

The selection of metal ions for specific biochemical roles is ultimately determined by differences in their physicochemical properties (Table 6.3). With the exception of Mg^{2+}, which may occasionally coordinate nitrogen ligands,[1] the alkali (Na^+, K^+) and alkaline earth (Mg^{2+}, Ca^{2+}) metal ions bind to oxygen ligands and adopt octahedral coordination. Potassium and calcium commonly expand their coordination numbers and adopt irregular geometries. Ligand exchange rates for Na^+, K^+, and Ca^{2+} approach the diffusion limit ($\sim 10^9$–10^{10}

Table 6.1 Distribution of Intracellular and Extracellular Ions in and Around a Typical Mammalian Cell System

Ion	$[M^{n+}]_{in}$ (mM)	$[M^{n+}]_{out}$ (mM)
Na$^+$	10	145
K$^+$	140	5
Mg^{2+}	30	1
Ca^{2+}	1	4
H$^+$	5×10^{-4}	5×10^{-4}
Cl$^-$	4	110

Table 6.2 Summary of the Common Biological Roles for Alkali and Alkaline Earth Metals

Ion	Intracellular	Transmembrane Function	Extracellular
Na$^+$		Osmotic balance	Electrolyte
K$^+$	Ribosomes	Osmotic balance	
	Enzyme activation (structural)		
Mg^{2+}	Enzyme activation (catalytic/structural)		
	Ribosomes		
	Chelate with NTPs^a		
Ca^{2+}	Second messenger		
	Muscle activation	Enzyme activation	
	Skeletal mass	(catalytic/structural)	

^{a} Nucleotide triphosphates.

Table 6.3 Physicochemical Properties of Group IA and IIA Ions

	Ion	Ionic radii (Å)	Charge Density $(q^2/r)^a$	Approximate $k_{ex}(H_2O)$ s^{-1}	Coordination Numbers^b
IA	Li$^+$	0.60	1.67	10^8	4,6
	Na$^+$	0.95	1.05	10^{10}	6
	K$^+$	1.33	0.75	10^{10}	6–8
	Rb$^+$	1.48	0.68	10^{10}	6–8
	Cs$^+$	1.67	0.60	10^{10}	6–8
IIA	Be^{2+}	0.31	12.90	10^3	2,4
	Mg^{2+}	0.65	6.15	10^6	6
	Ca^{2+}	0.99	4.04	10^9	6–8
	Sr^{2+}	1.13	3.54	10^{10}	6–8
	Ba^{2+}	1.35	2.96	10^{10}	6–8
	Ra^{2+}	1.43	2.80	10^{10}	—

^{a} Charge density calculated as $(Z^2)/$(ionic radius in Å).
^{b} Typically, alkali and alkaline earth ions are bound by oxygen ligands (carbonyl, carboxylate, phosphate, and occasionally alcohol or water).

s^{-1}), suggesting that they are highly mobile and are not trapped by extensive interaction with the large number of potential binding sites on proteins, nucleic acids, and membranes. In contrast to the divalent ions Mg^{2+} and Ca^{2+}, which possess respectable binding constants to a variety of ligand sites, the monovalent ions (Na^+, K^+, Cl^-) interact weakly with most biological ligands, and so are unlikely to be involved in the triggering of biological activity by direct binding. A combination of rapid-exchange kinetics and strong ligand binding suggests that Ca^{2+}, rather than Mg^{2+}, would make an effective trigger ion for the activation of biological reactions. For this reason, Ca^{2+} had to be excluded from much of the intracellular environment (Table 6.1).

Figure 6.1 shows that both magnesium and calcium are used to stabilize a variety of protein structures. Calcium is particularly suited to this role as a result of its ability to adopt unusual coordination geometries (e.g., thermolysin, a heat-stable peptidase that is stabilized by four Ca^{2+} ions). An important structural function of Ca^{2+}, which we shall note here and shall discuss more fully later, is its employment in skeletal matter due to the relatively low solubility of a wide range of calcium salts (e.g., phosphates and carbonates).

Approximately 90 percent of intracellular Mg^{2+} is bound to ribosomes (complexes of RNA and proteins that mediate protein synthesis) or polynucleotides. Although magnesium and potassium play important roles in the structural biology of nucleic acids, they are also activators of enzymes that regulate the biochemistry of nucleic acids. Typically, these ions will induce conformational change in the enzyme-active site (e.g., K^+ in yeast aldehyde dehydrogenase) or play a direct role in catalysis (e.g., Mg^{2+} in ribozyme activity). As a result of their high charge densities, both magnesium and zinc find use as Lewis acid catalysts. Zinc ion is the stronger Lewis acid and is typically employed in the hydrolysis of carbonyl functionality (i.e., esters and amides). The harder magnesium ion is more frequently associated with phosphate ester hydrolysis and phosphoryl transfer. Biology, therefore, has access to both strong (Zn^{2+}) and weak (Mg^{2+}) Lewis acids. Calcium is infrequently found in a catalytic role. The examples of staphylococcal nuclease and phospholipase A_2 described in Section 6.4 are atypical.

The monovalent alkali metals (including H^+ and Cl^-) control transmembrane potentials and regulate the equilibrium of cellular electrolytes and osmotic pressures. These topics will be considered after we have introduced some key equations relating to bioenergetics. Given the brief preceding summary, it should be increasingly clear that we can deduce the likely biological function of the alkali and alkaline earth metals (and the transition metals) by considering their physicochemical properties. As the following specific examples well demonstrate, nature has made effective use of these functional differences to regulate cellular metabolism.

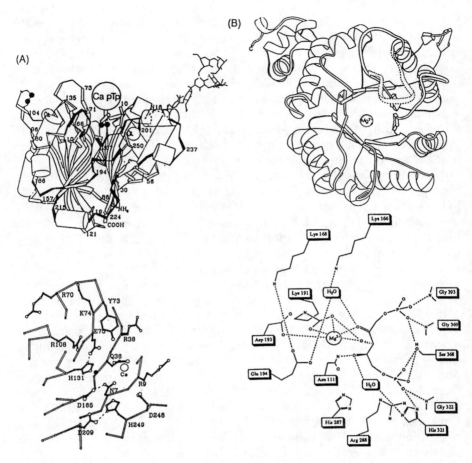

Figure 6.1 Examples of coordination sites in calcium- and magnesium-dependent enzymes. (A; upper and lower) The catalytic calcium site in deoxyribonuclease I. [Adapted from D. Suck and C. Oefner, *Nature, 321,* 620–625 (1986).] (B; upper and lower) The magnesium site in the C-terminal domain of ribulose bisphosphate carboxylase. [Adapted from T. Lundqvist and G. Schneider, *J. Biol. Chem., 266,* 12604–12611 (1991).]

6.2   Membrane Translocation

We have already seen several examples where the movement of protons across membranes drives chemical reactions (e.g., H^+-driven ATP synthesis in oxidative phosphorylation). Later in this chapter we shall see other examples of transmembrane ion movement that are coupled to the influx of essential nutrients, mechanisms for neurotransmission or cell signaling, and removal of toxic species. At the molecular level the mechanisms of transmembrane ion transport are extremely diverse, involving a bewildering array of proteins and ligands. However, Figure 6.2 shows that ion transport can be viewed in terms of a smaller

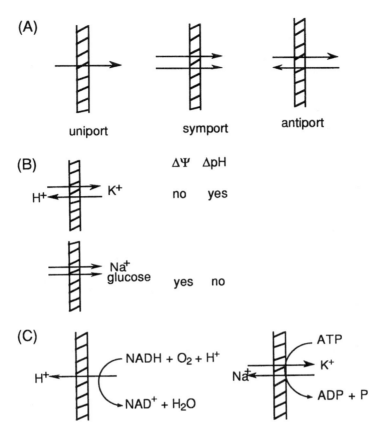

Figure 6.2 (A) General terms used to describe active and passive transport mechanisms for the movement of ions and neutral molecules across membranes. Uniports transport one species. Symports and antiports transport two species, which may travel in either the same or opposite directions, respectively. (B) These may occur either with or without a net change in electrical potential $\Delta\psi$ or solution pH. (C) Ion pumps are actively coupled to a metabolic pathway. In contrast, channels or ionophores simply serve as ion conduits for the passive transport of minerals, nutrients, or waste products to or from the cell and are not coupled to any metabolic reaction.

number of fundamental pathways that can be categorized under the general headings of *active* and *passive* transport. Molecules or ions transported by a passive mechanism move in the direction of an existing electrochemical gradient. Energetically, the pathway is "downhill," and so no energy is expended. Transport of metal cations by carrier ligands (ionophores) or through channels tends to occur by a passive mechanism. In active transport, a molecule is forced to move against an electrochemical potential gradient (e.g., translocation of Na^+ from a region of low to high concentration). By necessity this must be coupled to an energy source, which is frequently the hydrolysis of ATP. The movement of ions by an active mechanism may be coupled to the simultaneous translocation of other molecules (charged or neutral). If these species move in the same direction, the channel is termed a *symport*. If they move in the opposite direction, it is termed an *antiport*. Figure 6.2 shows that the movement of these other species may be either active or passive. Although several transmembrane proteins involved in the translocation of metal ions have now been identified (summarized previously in Table 3.7), the molecular mechanisms underlying these transport processes are not well defined.

Summary of Sections 6.1–6.2

1. Alkali and alkaline earth metals perform specific biochemical roles (Table 6.2) as a result of their distinct physicochemical properties (Table 6.3).
2. Passive transport, where metals migrate along a concentration gradient, and active transport, where metals migrate against a concentration gradient, are the two limiting mechanisms. Transport of two or more ions or substrates may be coupled. Active transport, in particular, may be coupled to an energetically favored passive transport mechanism or may be driven by ATP hydrolysis through an ion pump.

6.3   Alkali Metals and the Regulation of Membrane Potentials

Cells must receive a constant influx of nutrients, ions, and important metabolic substrates if they are to maintain their metabolism. In turn, there must be a mechanism for the efflux of waste or toxic products. Chapter 3 outlined how the rapid exchange of charged solutes or large neutral substrates through a hydrophobic membrane is a difficult process, requiring specific carrier molecules or transmembrane proteins. Energy is expended, and so transport phenomena are intimately coupled to the bioenergetics of the cell. In this section we shall try to understand some basic thermodynamic principles that underpin all transport processes and, then, consider some specific examples that illustrate the role of sodium and potassium gradients in the regulation of cellular events.

6.3.1   Thermodynamic Considerations

A detailed discussion of cellular energetics and transport phenomena requires an understanding of irreversible thermodynamics. Such a task is far beyond the scope of this book, and so we shall simply establish some important working relationships.[2] The electrochemical potential (μ_x) of n moles of a species (x) [equation 6.1] is a measure of its free energy under specified solution conditions [concentration ([x]), electrical potential (ψ), and pressure (P)],[3]

$$\mu_x = \mu_x^\circ + nRT \ln[x] + nZ_xF\psi + V(P - P_0) \tag{6.1}$$

where μ_x° is the electrochemical potential under ideal conditions, Z_x is the charge on species x, V is the volume, and $(P - P_0)$ is the pressure difference relative to $P_0 = 1$ atm. As a good approximation, constant atmospheric pressure can be assumed for solution studies. Of course, if a gas is taken up or evolved, the story changes! Equation 6.2 defines the free energy difference (ΔG) for a solution species x separated by a membrane,

$$\Delta G = \mu_{out} - \mu_{in} \tag{6.2}$$

Since $\mu_{out}^\circ = \mu_{in}^\circ$, we can write

$$\Delta G = nRT \ln ([x]_{out}/[x]_{in}) + nZ_x F \Delta\psi_x \tag{6.3}$$

From this equation it is clear that the movement of charged species across a biological membrane is dependent both on concentration gradients and on membrane potentials. This is a key equation for relating transport phenomena to the energy state of the cell. The flow of a charged species is driven by an electrochemical gradient, encompassing both a concentration gradient and a membrane potential. If $\Delta G > 0$, movement into the cell will be favored. If $\Delta G < 0$ the spontaneous direction of flow will be out of the cell. At $\Delta G = 0$ (equilibrium),

$$\Delta\psi_x = (RT/ZF) \ln ([x]_{out}/[x]_{in}) \tag{6.4}$$

which is the Nernst equation.

The resting potential of a cell is actually represented by a weighted sum of the electrochemical gradients of the three most abundant monovalent ions (K^+, Na^+, and Cl^-). The equation used to describe this potential is not the Nernst equation but a variation that accounts for the relative permeabilities of these ions through the cell membrane. The Goldman equation is defined next,

$$\Delta\psi = (RT/F) \ln\{(P_K[K^+]_o + P_{Na}[Na^+]_o + P_{Cl}[Cl^-]_i)/(P_K[K^+]_i + P_{Na}[Na^+]_i + P_{Cl}[Cl^-]_o)\}$$

where P_K, P_{Na}, and P_{Cl} are the relative permeabilities of the ions and the subscripts o and i represent outside or inside the cell, respectively. The effect of ion permeability on membrane potentials will be explained shortly in the section on Donnan potentials. From equation (6.4), we might deduce that, for an uncharged solute ($Z_x = 0$), transport is driven only by a concentration gradient. However, the flow of neutral metabolites is often coupled to the transport of ions (e.g., the Na^+-glucose transport system described later), and so the movement of neutral species may still be regulated indirectly by membrane potentials. We shall review specific examples shortly.

The approximate concentrations of intracellular and extracellular ions in a mammalian cell are listed in Table 6.1. The equilibrium ($\Delta G = 0$) concentrations of intracellular and extracellular K^+ represent the balance between a concentration gradient that tends to drive K^+ out of the cell and an opposing membrane potential (negative inside) that attempts to pull K^+ into the cell (Figure

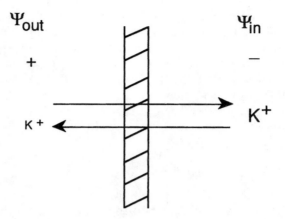

$$\Delta\mu = 0 = RT \ln ([K^+]_{in}/[K^+]_{out}) + F \Delta\psi = 8.314 \times 298 \times \ln(140/5)$$
$$+ 96{,}500 \times (-85.5 \times 10^{-3} \text{ V})$$

Figure 6.3 Balancing of transmembrane concentration gradients and electrical potential differences for K^+. The intracellular and extracellular potassium concentrations $[K^+]_{in}$ and $[K^+]_{out}$, respectively, are taken from Table 6.1. This concentration difference is balanced by a membrane potential ($\Delta\psi = \psi_{out} - \psi_{in}$) of -85.5 mV. A similar calculation can be made for any charged or neutral species. However, one should realize that the electrochemical potentials of many species are interdependent and must usually be accounted for by the inclusion of additional concentration or potential terms.

6.3). The magnitude of this potential clearly depends on a number of factors. For Na^+ ions, both the concentration gradient and the membrane potential tend to drive Na^+ into the cell. Since $[Na^+]_{out} > [Na^+]_{in}$, there must exist a mechanism to pump Na^+ out of the cell. This is provided by Na^+/K^+-ATPase, an antiport system that is driven by the hydrolysis of ATP to ADP and pumps two K^+ ions into the cell for every three Na^+ ejected. Figure 6.4 illustrates how the equilibrium concentrations of cellular ions and membrane potentials are created and maintained, with emphasis on alkali metal pumps that are coupled to the transport of other metabolites.

6.3.2 Na^+/K^+ ATPase

Very few transport systems have been studied in detail. Of these, Na^+/K^+ ATPase is one of the better characterized and was the topic of discussion in Chapter 3. This enzyme is an example of a *primary active transport* system that serves to establish a transmembrane concentration gradient of one or more species.[4] The electrochemical potential energy provided by this mechanism may be employed to transport other molecules against a concentration gradient.

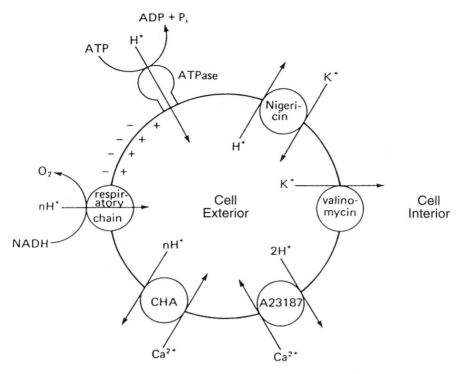

Figure 6.4 Model of the Ca^{2+}/H^+ antiport and the effects of ionophores in membrane vesicles of *E. coli*. In this case the vesicles (think of these as cell walls) are *inverted*, and so the orientation of the proton-motive force and other ion fluxes are reversed from those found in vivo. Primary proton pumps (ATPase or the respiratory chain) form a proton-motive force ($\Delta\psi$) with an acidic pH and positive potential in the interior. The antiport CHA couples the proton gradient to Ca^{2+} uptake. The ionophore A23187 mimics the operation of the natural antiporter. The antiport nigericin dissipates the pH gradient ΔpH and inhibits Ca^{2+} accumulation. In this example, the uniport valinomycin has no effect. (Adapted from B. P. Rosen and R. M. Brey, *Microbiology*, American Society of Microbiology, 1979, pp. 62–66.)

6.3.3 Na⁺-Glucose Transport

Figure 6.5 illustrates a glucose transport protein that drives glucose molecules into a cell by coupling this energetically unfavorable process to the passive transport of Na^+ ions into the cell. The use of preexisting ion gradients and membrane potentials to transport other solute molecules is termed *secondary active transport*.

6.3.4 Donnan Potentials

We have considered a variety of mechanisms that facilitate the movement of ions across membranes in response to a concentration or potential gradient.

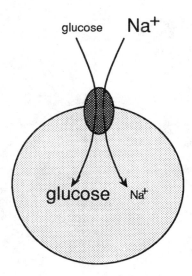

Figure 6.5   Glucose transport by way of a Na^+ symport.

These gradients are established even in the absence of carrier ligands, channels, or ion pumps because of variations in transmembrane ion diffusion rates. Biological membranes are normally impermeable to charged macromolecules within the cell. Small ions of high charge density (e.g., Na^+, Ca^{2+}, Mg^{2+}, and HPO_4^{2-}) are less permeable than larger ions, such as K^+ and Cl^-. The distinct permeabilities of each ionic species results in the generation of diffusion potentials that are named after Frederick Donnan. In 1911 Donnan reported his considerations of transmembrane potentials arising from differences in the membrane permeability of diffusible ions. In biology these diffusion-controlled electrochemical potentials arise from the many polyanionic macromolecules within a cell (of which the nucleic acids and lipids are the most common examples). As a result of their large size and high charge density, these cannot diffuse out of the cell, and so they must be charge-balanced by an appropriate number of counter cations. If we consider a starting situation where the cell is osmotically balanced with respect to the number of species both inside and outside the cell and is charge balanced, it is clear from Figure 6.6 that there are concentration gradients for both the polyanion and Cl^-. The polyanion cannot diffuse out of the cell, but Cl^- can diffuse into the cell. To retain charge balance, K^+ will also diffuse into the cell. The final concentrations of ions on both sides of the membrane can be calculated by considering the electrochemical potentials for K^+ and Cl^- (μ_K and μ_{Cl}, respectively). At equilibrium,

$$\Delta\mu_K = RT \ln([K^+]_{in}/[K^+]_{out}) + F\psi = 0 \tag{6.5}$$

$$\Delta\mu_{Cl} = RT \ln([Cl^-]_{in}/[Cl^-]_{out}) - F\psi = 0 \tag{6.6}$$

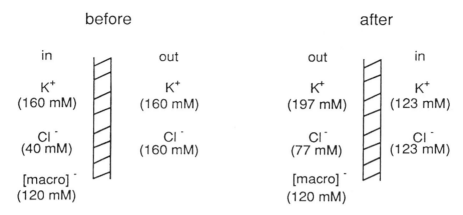

Figure 6.6 Diffusion potentials result from the selective permeability of an ion that leads to a concentration gradient and a transmembrane electrochemical potential. In the limit where one species (in this case a polyanionic macromolecule) cannot diffuse through the membrane, the resulting potential is called a *Donnan equilibrium potential.*

Adding (6.5) and (6.6), gives (6.7):

$$RT \ln([K^+]_{in}/[K^+]_{out}) = RT \ln([Cl^-]_{in}/[Cl^-]_{out}) \tag{6.7}$$

and so

$$[K^+]_{in}[Cl^-]_{in} = [K^+]_{out}[Cl^-]_{out}$$

To retain charge balance,

$$[K^+]_{in} = [Cl^-]_{in} + [poly^-]$$

$$[K^+]_{out} = [Cl^-]_{out}$$

and so

$$[Cl^-]_{in}^2 + [Cl^-]_{in}[poly^-] = [Cl^-]_{out}^2 \tag{6.8}$$

Because

$$[Cl^-]_{in} = 100 \text{ mM} + x$$
$$[Cl^-]_{out} = 200 \text{ mM} - x$$
$$[poly^-] = 100 \text{ mM}$$

the value of x can be readily determined by substituting these equations in (6.8). The final concentrations are shown in Figure 6.6. Clearly, the number of particles on each side of the membrane is different, and so the osmotic pressure within the cell increases. Although there is a tendency for water to diffuse into the cell to dilute the intracellular contents, this does not result in an increase of the cell volume, since most cells possess an outer wall (called the peptidoglycan layer, discussed in Section 6.6) that prevents such an expansion. Equilibrium is established when the downhill gradient for Cl^- migration into the cell is balanced by the uphill concentration gradient for movement of K^+. The resulting membrane potential can be evaluated from equation (6.5), where $\mu_K = 0$, to give $\psi \sim -12$ mV. These Donnan potentials arise from the impermeability of cellular membranes to macromolecular polyanions and are present irrespective of ion channels or ion pumps. Indeed, one might view the role of ion channels and pumps as mechanisms for regulating the background Donnan potential.

The example described earlier and shown in Figure 6.6 is clearly oversimplified. In particular, the presence of other solutes and ions has been neglected. In most cases, the concentrations of many species will be intimately connected. (Consider again the Na^+/K^+ antiport and the Na^+/glucose symport described earlier.) The example does, however, clearly establish the ideas of electrolyte potentials resulting from Na^+ and K^+ gradients and osmotic pressures. A combination of the two is often referred to as electro-osmosis. These transmembrane potentials underly many important pathways in biology (bioenergetic, neurochemical, and regulatory).

Summary of Section 6.3

1. Membrane transport processes can be understood in terms of the electrochemical potential μ_x for each species x. At equilibrium, $\mu_{out} = \mu_{in}$ for x.
2. By coupling together several species of ion or substrate, equation (6.1) can be used to explain active and passive transport phenomena, and the origin of Donnan equilibrium potentials.

6.4   Enzyme Activation

The common alkali and alkaline earth ions serve as enzyme activators. However, the relative importance of this function and the mechanism of activation varies according to the metal (Table 6.4).

6.4.1   Sodium and Potassium

Since the major functions of Na^+ relate to concentration gradients across membranes, the co-transport of solutes, and preservation of pH (Na^+/H^+ transport), no time will be spent discussing its rather limited chemistry as an enzyme acti-

Table 6.4   Alkali and Alkaline Earth Metals as Enzyme Activators

Metal	Frequency of Use	Function	Metal	Frequency of Use	Function
Na^+	Rare	Structural	Ca^{2+}	Moderate	Catalytic/structural
K^+	Moderate	Structural	Mg^{2+}	Common	Catalytic/structural

vator. Potassium, on the other hand, although serving principally as a counter-ion for negatively charged solutes and nucleic acids, is also required for the activation of a large number of enzymes (Table 6.5).[5] Of the monovalent alkali metals, only K^+ has been shown to serve as a stoichiometric cofactor. Pyruvate kinase (Figure 6.7) provides one of the few examples where the role of the potassium cofactor is well characterized. The low charge density precludes Lewis acid activity, and so K^+ helps to orient pyruvate in the active site by bridging the enzyme and substrate. A variety of monovalent cations may replace K^+, with the relative activities $Tl^+ > K^+ > Rb^+ > Cs^+ > Na^+ > Li^+$.[6] Ammonium ion ($NH_4^+$) does not normally compete with K^+. However, Ba^{2+} may block K^+ sites, since these ions are of the same size but the electrostatic attraction for the divalent ion is greater. Thallium(I) is a particularly good probe for K^+ sites and will be discussed in Chapter 9.

6.4.2 Magnesium and Calcium

A survey of any molecular biology catalog will provide ample evidence for the importance of Mg^{2+} as an activator of enzyme activity, especially those enzymes that act on nucleic acids. Restriction nucleases, ligases, and topoisomerases are among the many enzymes that are most effectively stimulated by divalent magnesium. Magnesium plays a role in enzymatic reactions in two general ways. First, an enzyme may bind the magnesium-substrate complex. In this case, the enzyme interacts principally with the substrate and shows little—or at best, weak—interaction with Mg^{2+} [e.g., MgATP (kinases), Mg isocitrate (isocitrate lyase)]. Alternatively, Mg^{2+} binds directly to the enzyme and alters its structure and/or serves a catalytic role. Although other divalent metal ions may also activate these enzymes, this is frequently accompanied by a reduction of enzyme efficiency and/or substrate specificity.[7] Magnesium binds weakly to proteins and enzymes ($K_a \leq 10^5$ M^{-1}), and magnesium-activated enzymes are not necessarily isolated in the metal-bound form. Magnesium must be added to the enzyme solution for in vitro reactions, whereas in vivo background magnesium concentrations are of the order of several millimolar. The chemistry of calcium enzymes has been more thoroughly developed as a result of the availability of crystallographic data for several calcium enzymes with clear definition of the Ca^{2+} binding site. Calcium enzymes, such as staphylococcal nuclease and deoxyribonuclease I, bind Ca^{2+} tightly and are isolated with the metal in situ. Binding constants for Ca^{2+} vary from the high-affinity ($K_a > 10^6$ M^{-1}) regulatory proteins (e.g., calmodulinin and troponin c are discussed in the next chapter) to the low-affinity structural or storage proteins ($K_a \sim 10^3$ M^{-1}) noted in Table 6.6.

Table 6.5 Examples of Enzymes Activated by Potassium Ion

Adenylate cyclase	Serine deaminase	Na^+/K^+-ATPase
Aspartate kinase	Tryptophan synthase	Aldehyde dehydrogenase
DNA polymerase	Pyruvate kinase	

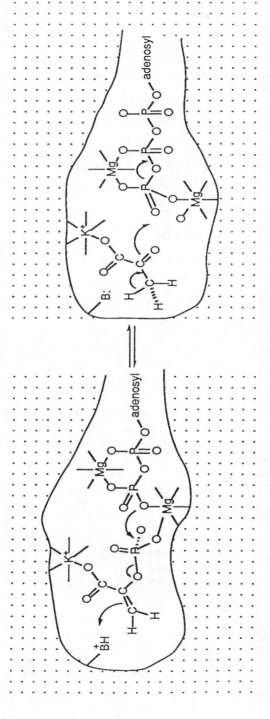

Figure 6.7 In glycolysis (see Figure 5.13), pyruvate kinase catalyzes reversible phosphoryl transfer from phosphoenolpyruvate to ADP, yielding pyruvate and ATP. Potassium ion serves to stabilize the enzyme–pyruvate complex. One Mg^{2+} forms a chelate complex with ATP, and the other may help to stabilize the enzyme–(Mg nucleotide phosphate) complex. The specific roles shown for the two enzyme-bound metals have yet to be firmly established.

Table 6.6 Calcium-Binding Proteins

	Ca^{2+} Sites	$K_a\,(M^{-1})$
EF-hand proteins		
Parvalbumin	2	10^9
Calbindin	2	10^6
Calmodulin	4	10^6
Troponin C (skeletal)	4	10^6
Extracellular digestive enzymes		
Staphylococcal nuclease	1	10^5
Phospholipase A_2	2	10^5, 10^3
Trypsin	1	10^4
Structural/storage proteins		
Thrombin	Many	10^3
Phosphodentine (material in teeth)	Many	10^3
Calsequestrin (Ca storage in the sarcoplamic reticulum)	40	10^3

Very few magnesium binding sites on enzymes have been crystallographically characterized. The binding pockets for the magnesium-dependent enzyme ribulose bisphosphate carboxylase (RuBISCO) are compared with the calcium enzyme deoxyribonuclease I in Figure 6.1. There are many other enzymes where magnesium is known to bind, but no role has been identified. For example, Mg^{2+} binds at the interface of the subunits of ribonucleotide reductase and presumably stabilizes the interfacial domain. It is important to note that both polynucleotides and the important enzymes in nucleic acid biochemistry that bind and utilize Mg^{2+} are intracellular species. In contrast, enzymes that have an absolute requirement for Ca^{2+} are typically extracellular digestive enzymes. This is a safety device to prevent premature activation by Mg^{2+} after synthesis in the cell. We shall now take a closer look at two of these calcium-dependent digestive enzymes.

Phospholipase A_2 is a calcium-dependent enzyme located in the outer membrane and catalyzes the hydrolysis of the 2-acyl ester bond in 1,2-diacylglycero-3-phospholipids, releasing free fatty acids. In mammals, fatty acids are transported through the bloodstream to the liver for further digestion in the mitochondrial energy cycle. Snake venom phospholipase preferentially hydro-

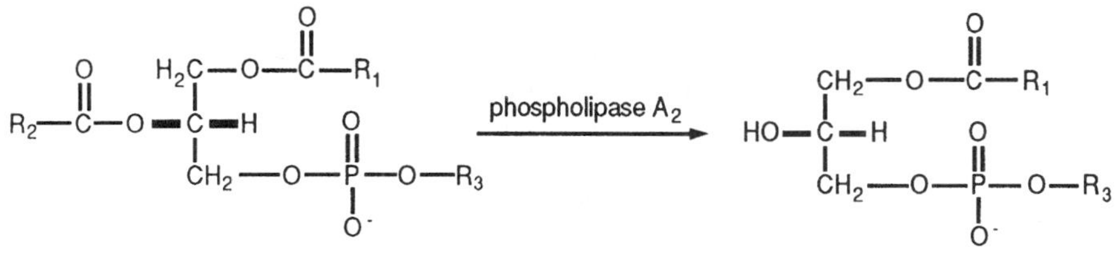

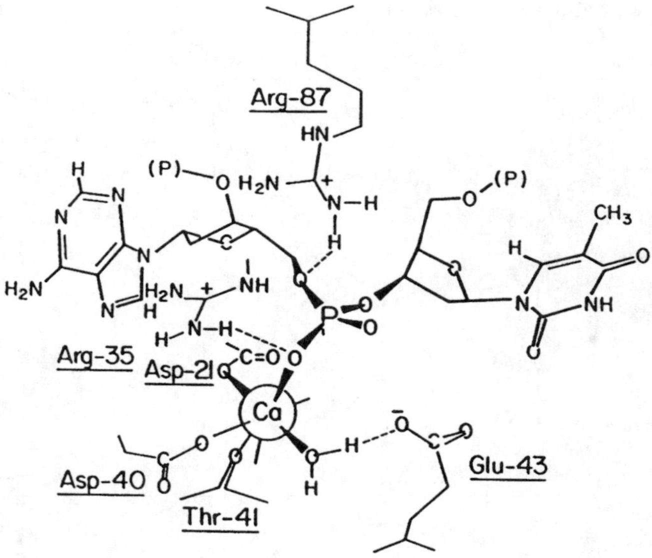

Figure 6.8 A putative model for the hydrolytic activity of staphylococcal nuclease (M_r ~ 16,900), an enzyme that cleaves DNA and RNA to form 3'-phosphomononucleotides. Calcium ion is bound by carboxylate residues (Asp-21 and Asp-40) in the active site and also serves as a Lewis acid by coordinating to the phosphate and activating it toward nucleophilic attack. Glu-43 is a general base that deprotonates H_2O to generate a more effective nucleophile (HO^-). [Reproduced with permission from Weber et al., *Biochemistry 33*, 8017–28 (1994)]

lyzes neutral lecithin molecules (Figure 3.3), whereas the mammalian pancreatic enzyme prefers negatively charged phospholipids, such as phosphatidyl glycerol and phosphatidic acid (Figure 3.3). Two positively charged lysine residues in the binding domain for the pancreatic enzyme assist in the selectivity for negatively charged substrates. The calcium binding site is heptacoordinate and does not readily bind Mg^{2+}, which prefers a regular octahedral geometry. Again, Ca^{2+} binds to the phosphate group and stabilizes the increased negative charge that arises in the transition state.

 Staphylococcal nuclease is an extracellular 5'-phosphodiesterase from the yeast *Staphylococcus aureus* that catalyzes the hydrolysis of DNA or RNA to give 3'-mononucleotide and dinucleotide products. The enzyme has been crystallographically characterized as the ternary enzyme–Ca^{2+}-inhibitor complex (Figure 6.8). The inhibitor (thymidine-3',5'-bisphosphate) models substrate binding at the active site. Calcium binds to the phosphodiester group and stabilizes the leaving group after hydrolysis by a calcium-bound water molecule.

6.5 Complexes with Nucleic Acids

Since the intracellular concentrations of Na^+ and Ca^{2+} are low, the metal-binding chemistry of the nucleic acids in vivo is dominated by K^+ and Mg^{2+}. Ribosomes contain a major fraction of these ions, which are bound to ribosomal RNA and proteins. Divalent magnesium binds with a greater affinity as a result of its higher charge. The binding of positively charged counterions by nucleic acids is a natural consequence of the polyanionic sugar–phosphate backbone. Metal ions alleviate electrostatic repulsion between phosphates, thereby stabilizing base-pairing and base-stacking. This is most clearly evidenced by the increase in melting temperature (T_m) and hypochromism, respectively (Figures 6.9 and 6.10). In contrast, transition metals that have a higher affinity for heteroatoms on the bases tend to inhibit base-pairing and stacking, and so destabilize the double helix. The stabilization of nucleic acid structure by alkali and alkaline earth ions is of great importance, but poorly characterized. Three illustrative examples are described next.

6.5.1 Ribozymes

One of the most exciting discoveries in nucleic acid chemistry has been the finding that RNA possesses catalytic activity and is able to promote hydrolysis

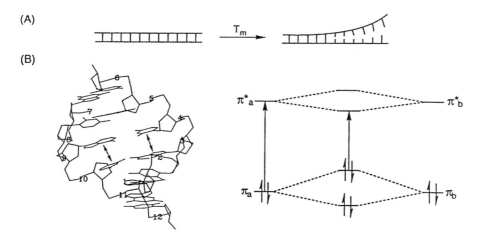

Figure 6.9 (A) The melting temperature (T_m) is the temperature where the two strands in a double-strand helix are half-dissociated. This can be determined by following the change in absorbance at 260 nm as a function of temperature (see Figure 6.10A). As the strands separate, the interactions between bases disappear, with a resulting loss of hypochromism (increase in absorbance). (B) Hypochromism arises from the close interactions between the π-orbitals of the stacked bases (left). The net result is to decrease (and red-shift) the main absorption bands resulting from π–π transitions, relative to the isolated bases (right). Metal binding reduces repulsion between backbone phosphates, favoring closer stacking of the bases and enhanced hypochromism (see Figure 6.10B).

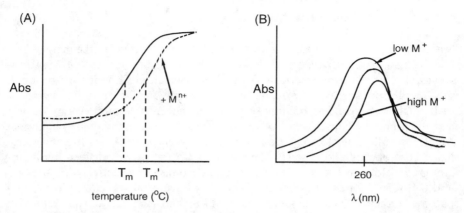

Figure 6.10 Influence of increasing metal ion concentration on (A) DNA melting profiles and (B) absorption spectra (see Figure 6.9). In (A), alkali or alkaline earth metals neutralize repulsive interactions between the negatively charged ribose–phosphate backbones and stabilize the interactions between base pairs. This further stabilizes the hydrogen bonds that pair the two strands, resulting in higher melting temperatures (A), and leads to enhanced base stacking with increased hypochromism and a decrease in absorbance (B).

and ligation of the RNA phosphodiester backbone. Such RNA enzymes have been termed *ribozymes*. Many ribozyme sequences are part of much larger stretches of messenger or ribosomal RNA molecules. Cleavage and ligation reactions are part of the post-transcriptional modification process that ultimately leads to the mature sequence. It has been possible to identify the shorter sequences required for ribozyme activity, and to isolate and study these as discrete molecules. One of the best studied and simplest is the "hammerhead" ribozyme shown in Figure 6.11, so-called because of its resemblance in shape to the head of a hammer! The three dimensional structure of this ribozyme and its RNA substrate have been determined (Figure 6.11), and the positions of five critical magnesium ions have been identified. Several of these serve important structural roles, helping to stabilize the complex secondary and tertiary structure of the ribozyme-substrate complex. These ions retain their solvation shell, binding as $Mg(H_2O)_6^{2+}$ by forming hydrogen bonds from the waters of solvation to oxygen and nitrogen atoms on the bases, phosphates, and sugar rings. One other magnesium ion has been located at the catalytic site and Figure 6.11B illustrates a putative transition state or intermediate on the reaction pathway. Here a magnesium-bound hydroxide abstracts a proton from the 2'-OH, which can subsequently attack the phosphate ester, forming a cyclic phosphate ester intermediate with elimination of the 5'-OH from the other side. In this model, the magnesium ion also serves as a Lewis acid by binding to the pro-R phosphate oxygen. Several other families of ribozymes have been characterized that differ in their reaction mechanisms. Nevertheless, the hammerhead is a convenient paradigm for understanding the essential molecular features.

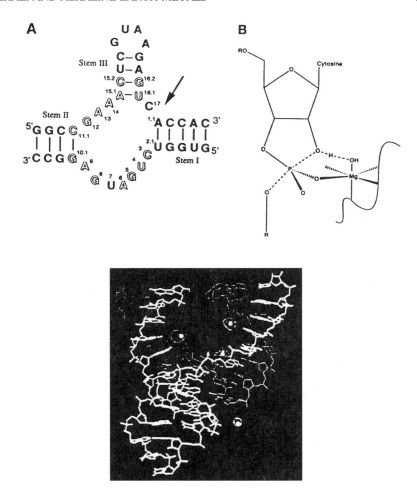

Figure 6.11 (A) Illustration of the secondary structure of the ribozyme-substrate complex. This represents part of an internal domain in a naturally occurring, self-cleaving RNA molecule. The arrow indicates the point of strand cleavage on the larger 25-nucleotide substrate unit. Absolutely conserved bases are outlined. (B) A proposed transition state or intermediate during ribozyme catalyzed hydrolysis, showing the magnesium ion in the active site. An outline of the three-dimensional structure of the ribozyme-substrate complex is shown below. [Reprinted with permission from W. S. Scott et al., *Cell 81*, 991–1002 (1995)].

6.5.2 B- and Z-DNA

Double-stranded DNA normally adopts a B-conformation; however, under certain solution conditions, transitions to other conformations can be observed. The B → Z transition can be induced by high salt concentrations (2.5 M Na$^+$, 0.7 M Mg^{2+}, or 0.04 M Co(NH$_3$)$_6^{3+}$). Figure 6.12 illustrates some crystallo-

(A)   (B)

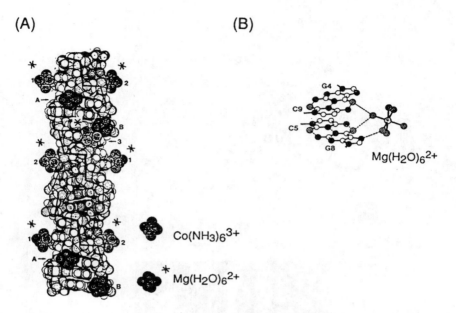

Figure 6.12 Conformational transitions induced by metal coordination. (A) Z-DNA is stabilized by complexation with $Mg(H_2O)_6^{2+}$ or $Co(NH_3)_6^{3+}$. Magnesium binding sites are indicated with (*). (B) The interaction of one $Mg(H_2O)_6^{2+}$ center with Z-DNA. A water molecule binds to O-6 of guanines 4 and 8 (G4 and G8), and another binds to N-7 of guanine 8. [Reprinted with permission from R. V. Gessner et al., *Biochemistry, 24,* 237–240; copyright (1985) American Chemical Society.]

graphically characterized metal-binding sites on Z-DNA. It should be noted that only certain sequences may adopt a Z-conformation [e.g., poly(dG–dC), poly(dA–dC), poly(dT–dG)].[8] Alternating pyrimidine–purine bases are required. Ultimately, the preferred backbone conformation is controlled by the stereochemistry of the ribose rings and orientations of base units that minimize steric and torsional strains (see, for example, Figures 1.21 and 1.22).[9]

6.5.3 Telomeric DNA

Telomeres are sequences of DNA that occur at the ends of chromosomes, are essential for chromosomal stability, and mediate the association of chromosomes during cell replication. Telomeric DNA is characterized by a 3'-overhang of two repeating sequences that are capable of dimerization by forming parallel guanine quadruplexes. Figure 6.13 summarizes this structural chemistry and illustrates the role of K^+ in stabilizing the telomeric complex. Potassium is particularly suited for the octacoordinate ligand environment formed from the O-6 atoms (oxygen at carbon-6) of eight guanine bases.

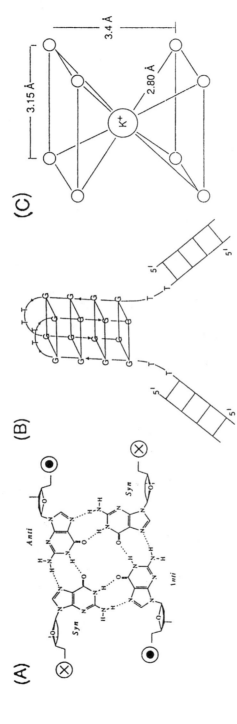

Figure 6.13 Stabilization of telomeric DNA by K^+. (A) A guanine tetrad within an antiparallel quadruplex. Strand orientations (5' to 3') are represented by $\otimes$ or ($\bullet$), respectively. Structural requirements at the ribose rings are indicated (see Figure 1.22). (B) A proposed structure for the overlapping ends of telomeric DNA. (C) An octacoordinate potassium chelation cage created by the O-6 oxygen atoms of the stacked guanine bases. Interatomic distances are derived from fiber diffraction studies and model building. [Reprinted by permission from W. I. Sundquist and A. Klug, *Nature, 342,* 825–829; copyright (1989) Macmillan Magazines Limited.]

Summary of Sections 6.4–6.5

1. The roles of Na^+, K^+, Mg^{2+}, and Ca^{2+} as enzyme activators are summarized in Table 6.4. Mg^{2+} and to a lesser extent K^+ are the major cofactors.
2. Mg^{2+} may bind to substrates and/or enzymes, serving either a catalytic or structural role.
3. Intracellular Mg^{2+} and K^+ are extensively complexed with nucleic acids. These cations stabilize the negatively charged ribose-phosphate backbone and specific structural motifs. Examples include RNA tertiary structure, DNA backbone conformation, and telomeric DNA in chromosomes.

Review Questions

- In the crystallographic study illustrated in Figure 6.11 one of the 2′-hydroxyls of the RNA substrate was methylated, forming a 2′-OCH_3 derivative. Why was this done?
- Draw a mechanism for formation and subsequent hydrolysis of the cyclic phosphate ester intermediate. Do you expect a high or low concentration of the putative Mg-hydroxide intermediate to be present in solution?
- The designation of the phosphate oxygen as pro-R implies a stereochemical distinction. How might you obtain supporting evidence that the pro-R rather than pro-S oxygen interacts with the hydrated magnesium?

[Scott, W. S. et al., *Cell 81*, 991–1002 (1995); Dahm, S. C. et al., *Biochemistry 32*, 13040–13045 (1993)]

6.6   Cell Walls and Membranes

At the end of Chapter 1 we learned that eukaryotic cells possess internal structures or compartments (mitochondrion, golgi apparatus, peroxisomes, and so on) that isolate the various functions essential to normal cell metabolism. The membranes that envelope these inner compartments and form the outer walls of the cell are composed of a variety of proteins, polysaccharides, and lipids. The organization of cell walls and some of the lipids used in their construction were described previously in Chapter 3 (Figures 3.2 and 3.3). The peptidoglycan layer found in the outer walls of bacterial cells is composed of a disaccharide repeat unit containing N-acetylglucosamine (NAG) and N-acetylmuramic acid (NAM). Figure 6.14 shows the structures of these units and how they are connected by short peptide chains made up from D-amino acids. The outer layer of gram-positive bacteria also contains teichoic acid polymers, of which there are several structural types.[10] Many of these are based on phosphodiesters of glycerol with alternating NAG and D-alanyl units on the central hydroxyl. One end of the polymer is attached to a lipid that forms part of a contiguous lipid membrane. It should be clear that many of these surface polymers are polyanionic

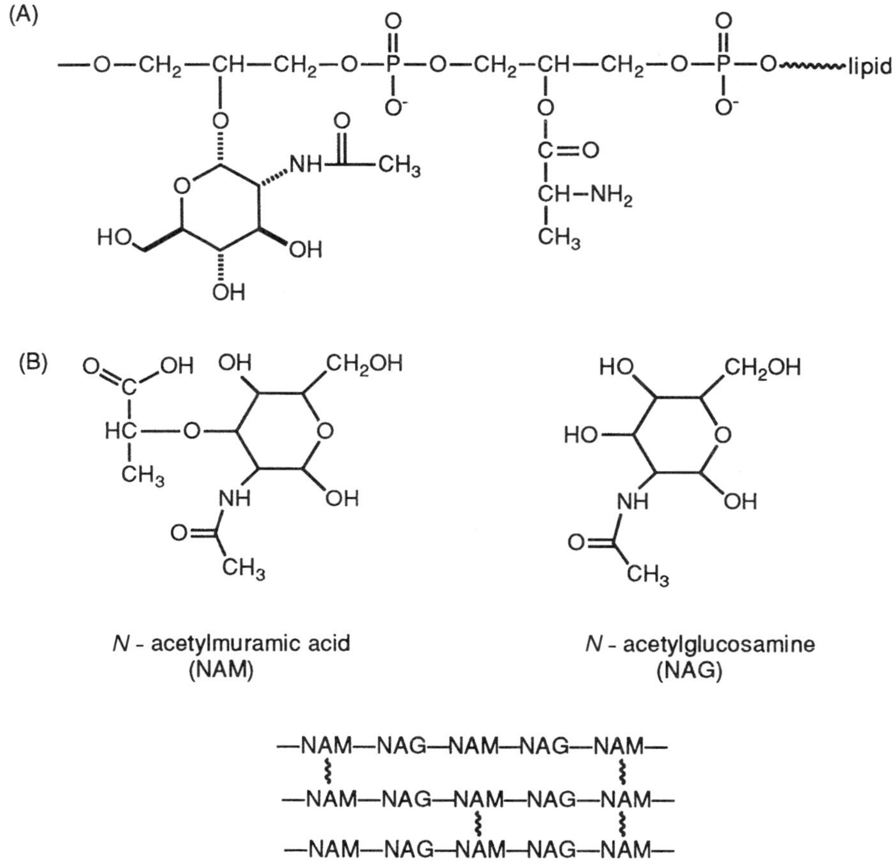

(A)

(B)

N - acetylmuramic acid
(NAM)

N - acetylglucosamine
(NAG)

—NAM—NAG—NAM—NAG—NAM—
—NAM—NAG—NAM—NAG—NAM—
—NAM—NAG—NAM—NAG—NAM—

Figure 6.14 The outer cell wall of certain bacteria is strengthened by a layer of peptidoglycans and teichoic acids. (A) A typical teichioc acid chain, showing units of N-acetyl glucosamine and D-alanyl linked alternately to a phosphoglycerol backbone. For gram-positive bacteria a terminal lipid facilitates anchoring to the cell membrane.[10] (B) The peptidoglycan layer is composed of a repeating disaccharide (–NAM–NAG–) that is cross-linked (wavy lines) by short peptides of D-amino acids.

carboxylates or phosphates. Both magnesium and calcium are used to cross-link and stabilize these structures, and so chelating agents that bind these metals strongly (e.g., $EDTA^{4-}$) will disrupt cell membranes.

6.7 Biominerals

Many organisms use biological minerals to provide internal or external *structure* (e.g., bones, shells), as *sensors* to probe the environment (navigational), or for convenient *storage* of metal ions as salt deposits until required (e.g., ferritin, metallothionein). A large number of inorganic salts and structural forms are known, however. Table 6.7 documents only some of the more important forms of biological minerals and their distribution in a variety of organisms. Our attention will focus on calcium and iron salts.

6.7.1 Skeletal Mass

Calcium is the major metal ion in skeletal minerals, constituting more than 20% by weight of most hard tissue. X-ray diffraction studies demonstrate a close structural relationship between these biological minerals and apatites (calcium phosphates). Hydroxyapatite [$Ca_5(PO_4)_3OH$] is a reasonable chemical prototype for biological apatites. Crystallographic studies show a hexagonal arrangement of Ca^{2+} and PO_4^{3-} ions around columns of OH^-. The use of fluoride ion in toothpastes reflects the high affinity of this ion for the hydroxide sites in the mineral lattice of hydroxyapatite. Fluoroapatite is less sensitive to acid degradation (fluoride is a weaker base than hydroxide); consequently, it strengthens the mineral coating surrounding teeth. Other substitutions of the Ca^{2+}, PO_4^{3-}, and OH^- sites are possible, and contaminant ions are commonly found in biological apatites. Examples include CO_3^{2-}, HCO_3^-, F^-, Cl^-, Mg^{2+}, Na^+, and K^+ that may act as direct replacement ions (e.g., CO_3^{2-} for PO_4^{3-}) or, more typically, lie in distinct interstitial sites in the lattice. Magnesium ion [$Mg(H_2O)_6^{2+}$] does not readily substitute for calcium, since the ion is small and

Table 6.7 Common Minerals Found in Living Organisms[a]

Cation	Anion	Formula	Mineral	Distribution and Function
Calcium	Carbonate	$CaCO_3$	Calcite Aragonite Vaterite	Exoskeleton in plants, balance sensor in animals, calcium store, eye lens
	Phosphate	$Ca_{10}(PO_4)_6(OH)_2$	Hydroxyapatite	Skeletal matter, calcium store in shells, bacteria, bones, and teeth
	Oxalate	$Ca(COO)_2 \cdot 2H_2O$	Weddellite	Calcium store in plants
	Sulfate	$CaSO_4 \cdot 2H_2O$	Gypsum	Balance sensor in plants, S store, Ca store
Iron	Oxide	Fe_3O_4	Magnetite	Magnetic sensor in bacteria and animals
	Hydroxide	$Fe(OH)_3$	Ferritin	Iron store in eukaryotes and prokaryotes
Silicon	Oxide	SiO_2	Amorphous	Skeletal matter in sponges and protozoa
Magnesium	Carbonate	$MgCO_3$	Magnesite	Skeletal matter in corals

[a] See H. A. Lowensteam and S. Weiner, in *Biomineralization and Biological Metal Accumulation* (Eds. P. Westbroek and E. W. de Jong), Reidel, pp. 191–203, for a comprehensive listing of minerals and their distribution in extant organisms.

does not readily incorporate into the apatite lattice structure during crystal growth. The lattice energy for this particular mineral form is large (see Chapter 1, Section 1.2), and so magnesium tends to bind at the mineral surface. Many of these ion replacements require omission of normal constituent cations or anions in the lattice to achieve charge neutrality. For example, substitution of tetrahedral PO_4^{3-} by planar CO_3^{2-} leaves a void that can be occupied by a monovalent anion, such as F^- or OH^-, to preserve electroneutrality and structural packing. Alternatively, a neighboring Ca^{2+} may be replaced with Na^+ or K^+. In any event, such changes can lead to structural variations in the mineral lattice.

Bones actually contain large amounts of fibrous protein filaments called *collagen*. This provides a framework around which calcium phosphate is deposited and defines the shape of the bone. Collagen fibers are composed of cross-linked tropocollagen molecules, and the gaps between these molecules provide nucleation sites for calcium deposition (Figure 6.15). Two types of cell that are involved in the regulation of bone formation and shaping are called *osteoblasts* and *osteoclasts*. Osteoblasts secrete collagen fibers used to make the bone matrix, whereas osteoclasts are mobile cells that secrete enzymes that hydrolyze collagen and lead to the breakdown and dissolution of the hydroxyl apatite matrix. These cells contain many lysosomes that carry out the final intracellular digestion of the organic fragments.

6.7.2 Biological Sensors

Biological minerals are frequently used as devices to position or orient cells relative to a chemical (possibly a food) source or a directional axis. Magnetic and gravity sensors are two common examples of how cells use crystals of inorganic solids as navigational devices. In vertebrates (including mammals and fish), calcium phosphate is the major calcium-containing mineral; however, the balance organ located in the inner ear is composed of calcium carbonate.[11] The crystal form of this organ varies according to the species (calcite, mammals; aragonite, fish). The balance organs of other species may contain distinct mineral forms, such as barium sulfate (baryte). Other types of sensors are used to guide growth.

The mechanism by which these systems function is poorly understood. One example that has been studied in some detail, which we shall outline even though it is an iron-based sensor, is the mineral magnetite. This also gives us the opportunity to develop further some important ideas in magnetism. Magnetotactic

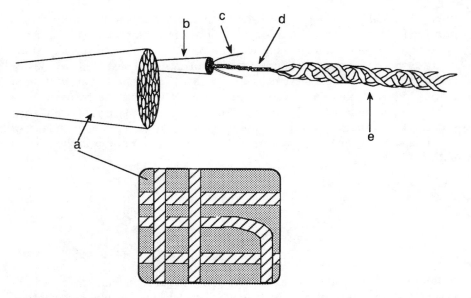

Figure 6.15   Intertwining collagen fibers (a) form the matrix around which calcium is deposited (as apatite). Each collagen fiber is made up from *macrofibrils* (b), which in turn consist of a collection of *microfibrils* (c). Each microfibril is made up from a number of tropocollagen helices (d), composed of three intertwining polypeptide chains (e).

bacteria contain Fe_3O_4 crystals in vesicles that lie close to the cell membrane. To understand how this works, we shall make a brief diversion to consider the magnetic properties of minerals. A disordered array of ions that possesses a

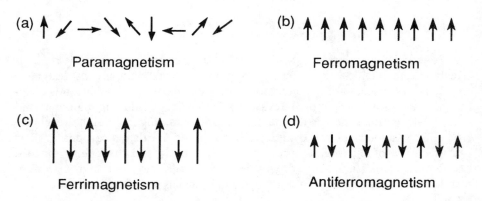

net magnetic moment is an example of a paramagnetic material (a). If such a sample is placed in an external magnetic field, the net magnetic moment from the sample will oppose the external field. Below a critical temperature, there may be a coupling of the individual moments that aligns them in the same direction and produces a permanent magnetic moment. Such a substance is termed *ferromagnetic* (b) and the critical temperature for spin alignment is called the *Curie temperature* (T_c). If some moments are aligned in the opposite direction but there is still an overall net spin, a ferrimagnetic material (c) may be formed. If the moments are aligned in an antiparallel fashion, to give no net moment, the material is antiferromagnetic (d). In this case, the critical temperature for alignment is the Néel temperature (T_N). Iron oxides, formed from ferric ions that are linked by oxide anions, show antiferromagnetic behavior because the overlap of orbitals, and the alignments of electrons within those orbitals tends to orient the spins on the iron centers in opposite directions. The mineral that is found in magnetotactic bacteria (magnetite, Fe_3O_4) is a ferrimagnetic material

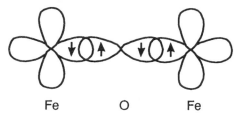

that typically adopts a spinel structure. Ions in the O_h holes have their spins aligned, whereas ions in T_d holes are oriented in the opposite direction. This results in antiferromagnetic coupling of the Fe(III) spins and a net ferrimagnetism from Fe(II). Magnetotactic bacteria, therefore, possess an internal magnetic moment (a "compass") that can be used to orient the cell in an external applied field.

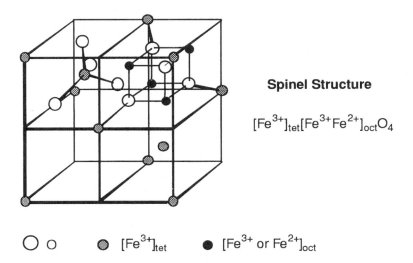

Spinel Structure

$[Fe^{3+}]_{tet}[Fe^{3+}Fe^{2+}]_{oct}O_4$

$\bigcirc$ O $\circledcirc$ $[Fe^{3+}]_{tet}$ $\bullet$ $[Fe^{3+}$ or $Fe^{2+}]_{oct}$

Figure 6.16 Key ionic centers in part of a spinel unit cell are shown to illustrate the coordination geometries of the tetrahedral and octahedral iron centers.

6.7.3   Storage

Skeletal materials constitute the major calcium reserve in many cellular systems. The role of skeletal minerals in calcium homeostasis is more complicated, however, than simple passive diffusion of Ca^{2+} to and from the mineral phase. Bone mineral cannot by itself readily exchange Ca^{2+} with extracellular fluids, since this is protected by a collagenous matrix that does not allow facile access to water or solvated ions. Bone tissue is in a continual state of flux, with constant formation and resorption of the protein–mineral complex.[12] Release of Ca^{2+} to the fluid phase requires that the bone tissue (protein and mineral) be enzymatically broken down. This occurs under the tight control of osteoclasts, and so Ca^{2+} uptake and release are under strict metabolic regulatory control.

Ferritin (discussed in Chapter 3) also offers an excellent example of the uses of a biological mineral for long-term storage. The inner core of ferritin is composed of iron phosphate (see Section 3.3), with the iron ions being transported by the carrier protein transferrin. From these two examples, it should be clear that biological minerals play a major role in the regulation of cation and anion concentrations in cellular systems, particularly for calcium, iron, and phosphate.

Summary of Sections 6.6–6.7

1. Cell membranes are composed of polysaccharides and lipids with negatively charged headgroups. Both Mg^{2+} and Ca^{2+} are used to cross-link and stabilize these structures.
2. Calcium forms relatively insoluble salts with phosphates and carbonates. These are the major inorganic components in a variety of skeletal materials in mammals, fish, insects, and so on. Bone in mammals is composed of intertwining protein fibers (collagen) that provide the skeleton for a calcium phosphate matrix (hydroxyl apatite).
3. Skeletal matter constitutes one of the major reserves of Ca^{2+} and phosphate. The release of Ca^{2+} is under strict cellular control. Specific cells (*osteoclasts*) secrete enzymes that break down the collagen framework and release Ca^{2+} into solution.
4. Calcium and magnesium salts are extensively used in the balance organs of mammals and fish. The most thoroughly studied navigational sensor is the mineral magnetite (Fe_3O_4) that is located in magnetotactic bacteria.

Notes

1. Chlorophyll, a pigment molecule in green plant photosynthesis, is a magnesium derivative of chlorin. The magnesium ion lies in the center of the chlorin plane coordinated to four pyrrolic nitrogens (see Figure 5.11).
2. Good discussions may be found in the articles by H. Rottenberg [*Biochim. Biophys. Acta, 549,* 225–253 (1979)] and H. V. Westerhoff and K. van Dam [*Curr. Topics Bioenergetics, 9,* 1–62 (1979)].

3. Concentration terms are more properly written as activities $a = \gamma[x]$, where γ is the activity coefficient (see note 8, Chapter 1). The symbol ψ is commonly used to denote membrane potentials. Elsewhere in the text, we have used the symbol $E°$ to denote the reduction potential of a specific prosthetic site in discussions of redox thermodynamics.
4. Primary active transport is directly coupled to ATP hydrolysis. Secondary active transport is coupled to preexisting gradients of ions or other solutes.
5. Most intracellular K^+ and Mg^{2+} are bound to ribosomes (complexes of RNA and protein).
6. Note that Tl^+ is a better activator than K^+. The reason that it is not the natural cofactor is related to the limited availability of this ion.
7. Mn^{2+} has been extensively used as a substitute for Mg^{2+}. Chapter 9 details the use of this high-spin d^5 ion as a spectroscopic probe.
8. This nomenclature denotes one strand having the sequence 5'-dG–dC–dG–dC–dG–dC–...-3', with a complementary 3'-dC-dG-dC-dG-dC-dG-...-5' strand. Similarly for the other polynucleotide sequences. The prefix d in dG, dC, and so on, denotes *deoxy*.
9. The interested reader should refer to W. Saenger, *Principles of Nucleic Acid Structure*, Springer-Verlag, 1984.
10. The classification as *gram-positive* or *gram-negative* bacteria is made according to whether the cell wall reacts with the gram stain (a colored complex of a dye, crystal violet, and iodine). In gram-positive bacteria, which take up the stain, the outer membrane is much thicker (about 50 nm) than that found in gram-negative bacteria. The cell walls, therefore, possess quite distinct chemistry.
11. The balance organs of mammals are called *otoconia*; those of fish are termed *otoliths*.
12. Contrast this with the mineral components of teeth. Dental enamel and dentin are formative processes only; that is, the matrix proteins (calcium-binding phosphodentine) are degraded as more enamel apatite is formed in the outer regions of the extracellular matrix. As a result, enamel mineralization is more akin to invertebrate mineralizations (i.e., the formation of shells and other protective coatings over snails, cockles, mussels, and so on) rather than the collagen-based mineralization found in mammals.

Further Reading
Bioenergetics and Membranes

Harrison, P. M., ed. Structure and Mechanism of (Na^+, K^+)- and (Ca^{2+})-ATPases, M. Forgac, G. Chin, *Metalloproteins,* Part 2, Macmillan, 1985, pp. 123–148.
Hille, B. *Ionic Channels of Excitable Membranes*, 2nd ed. Sinauer, 1992.
Neher, E. Ion channels for communication between and within cells. *Science 256*, 498–502 (1992).
Neher, E. and B. Sakmann. The patch-clamp technique. *Sci Amer. 266*, 44–51 (1992).
Nicholls, D. G. and S. F. Ferguson. *Bioenergetics*, 2nd ed., Academic, 1992.
Poole, R. K., and G. M. Gould, Eds. *Metal Microbe Interactions*, Special Publications of the Society for General Microbiology, vol. 26, IRL Press, 1989.
Williams, R. J. P. The biochemistry of sodium, potassium, magnesium, and calcium, *Quart. Rev. Chem. Soc., 24*, 331–360 (1970).

Enzyme Activation

Black, C. B., H-W. Huang, and J. A. Cowan. Biological coordination chemistry of magnesium, sodium, and potassium Ions. Protein and nucleotide binding domains. *Coordn. Chem. Rev. 135–136*, 165–202 (1994).
Cowan, J. A. *Biological Chemistry of Magnesium*, VCH, New York, 1995.
Suelter, C. H. Monovalent cations in enzyme-catalyzed reactions, *Metal Ions in Biological Systems,* Vol. 3, Dekker, 1974.

Nucleic Acids

Draper, D. E. The RNA folding problem. *Acc. Chem. Res. 25*, 201–207 (1992).

Eichorn, G. L. and L. G. Marzilli, eds. Metal Ions in Genetic Information Transfer, *Adv. Inorg. Biochem., 3* (1981).

Gessner, R. V. et al. Structural basis for stabilization of Z-DNA by cobalt hexaamine and magnesium cations, *Biochemistry, 24*, 237–240, 1985.

Klug, A., and D. Rhodes. Zinc Fingers, *Trends in Biochem. Sci., 12*, 464–469 (1987).

Mildvan, A. S., and C. M. Grisham. The role of divalent cations in the mechanism of enzyme-catalyzed phosphoryl and nucleotidyl transfer reactions, *Structure and Bonding, 20*, 1–21 (1974).

Pyle, A. M. Ribozymes: A distinct class of metalloenzymes. *Science 261*, 709 (1993).

Scott, W. S. et al. The crystal structure of an all-RNA hammerhead ribozyme: a proposed mechanism for RNA catalytic cleavage. *Cell 81*, 991–1002 (1995).

Spiro, T. G., Ed. Nucleic Acid–Metal Ion Interactions, *Metals in Biology*, Wiley, 1980.

Biominerals

Spiro, T. G., Ed. *Calcium in Biology*, Wiley-Interscience, 1983.

Problems

1. The solvent exchange rate for Mg^{2+} (aq) is around 10^5 s^{-1}. A novel magnesium channel is isolated that transports Mg^{2+} at a rate of 10^8 s^{-1}. Comment on this result and its implications for the transport mechanism.

2. (a) Calculate the ATP/ADP ratio that can be obtained from a $\Delta\mu_{(H^+)} = 5$ kcal per mole of H$^+$ translocated, if the H$^+$/ATP ratio required for ATP synthesis is one. Assume $\Delta G°$ for ATP synthesis is 8 kcal/mole and [phosphate] = 10 mM. What is the implication for the reversibility of ATPase activity?

   (b) GABA (γ-amino butyric acid) binds to the GABA receptor and opens chloride channels. Interestingly, the drug valium, which is used in the treatment of stress and anxiety, acts by blocking GABA binding to its receptor. If the extracellular and intracellular concentrations of Cl$^-$ are 123 mM and 4 mM, respectively, determine the direction of flow of chloride ion when the transmembrane potential is -120 mV and $+30$ mV.

   (c) How much free energy does it take to pump Ca^{2+} from a cell if the cytosolic concentration is 0.4 μM and the extracellular concentration is 1.5 mM, with a membrane potential of -60 mV?

3. Briefly describe three distinct biochemical roles for Ca^{2+} ion. In each case, explain why the physiochemical properties of Ca^{2+} are particularly suited to that role.

4. MinK is a 130 amino acid, voltage-gated potassium channel. It contains a central, putative, transmembrane spanning region of 32 amino acids. This shorter sequence has been chemically synthesized with fluorescent labeling and studied.

(a) Predict the secondary structure of the peptide in methanol solution from its CD spectrum.

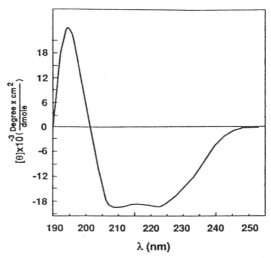

(b) The fluorescence spectra of the carboxyfluorescein-labeled peptide shows shifts in wavelength and intensity that depend on the solution environment: (-··-·-) in aqueous buffer; (——) in the presence of phosphatidyl choline vesicles. Spectrum (. . . . .) is a control obtained for carboxyfluorescein-labeled ethanolamine in aqueous buffer. Provide an explanation for these data.

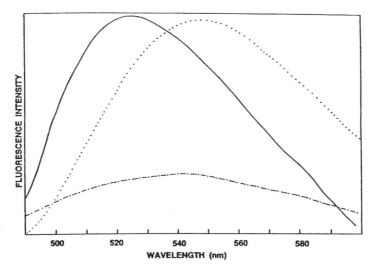

[Ben-Effraim, I. et al., *Biochemistry 32*, 2371–2377 (1993), Figure is reprinted with permission]

5. Divalent manganese will bind to the surface of artificial membranes that are composed of a variety of lipid molecules. Account for the relative binding affinities of Mn^{2+} to the following membrane surfaces: phosphatidyl choline ($9.1 \ M^{-1}$), phosphatidyl ethanolamine ($150 \ M^{-1}$), phosphatidyl serine ($1.8 \times 10^3 \ M^{-1}$), phosphatidic acid ($5.8 \times 10^4 \ M^{-1}$).

[Puskin, J. S. *J. Membrane Biol. 35,* 39–55 (1977)]

6. The ionic radius of Tl^+ is very similar to that of K^+, and both carry the same charge. However, the former cannot be transported by potassium selective ionophores. Suggest a reason why.

7. Li^+ and Mn^{2+} are both used as probes of Mg^{2+} chemistry.
   (a) Rationalize the use of each of these two metal ions in terms of their physicochemical properties relative to Mg^{2+}. Comment also on potential problems with their use.
   (b) What other methods can be used to investigate the chemistry of divalent magnesium?

[*The Biological Chemistry of Magnesium,* Chap. 1, Cowan, J. A., ed., VCH, New York, 1995]

8. You have a fragment of DNA. You do not know if it is single- or double-stranded. Describe a simple experiment to determine which you have, using equipment that would be readily available in most chemistry laboratories.

9. The dynamics of RNA substrate binding to a ribozyme has been studied by fluorescence-detected, stopped-flow methods. The reaction scheme under investigation can be summarized as,

$$\text{free} \underset{k_{-1}}{\overset{k_1}{\rightleftharpoons}} \text{bound} \underset{k_{-2}}{\overset{k_2}{\rightleftharpoons}} \text{complex}$$

$$\text{step 1} \qquad \text{step 2}$$

The kinetics of these steps were evaluated by monitoring the change in fluorescence intensity of a pyrene molecule that was covalently attached to the substrate molecule.
   (a) Why does the fluorescence of pyrene change? Explain the blue shift in emission wavelength that is observed after complex formation with the ribozyme.
   (b) Write equations for the dissociation constants K_{d1}, K_{d2} and the overall binding constant K_D in terms of the rate constants k_1, k_{-1}, k_2 and k_{-2}.
   (c) Assume $k_1 = 3.9 \times 10^6 \ M^{-1} \ s^{-1}$; $k_{-1} = 0.5 \ s^{-1}$; $k_2 = 2.5 \ s^{-1}$; $K_D = 1.1 \ nM$. Calculate K_{1d}, K_{2d}, and k_{-2}.
   (d) Determine the free energy of stabilization in the final complex from tertiary interactions.
   (e) Binding of a complementary base sequence to the substrate gives $k_1 = 55 \times 10^6 \ M^{-1} \ s^{-1}$ and $k_{-1} = 6.6 \ s^{-1}$. Compare and contrast with substrate binding to the ribozyme.

[Bevilacqua, P. C. et al., *Science 258,* 1355 (1992)]

10. The variation of substrate cleavage rates by the hammerhead ribozyme is

shown as a function of metal ion concentration. In this study, the substrate was labeled with sulfur as a replacement for oxygen at the pro-R_P position of the reactive phosphate. (A) shows the variation of rates as a function of Mn^{2+} (▲) and Mg^{2+} (□) concentration. (B) shows the variation of cleavage rate with Mn^{2+} concentration in the presence of 10 mM Mg^{2+}. Provide an explanation for these data. Why was the control experiment shown in (B) necessary?

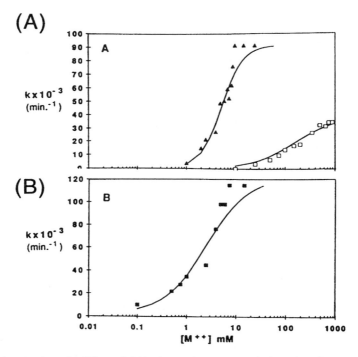

The kinetic data in (C) and (D) show the extent of cleavage (in a semi log plot) for normal substrate in the presence and absence of divalent Mg^{2+} (10 mM for C). Part (D) also shows a portion of an autoradiogram (an X-ray film of an SDS-PAGE gel on which radio labeled substrate and products were identified) used to monitor the cleavage kinetics. Use this raw data to calculate the rate constants for these reactions.

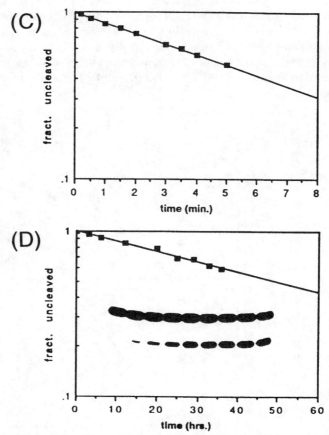

[Dahm, S. C. and Uhlenbeck, O. C. *Biochemistry 30,* 9464 (1991), Figure is reprinted with permission]

CHAPTER

7

Metals in the Regulation of Biochemical Events

In previous chapters we considered the structures and reactivities of discrete biological molecules. Each represents a challenging problem that may satisfy a lifetime of study. We must remember, however, that the chemistry of enzymes, hormones, nucleic acids, minerals, and membranes forms part of a more elaborate series of interdependent biochemical pathways that constitute the complexity of cellular metabolism. In multicellular organisms there is yet another level of biochemical function represented by molecular signaling between cells, cell differentiation, and replication. It should come as no surprise that the biochemistry of life processes is accompanied by a set of sophisticated control mechanisms that regulate the chemistry within cells and the communication networks between cells and their surroundings. This may sound a little vague. It is important to bear in mind that every aspect of life can be described in detail at the molecular level using the language of chemistry. We do not yet understand all the chemistry that underlies biology. That is where the challenges of tomorrow lie.

Metal ions play an important role in many regulatory pathways. In this chapter we shall review a representative selection of these mechanisms. First, we focus on the role of calcium as a secondary messenger. Then, we consider the role of ion gradients as triggers of cellular response. Finally, we examine the chemistry of gene regulation at the level of metal–nucleic acid and metalloprotein–nucleic acid interactions.

(Courtesy of R. Michell and *Trends in Biochemistry,* Elsevier)

7.1   Calcium as a Secondary Messenger

The main difference between a primary and a secondary messenger lies in their locus of operation (Figure 7.1). A primary messenger acts as an extracellular agent that either effects communication between cells or initiates a response to a change in the extracellular environment. Examples of primary messengers include hormones (e.g., insulin) and metabolites (e.g., glucose). In contrast, secondary messengers, such as cyclic nucleotides (cAMP, cGMP) and free calcium ion, function within the cell in response to an extracellular signal.

7.1.1   Calcium-Dependent Metabolism

The action of a secondary messenger relies on a rapid response to an external stimulus, and so calcium must satisfy certain functional criteria to serve this role:

1. There must be target molecules (effectors) within the cell that bind Ca^{2+} and lead to a cellular response. One can think of an effector as a molecule (protein, hormone, and so on) that can bind Ca^{2+} and is thereby activated to carry out a specific task. This molecule must remain passive until needed. Effector molecules for calcium are normally proteins that either initiate an immediate cellular response after binding Ca^{2+} (e.g., calcium binding to troponin C results in muscle contraction) or activate a target protein (e.g., cal-

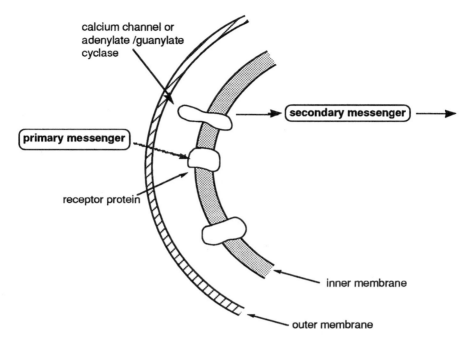

Figure 7.1 *Primary messengers* (e.g., insulin or glucose) bind to cell surface receptors and initiate a response that is transmitted to the interior of the cell through *secondary messengers* (Ca^{2+} or cyclic nucleotide monophosphates). The immediate effect of the primary messenger is to increase intracellular calcium levels or activate the synthesis of cAMP or cGMP.

modulin) that subsequently initiates further cellular responses. These protein targets in mammalian cells are localized in the cytosol and membrane surfaces open to the cytosol (Figure 7.2).

2. The concentration of calcium ion must remain low in an unstimulated cell to avoid interaction with effector proteins. These proteins must be highly selective for Ca^{2+}, since they must recognize this ion in the presence of high intracellular levels of free Mg^{2+} (~0.5 mm) and K^+(~100 mM). Typically, they will bind Ca^{2+} with affinities of $>10^6$ M^{-1}, whereas Mg^{2+} ($< 10^3$ M^{-1}) and K^+ ($< 10^1$ M^{-1}) bind more weakly. Moreover, the conformations of the apo-protein (protein lacking an essential cofactor) and its native calcium or magnesium forms may be different, allowing specific interactions with target enzymes or structural proteins.

3. During stimulation, the intracellular calcium levels must increase rapidly. Subsequently, the calcium levels must be reduced to normal submicromolar concentrations.

Table 6.6 in the previous chapter summarized some important binding data for extracellular and intracellular proteins and enzymes. Certain hydrolytic en-

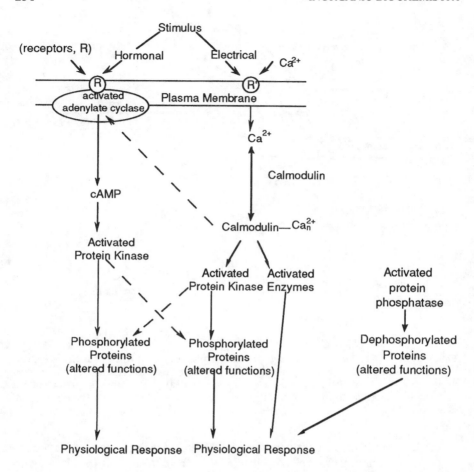

Figure 7.2 Secondary messengers (Ca^{2+} or cAMP) are released inside the cell in response to an external stimulus from a primary messenger. These bind to and activate other proteins and enzymes that regulate important pathways in cellular metabolism. Calmodulin-activated enzymes play a key role in the regulation of cellular responses to external stimuli. Kinases and phosphatases phosphorylate or dephosphorylate residues on proteins, respectively, thereby altering the structure and function of that protein. Interdependencies between the two common secondary messenger systems are indicated by the broken lines. [Adapted from P. Cohen, *Nature, 296,* 613 (1982).]

zymes [e.g., phospholipase A_2 and staphylococcal nuclease (see Figure 6.8)] show a high degree of specificity for Ca^{2+}. If these enzymes were readily activated by Mg^{2+}, it would be more difficult to regulate their intracellular activity. In keeping with the flexible coordination demands of Ca^{2+}, the coordination shell defined by protein residues at the metal-binding site is often expanded to seven ligands. This inhibits Mg^{2+} binding, which prefers to adopt a more regular octahedral coordination geometry.

Intracellular calcium ion is distributed between the cytosol (10^{-7} M) and internal compartments, such as the endoplasmic and sarcoplasmic reticula, and the mitochondrion (all 10^{-3} M). A cell responds to an external stimulus by mechanisms that provide for a rapid influx of calcium into the cytosol (Figure 7.2). The release of Ca^{2+} from internal compartments, such as the sarcoplasmic reticulum, is triggered by a small increase of $[Ca^{2+}]$ in the cytosol. These ions bind to proteins that activate calcium channels. After the release of Ca^{2+} into the cell, there must be rapid binding to an effector molecule. In comparison with Mg^{2+}, the rapid exchange rate makes Ca^{2+} an ideal choice as a trigger ion $[k_{ex} (Ca^{2+}) \sim 2 \times 10^8 \text{ s}^{-1}; k_{ex} (Mg^{2+}) \, 7 \times 10^5 \text{ s}^{-1}]$.

7.1.2 Muscle Contraction—Troponin C

Since Ca^{2+} may bind to many molecules in the cytosol, it is important that the effector protein competes favorably and binds Ca^{2+} tightly. For example, troponin C (Tn C) is an important protein in the regulation of muscle action and binds Ca^{2+} with $K_a \sim 10^8$ M^{-1}. Figure 7.3 shows that a typical muscle fiber is composed of thick (myosin) and thin (actin) filaments. The thin filament is also illustrated in Figure 7.3 and consists of a protein complex termed *tropomyosin* that forms a helix around the actin filament. The actin filament is composed of a globular protein called G-actin that polymerizes to form the filamentous F-actin. Troponin, a complex of three subunits (Tn I, Tn T, Tn C), regulates the interaction between the thick and thin filaments. Muscle contraction results from the sliding of the thick and thin filaments in opposite directions. The thick filament is constituted mostly from the protein myosin. Head groups project from this filament, which interact with the thin filament after the troponin complex binds Ca^{2+}. Specifically, Ca^{2+} binds to Tn C, which serves to remove Tn I (the inhibitory subunit) from the actin filament.

The triggering of muscle action may be summarized as follows. In response to a nerve stimulus, calcium ion is released from the sarcoplasmic reticulum. The sarcoplasmic reticulum is a special form of the endoplasmic reticulum and surrounds the fibers within the muscle cell. This compartment contains a large store of Ca^{2+} (10^{-3} M), which minimizes the diffusional time between the site of Ca^{2+} release and its uptake by troponin C on the muscle fiber. Four Ca^{2+} ions bind to Tn C and induce a large conformational change that releases the inhibitory complex Tn I from actin and facilitates direct interactions between the thick and thin filaments, resulting in muscle contraction (Figure 7.3). The muscle relaxes when Ca^{2+} is removed from the cell. Again, it should be noted that a rapid response is facilitated by the favorable exchange kinetics of calcium ion. Excess Ca^{2+} is rapidly removed from the cell by the ion pump Ca^{2+}–ATPase (Table 3.7).[1]

There is an obvious need to maintain $[Ca^{2+}]_i$ at low levels. In particular, Ca^{2+} may have an inhibitory effect on many Mg^{2+}-dependent enzymes, and the low solubility of calcium salts poses the problem of precipitate formation. As a means of regulating $[Ca^{2+}]_i$, even in those cases where an increase is required in

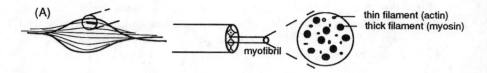

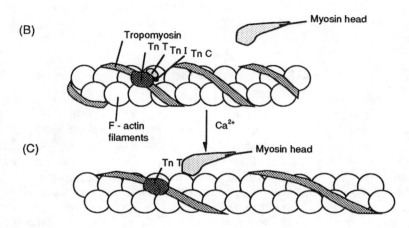

Figure 7.3 (A) Schematic illustration of a muscle fiber showing the network of thick and thin filaments. (B) Actin filaments are formed from G-actin, a globular protein that polymerizes to form the filamentous F-actin complex. Tropomyosin is a helical dimer that wraps around the actin filament. Movement of the tropomyosin chain, relative to the actin filament, is responsible for muscle contraction. The interaction between the myosin heads in the thick filament and actin is regulated by the trimeric troponin complex consisting of Tn T, Tn I, and Tn C. Calcium binds to Tn C, which results in loss of the inhibitory Tn I subunit and allows the myosin head to make intimate contact with the actin filament. The relative motion of the myosin and actin filaments results in a contraction. ATP phosphorylation of proteins provides the energy source for myosin head movement. After release of Ca^{2+} from the cell, the system reverts back to its original state.

response to an external stimulus, the cell makes use of intracellular buffering proteins, such as the parvalbumins, that have large binding affinities for Ca^{2+} ($K_a \sim 10^9\ M^{-1}$).

Whereas troponin C initiates a direct mechanical response following Ca^{2+} binding, other proteins, such as calmodulin, regulate cell metabolism by activating key enzymes. Before discussing the biochemistry of calmodulin, we should note that many of these regulatory proteins possess similar structural features. In particular, each Ca^{2+} is bound in a helix–loop–helix site named after the E and F helices that comprise part of the structure (Figure 7.4). The nomenclature derives from the notation used to describe the six α-helical segments of carp muscle parvalbumin B. Loops occurring between the helical se-

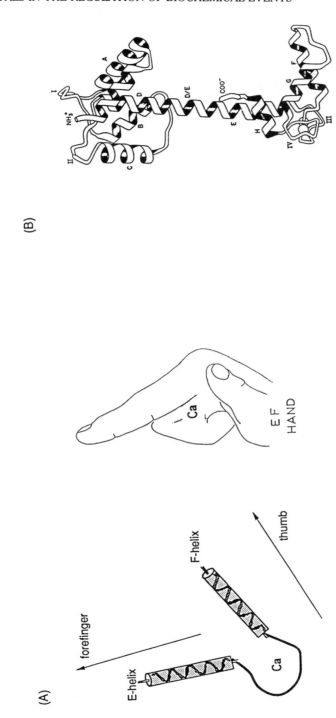

Figure 7.4 (A) The term EF-hand originates from the seminal crystallographic work on carp parvalbumin [R. H. Kretsinger, *Ann. Rev. Biochem.*, *45*, 239–266 (1976).] (B) Structural features of the calcium-binding regulatory protein troponin C, showing the EF-loops that form the metal sites. In this diagram, only the two lower sites are occupied. [Reprinted by permission from A. Herzberg and M. N. G. James, *Nature*, *313*, 653–658; copyright (1985) Macmillan Magazines Limited.]

quences were named according to the two helices defining the loop. Parvalbumin binds two Ca^{2+} ions in domains defined by the CD and EF loops. Only the latter is solvent accessible. The term *EF-hand domain* has been extensively used to define Ca^{2+}-binding domains that are structurally homologous to that identified in parvalbumin. Each protein contains one or two pairs of Ca^{2+} binding sites that display a considerable degree of cooperative binding. As expected, the ligands that bind Ca^{2+} are always oxygen atoms from side-chain carboxylates, backbone carbonyls, or water molecules. Since the EF-hand is a central structural feature of all the calcium-binding proteins involved in secondary messenger pathways, we will consider this topic further before going on to discuss calmodulin.

7.1.3   The Chemistry of EF-Hands

The chemical interactions of EF-hand Ca^{2+} binding sites is best described with reference to calbindin D_{9k} (M_r 8,500),[2] the smallest protein known to possess these sites. In all such proteins, the EF-hands are defined by two helices (labeled the E- and F-helices in the original structural work on carp parvalbumin) that are separated by a calcium-binding loop and always occur in pairs (Figure 7.5). Calcium binding can be described by either the macroscopic (K_1 and K_2) or microscopic (k_I, k_{II}, $k_{I,II}$, $k_{II,I}$) binding constants defined here.[3] The cooperativity

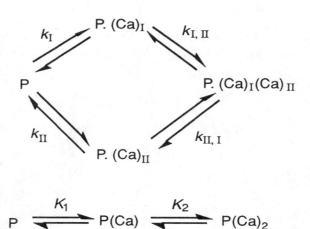

of Ca^{2+} binding is readily demonstrated by use of a fluorescent chelator as a competitive ligand for Ca^{2+}. Figure 7.6 shows the variation of fractional fluorescence intensity with total $[Ca^{2+}]$. The sigmoidal shape of the plot indicates a mildly cooperative interaction ($K_1 = 2.2 \times 10^8$ M^{-1}; $K_2 = 3.7 \times 10^8$ M^{-1}) where calcium binding at site I promotes tighter binding at site II.[4]

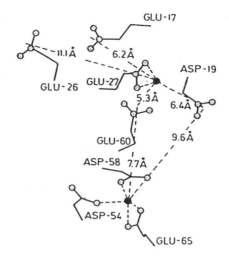

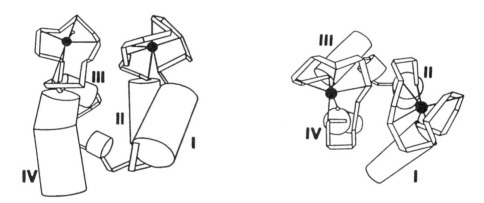

Figure 7.5 Schematic structure of calbindin D_{9k} showing the pair of EF-hands. [Adapted from D. M. E. Szenbenyi and K. Moffat, *J. Biol. Chem. 261,* 8761–8777 (1986), and reprinted by permission from S. Linse et al., *Nature, 335,* 651–652; copyright (1988) Macmillan Magazines Limited.] The coordination environment of Ca^{2+} in these EF-domains is approximately octahedral. This explains the greater affinity of these sites for Mg^{2+} relative to the Ca^{2+}-binding sites of staphyloccal nuclease and phospholipase A_2 (Figures 6.8 and 1.11).

7.1.4 Calmodulin and Protein Activation

In vivo it is not known how many Ca^{2+} ions must be bound to activate calmodulin. With this in mind, we shall represent the activated form as calmodulin–Ca_n^{2+}. Calmodulin–Ca_n^{2+} binds to and activates many proteins within the cell; however, we can consider only a few representative examples of these.

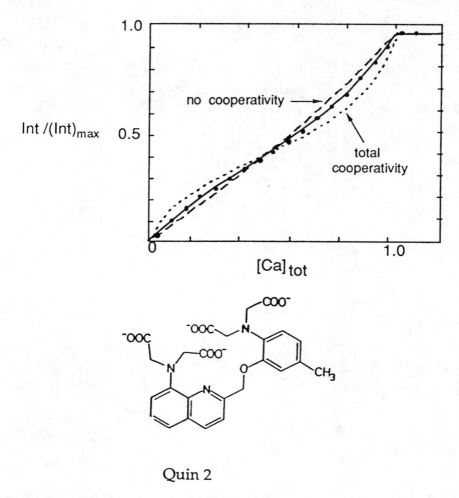

Quin 2

Figure 7.6 Plot of fluorescence intensity versus $[Ca^{2+}]$. Titration of calbindin D_{9k} with Ca^{2+} in the presence of an equivalent amount of the fluorescent chelator Quin 2 (a competing ligand for calcium). The fluorescent probe (structure shown) serves as a convenient reporter of Ca^{2+} binding to calbindin. (---) shows the best fit assuming two noninteracting sites. (····) shows the appearance of the curve assuming total cooperativity. (———) is a calculated curve assuming a strong site–site interaction ($K_1 = 2.2 \times 10^6$ M^{-1}; $K_2 = 3.7 \times 10^{10}$ M^{-1}). [Adapted from Linse et al., *Biochemistry*, *26*, 6723–6735 (1987).]

Kinases are an important class of enzymes that phosphorylate other target proteins within the cell and are regulated by calmodulin. It is the structural changes resulting from phosphorylation that lead to a physiological response. Figure 7.2 illustrates a typical scenario in neural (brain) cells. In addition to kinases, these cells possess cyclic nucleotide phosphodiesterase enzymes that catalyze the hydrolysis of the secondary messengers cAMP and cGMP. These enzymes are activated by calmodulin–Ca_n^{2+}, emphasizing the close relationship in the biochemistry of the two types of secondary messengers (Ca^{2+} and cyclic nucleotides).

Calmodulin is a key regulator of cellular mechanisms that remove Ca^{2+} from the intracellular space. It appears that calmodulin–Ca_n^{2+} binds to the calcium transport protein Ca^{2+}–ATPase, situated both in the plasma membrane (pumping outside the cell) and sarcoplasmic reticulum (pumping into internal compartments). Calcium binding by calmodulin, therefore, initiates Ca^{2+} flow away from the cytosol. Figure 7.7 summarizes how the initial stimulus that leads to the rapid increase in $[Ca^{2+}]_i$ also triggers a mechanism, via calmodulin–Ca_n^{2+}, that serves to return $[Ca^{2+}]_i$ to prestimulus levels, and so the cycle is carried full circle.

7.1.5   Calcium Channels as Regulatory Elements

Some of the most interesting problems in biological science lie in the field of neurochemistry. It has been found that Ca^{2+} is involved in the release of neurotransmitters from nerve cells, and an instructive connection can be made with

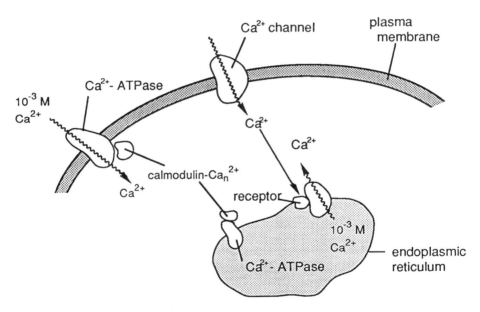

Figure 7.7   The increase in cytoplasmic calcium levels results from an external stimulus that triggers the release of Ca^{2+} from intracellular storage compartments (endoplasmic and sarcoplasmic reticula). Calcium binds to effector proteins that trigger enzymatic or cellular responses. One of these proteins (calmodulin), in turn, initiates a mechanism for removing Ca^{2+} from the cytosol by active transport through the Ca^{2+} pump Ca^{2+}-ATPase. The period between initial calcium release and return to prestimulus levels is referred to as the *response time*.

the chemistry of troponin C and muscle contraction discussed earlier. The relevant chemistry is illustrated in Figure 7.8. We have already seen (Figure 7.7) that an external stimulus results in an influx of Ca^{2+} to the cell. In response to a change in polarization at the axon terminal of a motor neuron (nerve cell), voltage-gated Ca^{2+} channels open and Ca^{2+} flows into the axon.[5] The higher calcium levels stimulate both the fusion of acetylcholine-containing vesicles in the axon with the presynaptic membrane and subsequent release of acetylcholine into the synaptic cleft. These neurotransmitters cross the synaptic cleft and bind to acetylcholine receptors. The change in receptor protein conformation allows the influx of Na^+ into the muscle cell, and the resulting depolarization of the cell interior triggers the release of Ca^{2+} from the sarcoplasmic reticulum into

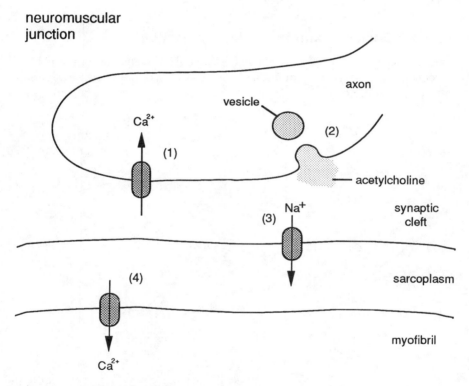

Figure 7.8 Schematic drawing of a neuromuscular junction. The axon is part of a nerve cell (*neuron*) that interacts with the muscle fiber. The region between the nerve and muscle cells is called the *synapse*. Signal transduction at the neuromuscular junction follows the following sequence of events. (1) Influx of Ca^{2+} to the axon of the motor neuron. (2) Increasing $[Ca^{2+}]$ results in release of acetylcholine following fusion of synaptic vesicles with the presynaptic membrane. (3) Influx of Na^+ to the muscle cell following binding of acetylcholine to a receptor protein on the sarcoplasmic reticulum. (4) Depolarization of the muscle cell membrane is communicated to the sarcoplasmic reticulum, triggering the release of Ca^{2+}, which ultimately activates muscle contraction as illustrated in Figure 7.3.

the cytoplasm (or sarcoplasm), which triggers muscle contraction as described previously in the section on troponin C.

Summary of Section 7.1

1. Primary messengers act outside of the cell. Secondary messengers function within a cell by binding to effector proteins that stimulate a further physiological response.
2. Ca^{2+} is a secondary messenger. Its function as a trigger is enhanced by rapid exchange rates and low intracellular concentrations relative to extracellular levels (10^{-7} M and 10^{-3} M, respectively).
3. Many calcium regulatory proteins bind Ca^{2+} ion in structural motifs termed *EF-hands*. These occur in pairs and demonstrate strong cooperative binding of Ca^{2+}.
4. Specific regulatory agents include troponin C in muscle control and calmodulin, which activates a variety of proteins and enzymes.

7.2   Regulation of Cellular Concentrations of Metal Ions

7.2.1   Introduction

This section addresses the interesting question of how a cell responds to changes in the cellular levels of essential metal ions, with particular emphasis on iron deficiency. Other examples (in particular, detoxification pathways for excess transition metals) will be discussed in the next chapter.

Iron is a truly multifunctional element, serving a diversity of roles that include O_2-transport, redox catalysis in enzymes, and DNA-cleaving antibiotics. When intracellular iron levels drop below a minimal level, this initiates a response that results in iron ion being transported into the cell. In Chapter 3 we saw that many bacteria release siderophore ligands to the environment to trap extracellular iron. The iron–siderophore complexes are then taken back into the cell by a set of specific receptor and transport proteins. In this section we shall examine the molecular mechanisms by which iron uptake is regulated, especially at the level of DNA expression. The biosynthesis of siderophores is carried out by a number of enzymes that are encoded by genes in the cellular DNA. Since these enzymes need not be expressed constantly but are produced in response to a specific need, a regulatory mechanism must exist that controls gene expression. One would think that this regulatory mechanism in iron metabolism might at some stage involve iron ion, and this is indeed the case. The genes encoding the enzymes and proteins required for siderophore synthesis and transport are grouped together. Such a cluster of genes is termed an *operon*. All the proteins may be expressed essentially under the control of one regulatory protein (Figure 7.9). It is the control mechanism that activates expression of this set of genes that concerns us here. The various stages of protein synthesis were pre-

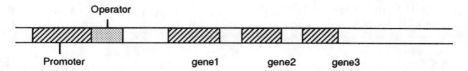

Figure 7.9 Structural arrangement of genetic elements. The *promoter* region is the initial binding site for RNA polymerase, and the *operator* binds transcriptional activators or repressors (see Figure 7.10). The DNA sequence enclosing the promoter, operator, and gene sites is termed the *operon*.

viously described in Section 1.13. Although expression may be inhibited at either the transcriptional or translational level, transcriptional control is most often employed. Transcription of DNA to mRNA is initiated by the binding of RNA polymerase (RNAP) to a sequence of DNA called the *promoter* region. The promoter is a defined segment of DNA to which RNAP initially binds and is located in front of the gene. The binding of RNAP can be prevented if another protein binds close to the promoter region. Such a protein is termed a *repressor* protein since it represses the binding of RNAP to the promoter sequence of DNA (Figure 7.10). The binding site for the repressor protein is called the *operator* site. Figure 7.10 also shows an alternative method of transcriptional control where RNAP may only bind to the promoter site in the presence of a *transcriptional activator*. Regulation of the iron uptake operon involves a repressor protein. DNA binding and release mechanisms for these proteins can generally be understood in terms of a change in protein conformation, resulting from the binding of molecular substrates or ions, that creates or destroys a favorable structural motif and binding interaction between protein side chains and the bases of DNA. In cases where this species is a metal ion, these are commonly referred to as *ion-responsive systems,* and the proteins are described as *metalloregulatory proteins.*

7.2.2 Regulation of Siderophore-Mediated Iron Uptake: The Aerobactin Operon

The aerobactin operon on plasmid ColV-K30 is one of the best-understood iron uptake systems (Figure 7.11). The operon contains three distinct domains: regulation (promoter and operator sites), biosynthesis (encoding enzymes required for aerobactin synthesis), and transport (encoding an outer membrane receptor protein for ferric aerobactin, which facilitates transport across the cell membrane). The biosynthetic pathway for aerobactin is summarized in Figure 7.11. The metalloregulatory repressor protein that binds to the operator site and inhibits transcription of the other genes in the operon is termed *Fur* (*f*erric *u*ptake *r*egulation). Divalent metal ions (in particular Fe^{2+}) must bind to Fur to promote binding to its DNA operator site. In vitro studies show that addition of chelating agents (EDTA or bipyridine) that remove Fe^{2+} leads to repressor

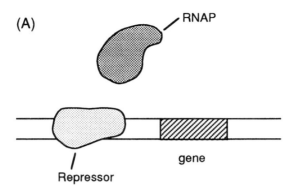

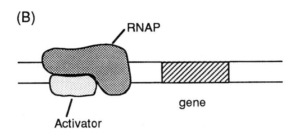

Figure 7.10 Transcriptional control of DNA expression by (A) repressor or (B) activator proteins.

release and the initiation of transcription. The DNA binding domain for Fur can be mapped out by high-resolution footprinting (Figure 7.12) using the methods described in Chapter 9.[6]

Many proteins or enzymes that bind to or carry out chemistry on DNA function as multimeric complexes of a basic unit. For example, Fur actually binds to operator DNA as a dimer in a fully cooperative interaction. Cooperativity implies that the binding of one protein (P) to DNA (D) enhances the binding of other protein molecules. In the limit of full cooperativity, it is assumed that either all the protein binds (DP_n) or none at all does (i.e., there are no DP_{n-1}, DP_{n-2}, ... species). This can be readily established by means of a Hill plot, the theory of which is now described. Consider the binding equilibrium between a piece of DNA (D) and n repressor proteins (P), defined by an equilibrium constant K_b (equations 7.1 and 7.2).

$$D + nP \overset{K_b}{\rightleftharpoons} DP_n \tag{7.1}$$

$$K_b = [DP_n]/[D][P]^n \tag{7.2}$$

Assuming $[D] = [D]_o - [DP_n]$, we can write

$$\log K_b + n \log P = \log [DP_n]/([D]_o - [DP_n]) \tag{7.3}$$

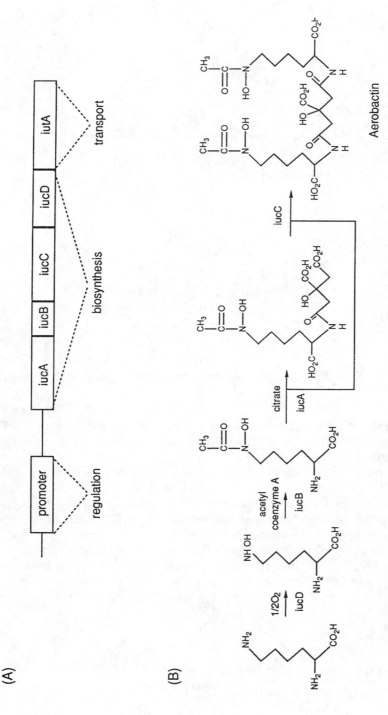

Figure 7.11 (A) Genetic organization of the *Escherichia coli* aerobactin operon and its promoters. The four *iuc* (*iron uptake chelate*) genes A, B, C, and D code for the biosynthetic steps from lysine to aerobactin illustrated in (B), and gene *iut* A (*iron uptake transport*) codes for an outer membrane receptor. [Adapted from J. B. Neilands, *Adv. Inorg. Biochem. 8*, 63 (1990).]

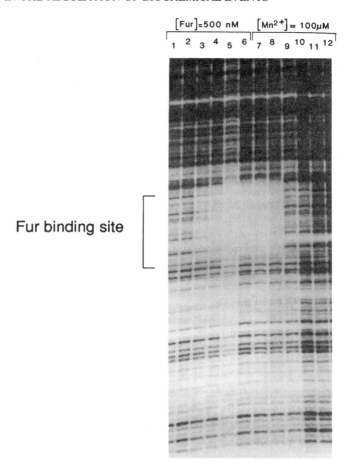

Figure 7.12 DNase I footprints (see Section 9.1) of the Fur binding site at the promoter of the *fur* gene. The autoradiograph was obtained from a 3'-labeled 250-bp fragment of DNA containing the binding domain. A divalent metal ion (in this case Mn^{2+}) is required for Fur to bind to the promoter DNA. From left to right, lanes 1–6 correspond to increasing Mn^{2+} concentrations at fixed [Fur] and lanes 7–12 to decreasing Fur concentrations at fixed [Mn^{2+}]. The DNA backbone is protected from cleavage when the protein binds, and the footprint is clearly visible when the [Mn^{2+}] and [Fur] are above critical concentrations. DNAse I is an enzyme that nicks the DNA backbone. [Adapted from V. DeLorenzo et al., *Eur. J. Biochem.*, *173*, 537–546 (1988).]

By dividing the numerator and denominator of the log term on the right by $[D]_o$, equation (7.4) is obtained, where $R = [DP_n]/[D]_o$. By plotting log $R/(1 - R)$ versus log P, the value of n may be obtained from the slope and

$$\log K_b + n \log P = \log R/(1 - R) \tag{7.4}$$

K_b from the intercept (Figure 7.13). In this way it was established that $n = 2$ for Fur binding. On a technical note, the concentration term [DP_n] at equilib-

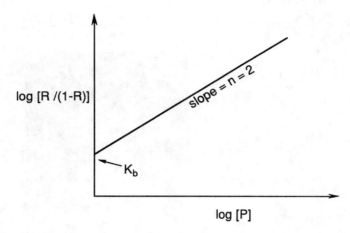

Figure 7.13 Hill plot showing cooperative binding of the Fur protein to the operator region of the aerobactin operon. The fraction of protein-bound DNA (R) is related to the levels of activity of β-galactosidase, an enzyme that is expressed under the control of the aerobactin operon (Figure 7.14).

rium was monitored by following the activity of β-galactosidase produced in a lac Z fusion to the aerobactin operon (Figure 7.14). The activity of β-galactosidase is therefore proportional to $([D]_o - [DP_n])$.

Other transport proteins required for iron utilization (e.g., *ton*B and *exb*B; see Section 3.2) are encoded by the chromosomal DNA and are not found on the plasmid.

7.2.3 Regulation of Protein-Mediated Iron Uptake and Storage

Siderophores are exclusively used by bacterial cells for the uptake of iron ion, but not by higher plants and animals. In Section 3.3 we saw that mammalian cells use transferrin and ferritin for the transport and storage of iron, respectively. Transferrin is required when the levels of cellular iron are low, whereas ferritin stores excess iron when the levels are high. Clearly it does not make sense for a cell to synthesize large amounts of transferrin and ferritin at the same time. We might expect that the expression of both transferrin and ferritin is tightly regulated and that the regulatory mechanism might depend on the local concentration of iron ion. We shall soon see that this is indeed the case. Moreover, the biochemical studies of this regulatory mechanism resulted in a surprising finding. The protein that regulates expression of transferrin and ferritin is a cytoplasmic aconitase: a relative of a mitochondrial enzyme in the citric acid (TCA) cycle, and one of a family of iron-sulfur proteins that we reviewed in Section 4.3!

How does it all work? Both transferrin and ferritin are regulated posttran-

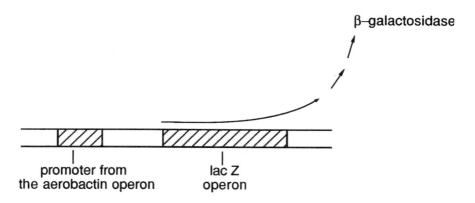

Figure 7.14 Summary of a method for quantitating repressor binding to promoter DNA. The *lac* Z gene encodes β-galactosidase, an enzyme that cleaves a Gal-Glu dimer in a reaction that can be readily monitored. By fusing the gene to the aerobactin operon, the *lac* Z gene comes under the strict control of the Fur repressor. *Lac* Z is not expressed when repressor binds, and so β-galactosidase activity, which reflects the amount of enzyme expressed, is proportional to $[D]_0-[DP_n]$.

scriptionally at the level of translation. This means that the mRNA is formed by transcription from the nuclear DNA, and then moves out of the nucleus to the ribosome, the site of protein synthesis (Figure 7.15A). Figure 7.15B shows that iron-response elements (IRE's) are regulatory sequences that are located just before the ribosomal binding site (rbs) and initiation codon of the mRNA (review Section 1.13 if you need a reminder of these terms). These IRE's are approximately 30 to 40 bases in length and possess a complex secondary and tertiary structure. Figure 7.16 shows that this IRE is complexed by an iron regulatory protein (IRP); formerly called an iron-response-element binding pro-

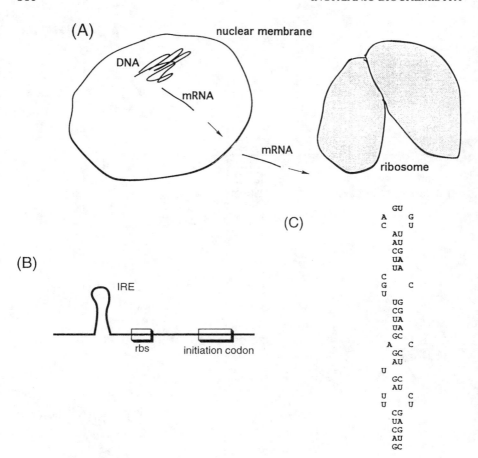

Figure 7.15 (A) Proteins are synthesized at the ribosome using the information encoded in the mRNA "blueprint". The mRNA is synthesized from the genomic DNA template in the nucleus of the cell by a process termed transcription. (B) An iron-response element is a structured region of mRNA that, in this case, lies just before the ribosomal binding site (rbs) and can, therefore, regulate the translation of the message and its stability. (C) An example of an IRE. This one comes from the mRNA encoding bullfrog ferritin H chain. The sequence and base pairing will vary slightly for ferritin and transferrin, and over different species.

tein (IRE-BP). When this protein was isolated and characterized it was found to be identical to the cytoplasmic aconitase! Apparently, we have a protein that has been designed to carry out two distinct activities, serving both as a hydro-lyase enzyme and as a translational regulatory protein. Enzyme activity requires an [Fe$_4$S$_4$] cluster, which can only form at high iron levels. When the cluster is present, the protein does not bind to any IRE, that is, it's function is purely enzymatic. The cluster does not form under conditions of low iron, and it is the

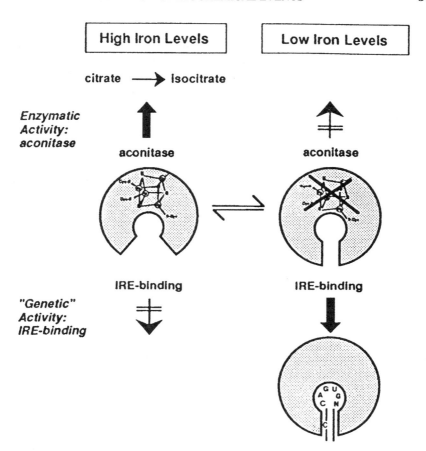

Figure 7.16 A schematic illustration of the iron-dependent switch between aconitase and gene regulatory activities. Under conditions of low "Fe" the apo-IRP lacks the [Fe$_4$S$_4$] cluster and binds to the IRE. In the case of the transferrin gene, the IRP serves to protect the IRE from nucleases and enhances initiation of protein synthesis. In the case of the IRE for ferritin, the IRP inhibits binding to the ribosome, and no ferritin synthesis is possible. Under conditions of high "Fe", the [Fe$_4$S$_4$] cluster is assembled, which prevents binding of IRP to IRE's. In the case of the transferrin gene, the mRNA is digested by nucleases. But, for the ferritin gene, the absence of the IRP allows binding to the ribosome, and translation of the gene can proceed. [Reprinted with permission from Gray, N. K. et al., *Eur. J. Biochem. 218*, 657–667 (1993)].

apo-IRP that is genetically active and regulates the synthesis of transferrin and ferritin.

At first glance, there might appear to be a dichotomy in these statements. Transferrin and ferritin should be expressed when the iron levels are low and high, respectively, and yet the IRE-binding protein functions as a regulatory protein only at low iron levels. In fact, the answer is simple. Regulation of transferrin translation is under positive control, but is negatively controlled for ferritin gene translation. Figure 7.16 illustrates what is meant by this statement. Under conditions of low iron, the IRP exists as the apo-protein and binds to the IRE. This binding stabilizes the IRE to nuclease digestion and enhances translational efficiency. The protein has a *positive* effect on protein synthesis. In contrast, the protein inhibits ferritin-encoding mRNA from binding to the ribosome, and no synthesis can occur. The protein has a *negative* effect on protein synthesis. Under conditions of high iron levels, the IRP contains an $[Fe_4S_4]$ center and is unable to bind to an IRE; rather it displays aconitase activity. (It is not yet clear if this aconitase activity is functionally relevant in the cytoplasm.) Transferrin mRNA is susceptible to nucleolytic digestion and is removed prior to binding to the ribosome. No transferrin is synthesized. In contrast, ferritin mRNA is free to bind to the ribosome and the ferritin protein is expressed.

Clearly, this last section on iron regulation has brought together many threads from earlier chapters. It is the synergy of these topics that most often brings the fullest gratification to workers in the field.

7.2.4   Regulation of Cellular Copper Levels

Copper homeostasis is controlled to a large degree by the copper-binding protein metallothionein. In yeast, metallothionein biosynthesis is regulated by a transcriptional activator called ACE1. This protein binds to the *c*opper *u*ptake *p*romoter (*CUP1*) locus of the metallothionein gene, inducing the binding of RNA polymerase, which then reads off the metallothionein gene sequence. Binding of ACE1 to its DNA sequence requires copper ion. It appears that the copper-binding domain of ACE1 is much like that of metallothionein, a cysteine-rich sequence that binds a cluster of Cu^+ ions [see Section 3.3.2 and Figures 2.13 and 3.19)]. Figure 7.17 summarizes the details.

Summary of Section 7.2

1. Proteins that bind metal ions and regulate cellular metabolism are called *metalloregulatory proteins,* or *metal-responsive systems.*
2. Several metalloregulatory proteins function at the level of gene transcription. Proteins that inhibit RNA polymerase binding to a promoter sequence are called *DNA-binding repressor proteins.* Proteins that are required for RNA polymerase binding to promoter DNA are called *transcriptional activators.*
3. Fur is an iron-binding repressor protein that functions as a dimer and regulates expression of genes encoding enzymes and proteins for the biosynthesis and transport of iron uptake siderophores. ACE1 is a transcriptional activator that mediates the biosynthesis of metallothionein.
4. The iron regulatory protein (IRP) contains a labile $[Fe_4 S_4]$ cluster that degrades when cellular iron levels are low. The apo-protein binds to a segment of mRNA called an iron response element (IRE) and regulates mRNA stability, and translation of iron uptake and storage proteins (transferrin and

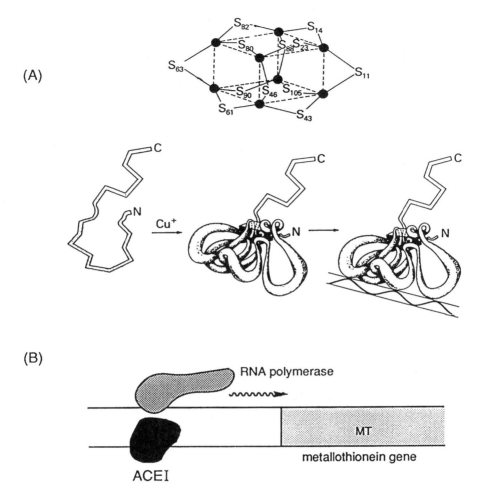

Figure 7.17 (A) The amino acid sequences for metallothionein and the metal-binding domain of ACE1 are similar. A putative cluster domain in ACE1 is shown (compare Figure 2.13). [Adapted from J. Imbert et al., *Adv. Inorg. Chem., 8,* 139–164 (1990).] (B) At high concentration, Cu^{2+} binds to ACE1 and changes the protein conformation such that binding to the metallothionein promoter is favored. RNA polymerase then binds and synthesizes mRNA from the metallothionein gene.

ferritin). Under conditions of high cellular iron, the iron-sulfur cluster assembles and displays aconitase activity.

Review Question

Note that the iron-responsive Fur protein regulates metal uptake to the cytosol, whereas the copper-responsive ACE1 protein regulates metal removal from the cytosol. Consider the relevance of this to the functional mechanism of each protein: Fur, a repressor, and ACE1, a transcriptional activator.

Notes

1. As an illustrative example, muscle stiffness after physical exercise is a direct result of the depletion of ATP levels in the cell. The Ca^{2+}–ATPase cannot function, $[Ca^{2+}]_i$ remains high, and the muscle cannot relax. Rigor mortis in a corpse is the "limiting case," where no ATP is generated and the muscles are firmly locked.
2. Previously called the intestinal calcium-binding protein (ICBP).
3. The notation I,II implies that site I is populated before site II, and vice versa for II,I. See Chapter 3 (Section 3.3.1) for another example of macroscopic versus microscopic binding constants.
4. The macroscopic and microscopic binding constants are related as follows: $K_1 = k_I + k_{II}$ and $K_2 = k_I k_{II,I}/(k_I + k_{II})$. It can be instructive to work through the derivation of these equations.
5. See Chapter 3 (Section 3.2) for a discussion of voltage-gated channels.
6. Section 9.1 introduces the footprinting method in some detail.

Further Reading
Calcium

Ikura, M. et al. Solution structure of a calmodulin-target peptide by multidimensional NMR. *Science 256*, 632–638 (1992).

McCubbin, W. D., and C. M. Kay. Calcium-induced conformational changes in the troponin-tropomyosin complexes of skeletal and cardiac muscle and their roles in the regulation of contraction-relaxation, *Acc. Chem. Res., 13*, 185–192 (1980).

O'Neil, K. T. and W. F. DeGrado. How calmodulin binds its targets: Sequence-independent recognition of amphiphilic α-helices. *Trends in Biochemical Sciences 15*, 59–64 (1990).

Spiro, T. G., ed. *Calcium in Biology*, Wiley-Interscience, 1983.

Metalloregulatory Proteins

Crichton, R. R. *Inorganic Biochemistry of Iron Metabolism*. Ellis Horwood, 1991.

Crichton, R. R. and R. J. Ward. Iron metabolism: New perspectives in view. *Biochemistry 31*, 11255–11264 (1992).

Eichorn, G. L. and L. G. Marzilli, eds. Metal-ion induced regulation of gene expression, *Adv. Inorg. Biochem.*, 8 (1990).

Hu, S., P. Furst, and D. Hamer. DNA and Copper Binding Functions of ACE 1, *The New Biologist, 2*, 544–555 (1990).

Rouault, T. A. et al. Structural relationship between an iron-regulated RNA binding protein (IRE-BP) and aconitase: Functional implications. *Cell 64*, 881–883 (1991).

Thiel, E. C. Iron Regulatory Elements (IREs): a family of mRNA non-coding sequences. *Biochem. J. 304*, 1–11 (1994).

Problems

1. Nitric oxide is a biological signaling molecule. Remarkably, the ·NO radical is a potent toxin that mediates its action through oxidative chemistry. The weak oxidant ·NO can be readily converted to the more powerful oxidant, peroxynitrite ($ONOO^-$), by reaction with superoxide (a by-product of cell metabolism that is discussed in Chapter 8). Protonation yields the conjugate acid, peroxynitrous acid (ONOOH).

$$O_2^{-\cdot} + \,^{\cdot}NO \rightarrow ONOO^-$$
$$H^+ + ONOO^- \rightleftharpoons ONOOH$$

(a) Peroxynitrous acid can readily oxidize lipids, thiols, and amino acid side-chains. Reaction with methionine (Met) follows two pathways, producing the sulfoxide derivative of methionine (MetO) by one, and $H_2C = CH_2$ (+ other products) by the other. Propose mechanisms for the production of MetO and ethylene.

(b) Peroxynitrous acid isomerizes to nitrite through an activated intermediate, ONOOH*.

$$H^+ + ONOO^- \overset{K_a}{\rightleftharpoons} ONOOH$$

$$ONOOH \overset{k_1}{\underset{k_{-1}}{\rightleftharpoons}} ONOOH^*$$

$$ONOOH^* \overset{k_N}{\rightarrow} H^+ + NO_3^-$$

Show that k_{obs}, for the decomposition of peroxynitrite, is given by

$$k_{obs} = \left(\frac{k_1 k_N}{k_N + k_{-1}}\right)\left(\frac{[H^+]}{K_a + [H^+]}\right)$$

[Pryor, W. A. et al., *Proc. Natl. Acad. Sci. USA 91*,11173 (1994)]

2. In Section 7.1 we examined the cooperative binding of two Ca^{2+} ions to the regulatory protein calbindin

$$\text{calbindin} + Ca^{2+} \overset{K_1}{\rightleftharpoons} \text{calbindin.}(Ca^{2+})$$

$$\text{calbindin.}(Ca^{2+}) + Ca^{2+} \overset{K_2}{\rightleftharpoons} \text{calbindin.}(Ca^{2+})_2$$

By use of site-directed mutagenesis (substituting Gln for Glu, and Asn for Asp) the contribution of charged surface residues to the binding free energy of the two Ca^{2+} ions can be determined. The table summarizes binding constants for each site and the total binding free energy $\Delta G_{tot} = -RT \ln (K_1 K_2)$ for the two ions. Figure 7.5 illustrates the positions of the surface charged residues relative to each calcium site.

Protein	Neutralized side chains	$\log_{10}K_1$	$\log_{10}K_2$	ΔG_{tot} (kJ mol^{-1})	$\|\Delta\Delta G\|_{min}$ (kJ mol^{-1})
M0	—	8.3	8.6	−97	4.7
M5	Glu 17	7.4	8.1	−89	7.7
M6	Asp 19	7.6	8.0	−89	5.8
M7	Glu 17, Asp 19	7.4	7.1	−83	1.7
M8	Glu 26	7.8	8.4	−93	6.4
M9	Glu 17, Glu 26	6.8	7.8	−83	9.2
M10	Asp 19, Glu 26	6.9	7.6	−83	7.8
M11	Glu 17, Asp 19 Glu 26	6.8	6.7	−77	2.9

(a) With reference to Figure 7.5, comment on the effect of the single, double, and triple mutations on the individual K_1 and K_2 values, and on ΔG_{tot}.

(b) Estimate the contributions of each charge to the binding free energy for two Ca^{2+} ions.

(c) The minimum value for the free energy of interaction between the sites can be calculated from the macroscopic binding constants.

$$|\Delta\Delta G|_{min} = -RT \ln (4K_2/K_1)$$

This parameter reflects the cooperativity for Ca^{2+} binding ($\Delta\Delta G < 0$ for positive cooperativity). With no cooperativity $\Delta\Delta G = 0$ kJ mole^{-1}. Comment on the values you obtain for the single, double, and triple mutant proteins.

(d) The actual value for $\Delta\Delta G$ is given by $-RT \ln (K_{I,II}/K_{II}) = -RT \ln (K_{II,I}/K_I)$, where $K_I = K_I + K_{II}$ and $K_1K_2 = K_IK_{II,I}$. Show that the expression in part (c) is in fact a minimum for $\Delta\Delta G$.

(e) Do you think that Gln and Asn are good choices for the substituting amino acids? Explain.

(f) Calbindin binds to Ca^{2+}-ATPase, a membrane protein that pumps Ca^{2+} out of a cell. Assuming that the binding interaction is mediated through complementary charges that neutralize the three surface carboxylates on calbindin, provide a rational explanation of the significance of the data in the table in terms of physiological function.

[Linse, S. et al., *Nature 335*, 651 (1988)]

3. You have isolated a new metalloregulatory protein that binds cadmium and regulates the expression of a cadmium storage protein. The enzyme is known to act as a monomer.

(a) The data below was obtained from a gel shift assay of a 350 bp oligo-nucleotide that binds to the regulatory protein. Assume the following equilibrium, where P = protein, D = DNA, DP = DNA-protein complex, D_0 is [DNA]$_0$, and P_0 is [protein]$_0$.

$$P + D \cdot \overset{K_a}{\rightleftharpoons} DP$$

Derive an equation for K_a in terms of the above constants that will permit you to determine K_a in the presence and absence of Cd^{2+}.

[Cd^{2+}] = 0 M, D$_0$ = 3.6 × 10^{-12} M				[Cd^{2+}] = 6 × 10^{-7} M, D$_0$ = 3.6 × 10^{-12} M			
P$_o$(M)	DP/D$_o$	P$_o$(M)	DP/D$_o$	P$_o$(M)	DP/D$_o$	P$_o$(M)	DP/D$_o$
2.1 × 10^{-11}	0.21	1.5 × 10^{-10}	0.66	4.3 × 10^{-5}	0.34	3.1 × 10^{-4}	0.79
3.2 × 10^{-11}	0.29	1.8 × 10^{-10}	0.70	6.8 × 10^{-5}	0.45	3.8 × 10^{-4}	0.82
4.4 × 10^{-11}	0.36	2.1 × 10^{-10}	0.73	9.0 × 10^{-5}	0.52	4.4 × 10^{-4}	0.84
5.5 × 10^{-11}	0.42	2.8 × 10^{-10}	0.78	1.2 × 10^{-4}	0.58	6.1 × 10^{-4}	0.88
7.7 × 10^{-11}	0.50	3.7 × 10^{-10}	0.82	1.6 × 10^{-4}	0.66	7.5 × 10^{-4}	0.90
1.2 × 10^{-10}	0.61			2.5 × 10^{-4}	0.75		

(b) Comment on the relative magnitudes of the K_as obtained in the cadmium-bound and free forms with reference to the likely mode of regulation.

(c) Describe, as fully as you can, how you might use a transcription assay to study the role of Cd^{2+} in promoting transcription by RNA polymerase.

[refer to O'Halloran, T. V. et al., *Cell 56*, 119–29 (1989) for a description of methodology]

4. Fluorescence from lanthanide ions can often be used to monitor kinetic and thermodynamic properties of metal binding sites on biological macromolecules, especially for calcium ion. Tb^{3+} luminescence has been used to evaluate the Tb^{3+} dissociation rate from the EF-hand motif of the *E. coli* galactose-binding protein. Release of Tb^{3+} is induced by addition of $EDTA(Na)_4$. The latter did not perturb the system and rates were independent of $EDTA^{4-}$ concentration.

(a) What purpose does the $EDTA^{4-}$ ligand serve?

(b) The luminescence intensity decreases when Tb^{3+} dissociates from the protein-binding site. The plots below show intensity versus time profiles for release of Tb^{3+} from wild type (Asn142), and mutant proteins (Gln142, Glu142), where N, Q, D and E are the one-letter abbreviations for Gln, Asn, Asp and Glu. Rationalize the observed trend.

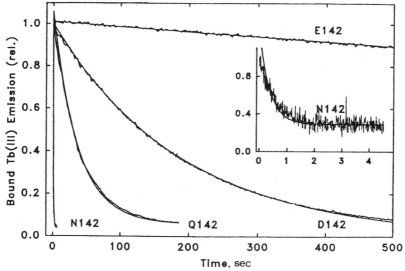

(c) The actual dissociation rate ($k_{off} \times 10^3$ s^{-1}) are 0.1, 2.4, 11, and 1200 for E, D, Q and N142 proteins, respectively. If the on rates (k_{on}, M^{-1}s^{-1}) are 0.36, 48, 73, and 2.4 $\times$ 10^4, respectively, calculate the dissociation constant (K_d) from the data provided, and comment on the association rates and metal binding constants for each mutant.

(d) Discuss the physiological relevance of these data in terms of a role for Ca^{2+} as a secondary messenger.

[Renner, M. et al., *Proc. Natl. Acad. Sci. USA 90,* 6493 (1993), figure is reprinted with permission; see also Drake, S. K. and Falke, J. J. *Biochemistry 35,* 1753 (1996)]

5. For calbindin, the macroscopic and microscopic binding constants for the two Ca^{2+} sites are related through the equations $K_1 = k_I + k_{II}$, and $K_2 = k_I k_{II}/(k_I + k_{II})$. Derive these relationships from first principles.

6. Nitrosyl oxide is an unusual example of a messenger system that functions by binding to heme centers. Draw the Lewis structures for NO, CO and O_2, and compare the coordination of each to a heme center. In each case, comment on the oxidation state of the iron. Suggest a reason why NO should be used in this capacity rather than the other two ligands.

[Traylor, T. G. and V. S. Sharma, *Biochemistry 31,* 2847 (1992)]

CHAPTER

8

Cell Toxicity and Chemotherapeutic

By evolutionary design, cells have developed remarkable defense mechanisms to deal with the reactive and potentially harmful by-products that arise from cellular metabolism. For example, oxidative chemistry often results in the formation of hydrogen peroxide (H_2O_2), a powerful oxidant that can cause oxidative damage to cellular infrastructure, proteins, and metabolites.

$$RH_2 + O_2 \rightarrow R + H_2O_2$$

In higher organisms, a complement of enzymes that carry out important oxidations and hydroxylations are isolated in intracellular compartments called *peroxisomes*. These compartments also contain a set of enzymes that quickly destroy reactive by-products as they are formed. A comparable situation is found for the lysosomes noted in Chapter 1, which contain intracellular digestive enzymes that are, by necessity, separated from the bulk contents of the cell to prevent random degradation of proteins and metabolites in the cytosol. Hazardous chemical species can also be absorbed from the extracellular environment (e.g., heavy metals, carcinogens, and neurotoxins) that must be met by an appropriate defense response.

Many of the natural defense mechanisms employed by prokaryotes and eukaryotes are well understood. By a combination of serendipity and rational design, it has proved possible to synthesize specific molecules that assist the recuperative and resistance power of cells when delivered as exogenous reagents. In Sections 8.1 through 8.3 we shall review three important areas related to cellular response to toxic species and disease states: (1) natural defense mechanisms developed by cells to neutralize reactive by-products from normal metab-

319

olism, (2) the chemistry underlying natural resistances to exogenous substances, and (3) the use of synthetic reagents as therapeutics.

8.1 Oxygen Toxicity

A brief review of Table 4.2 demonstrates that the reduced forms of molecular oxygen are reactive species that are readily formed and need to be rapidly eliminated by an in vivo defense mechanism. Hydrogen peroxide (H_2O_2) is a byproduct of enzymatic oxidation. Moreover, both H_2O_2 and the extremely reactive species hydroxyl radical ($HO^{\cdot}$) can result from the decomposition of superoxide ($O_2^{-\cdot}$), which is formed in respiring cells in small but not insignificant amounts.

$$O_2^{-\cdot} + H_2O \rightarrow HO^{\cdot} + HOO^-$$

The auto-oxidation of thiols, flavins, hydroquinones, and other redox centers leads to $O_2^{-\cdot}$ formation. Reduced ferredoxins and oxygen-binding proteins (myoglobin and hemoglobin) are also subject to one-electron oxidation by O_2 to form $O_2^{-\cdot}$. Note that dioxygen is relatively unreactive toward one-electron reduction (cf. Table 4.2), and so O_2 carriers are fairly stable toward oxidation. However, oxidation of ferrous hemoglobin to form $O_2^{-\cdot}$ and the ferric protein derivative (methemoglobin) is a common enough side reaction, so that many cells possess a methemoglobin reductase to regenerate the active reduced form of the protein. To combat the problem of reactive oxygen species, both prokaryotic and eukaryotic cells have evolved enzymes (superoxide dismutases) that catalyze the disproportionation of superoxide to oxygen and hydrogen peroxide.

$$2\,O_2^{-\cdot} + 2H^+ \rightarrow O_2 + H_2O_2$$

Catalase and peroxidase are both enzymes that catalyze the breakdown of product H_2O_2. However, they differ insofar as many peroxidases use the oxidative power of peroxide to oxidize organic substrates.

$$H_2O_2 + H_2O_2 \rightarrow 2H_2O + O_2 \qquad \text{(catalase)}$$
$$H_2O_2 + RH_2 \rightarrow H_2O + RHOH \qquad \text{(peroxidase)}$$

Both catalase and heme-peroxidase enzymes destroy the peroxide formed during oxidative chemistry in the intracellular peroxisomes. Catalase is also located in the cytosol, since H_2O_2 is a product from the catalytic reaction of the cytosolic enzyme superoxide dismutase. We will discuss the chemistry of each of these enzymes in turn, ending with a discussion of cytochrome P-450, an important enzyme in oxidative metabolism.

8.1.1 Superoxide Dismutase (SOD)

Superoxide dismutases isolated from eukaryotes contain both copper and zinc (CuZn-SOD). In mammals, superoxide dismutase is predominantly located in the liver, blood cells, and brain tissue. Prokaryotes possess iron (Fe-SOD) or manganese (Mn-SOD) enzymes that bind only one metal ion per subunit. Although cuprozinc superoxide dismutases have been extensively characterized, their function remains a topic of controversy. The turnover rate for reaction of $O_2^{-\cdot}$ with CuZn-SOD ($\sim 2 \times 10^9 \, M^{-1} \, s^{-1}$) is close to the diffusion limit and so the primary role of SOD may be to remove $O_2^{-\cdot}$ from the cytosol, serving as an antioxidant to inhibit aging and carcinogenesis. Also, mutations of SOD have been associated with amyotrophic lateral sclerosis (ALS)—Lou Gehrig's disease. It is also possible that the protein may serve as an intracellular repository for copper and zinc, with an active role in homeostatic control mechanisms for these elements. Bovine SOD is a dimeric enzyme that contains two identical 31,200-Da subunits. Figure 8.1 shows the overall topology of one subunit and a view of the active site, which contains the interesting and unique Cu–Zn pair bridged by a histidinate anion.

By variation of solution pH, addition of reducing agents, and dialysis against metal selective ligands, either copper or zinc may be removed and substituted with a variety of metal replacements. For example, (a) reduction at pH 6 allows release of Cu^{1+} by dialysis against CN^- [$Cu(CN)_2^-$ is a stable complex],

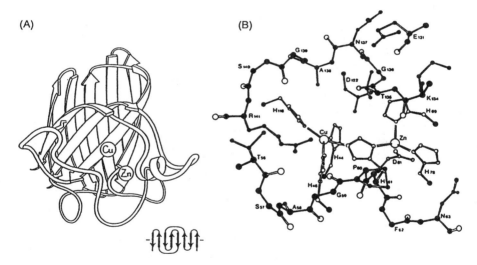

Figure 8.1 (A) Backbone schematic of a superoxide dismutase subunit showing the positions of the zinc and copper ions. The greek key topology of the β-barrel (so termed because of its resemblance in shape to a classical Greek key) is shown as an inset. [Adapted from J. A. Tainer et al., *J. Mol. Biol., 160,* 181–217 (1982).] (B) Residues of the active site viewed from the solvent channel. The Cu(II) coordination sphere includes His-118, 44, 61, and 46 and approximates to a square planar geometry. The geometry of Zn ligands His-61, 69, 78 and Asp-81 (clockwise from left) is approximately tetrahedral. Asp-81 lies behind the Zn and is buried. His-61 bridges the Cu and Zn. There is extensive hydrogen bonding from the His and Asp ligands to other residues in the cavity. [Reprinted by permission from J. A. Tainer et al., *Nature, 306,* 284–287; copyright (1983) Macmillan Magazines Limited.]

(b) dialysis against the chelating ligand EDTA at pH 3.8 removes both metals; and (c) dialysis at pH 3.8 selectively removes Zn^{2+}.

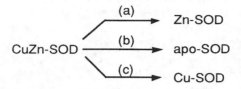

In this way, numerous metalloderivatives (M_1M_2-SOD) are formed that can be used to probe the coordination details at each site, the structural arrangement of active site residues, and the mechanistic chemistry of the enzyme. Some specific examples will be described shortly.

The copper coordination environment in bovine cuprozinc SOD is summarized in Figure 8.2 and differs in several important respects from the square planar geometry normally adopted by Cu(II). Specifically, one of the histidines is deprotonated and forms a bridge to a neighboring zinc ion, while an axial position is occupied by a solvent molecule, leading to a square pyramidal geometry at copper since the remaining axial site is not solvent accessible and is not ligated by a protein residue. The EPR g-values and hyperfine coupling constants for the copper site fully support the square pyramidal geometry determined from crystallographic studies (Figure 8.1). Substitution of Zn with Cu gives EPR parameters ($g_{\parallel}$ 2.312, $A_{\parallel}$ 97 G) that indicate tetrahedral coordination at this site. Larger values of $A_{\parallel}$ indicate a more tetragonal (or square planar) coordination environment, whereas smaller $A_{\parallel}$ values suggest tetrahedral character (refer to Table 5.4). All these conclusions are again in complete agreement with the crystallographic structure shown in Figure 8.1.

Ligand binding to SOD may be monitored in several ways. From NMR dispersion experiments the exchange rate for the axial H_2O has been estimated ($k_{ex} \sim 4 \times 10^6$ to 1×10^8 s^{-1}).[1] This is very fast, in keeping with a labile d^9 system. In spite of the rapid exchange, it turns out that exogenous ligands do not bind by displacing the axial H_2O. Clearly, $O_2^{-\cdot}$ itself cannot be used to probe the ligand-binding properties of SOD, since catalytic turnover would result. Instead cyanide (CN^-) or azide (N_3^-) are often used. Cyanide binding can be readily monitored by electron paramagnetic resonance spectroscopy (EPR), since $^{13}CN^-$ couples to the paramagnetic Cu^{2+} and the resulting splitting of the parallel signals (g_z) indicates coordination in the equatorial plane with loss of one of the bound His. We cannot tell which histidine is displaced from EPR data alone; however, this information can be directly obtained from ^{1}H NMR studies. In Chapter 2 we saw that the NMR resonances from protons adjacent to a paramagnetic center tend to be shifted to a less cluttered region of the spectrum. Unfortunately, the unpaired spin on Cu^{2+} relaxes relatively slowly and has the effect of rapidly relaxing nuclear moments on adjacent protons, which broadens the lines beyond detection.[2] Consequently, ^{1}H NMR studies of copper proteins seldom give detailed structural information concerning the

(A)

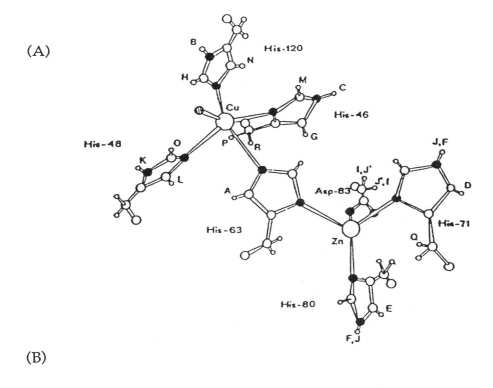

(B)

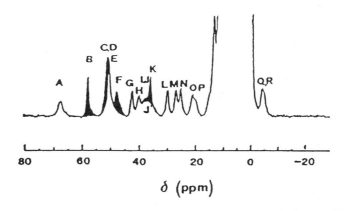

δ (ppm)

Figure 8.2 (A) The active site residues highlighting the assignments made in the NMR spectrum of the CuCo-SOD derivative shown in (B). (B) 300-MHz ¹H NMR spectrum of CuCo-SOD. The paramagnetically shifted resonances are easily resolved from the diamagnetic region. Exchangeable resonances (shaded) can be identified by comparing spectra run in H₂O and D₂O. In the latter solvent system, resonances from exchangeable protons are absent. [From L. Banci et al., in *NMR and Biomolecular Structure* (I. Bertini et al., eds.), VCH, 1991, p. 44.]

copper coordination environment. However, by substituting Zn^{2+} with a rapidly relaxing metal, such as Co^{2+} ($S = 3/2$), the relaxation of Cu^{2+}, which is electronically connected to Co^{2+} by the histidinate ligand, is increased and the line broadening of adjacent ^1H resonances is reduced. As a result, the resonances from residues adjacent to both metal sites can be observed, assigned, and subsequently used to probe ligand binding and the electronic properties of the coupled metal ions (Figure 8.2). In this way, it can be shown that His-46 is displaced by ligand binding, since the paramagnetically shifted C2-H resonance returns to the diamagnetic region of the spectrum.

In Chapter 1 we saw how site-directed mutagenesis can be a powerful tool in the evaluation of structure–function relationships in proteins and enzymes. In particular, we can use this technique to investigate the functional role of specific amino acid residues in the active site of SOD. For example, residue Arg 143 flanks the entrance to the ligand-binding site on Cu^{2+} (Figure 8.3) and we wish to determine whether this contributes predominantly to substrate binding or catalysis (i.e., lowers the transition state energy). By mutating Arg 143 to Lys, Ile, or Glu (positive, neutral, and negatively charged residues, respectively), it has been deduced that this residue influences ligand- or substrate-binding constants (K_d or K_m) but has no effect on the catalytic rate constant (k_{cat}). Table 8.1 summarizes the pertinent data. Note that the relative K_m values had to be inferred from binding constants for nonsubstrate ligands, such as azide (N_3^-), since the turnover rate is too fast (diffusion-controlled) to be measured at saturating substrate concentrations. These binding constants can be readily monitored by following the changes in chemical shift of the C2-His proton accompanying ligand binding (Figure 8.4).

Cuprozinc SOD displays an interesting electrochemistry (Figure 8.5). A midpoint reduction potential of $+0.42$ V can be estimated from optical titration experiments using a ferro/ferricyanide redox couple to control the solution potential, while the pH dependence for reduction of Cu^{2+} to Cu^{1+} indicates uptake of one H^+ between pH 5 and 8.4.[3] Reduction is accompanied by cleavage and protonation of the bridging histidinate ligand.

A working model for the reaction is illustrated in Figure 8.3. Since ligands apparently bind at the redox active copper site, it is reasonable to think of the enzyme as possessing a structural zinc site and a catalytic copper site. The overall reaction scheme for superoxide dismutation may be written as follows:

$$E(H^+)Cu^{1+} + O_2^{-\cdot} + H^+ \rightarrow H_2O_2 + ECu^{2+}$$

$$ECu^{2+} + O_2^{-\cdot} + H^+ \rightarrow O_2 + E(H^+)Cu^{1+}$$

(A)

Arg 143

Thr 137

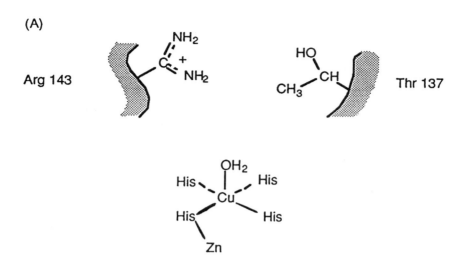

(B)

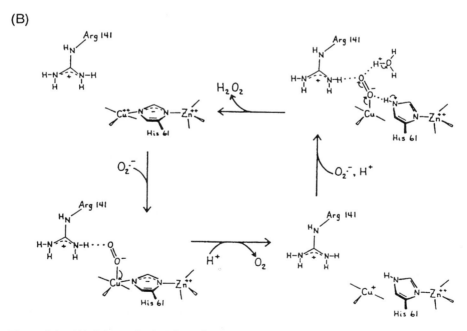

Figure 8.3 (A) Schematic drawing of the residues flanking the entrance to the copper site. (B) A proposed catalytic mechanism. The difference in residue numbering reflects the comparison of different, but homologous, SOD enzymes. [Reprinted by permission from J. A. Tainer et al., *Nature, 306,* 284–287; copyright (1983) Macmillan Magazines Limited.]

Table 8.1 Kinetic and Binding Data for Mutant SODs

Residue-143	$K_a(N_3^-)$ M^{-1}	Relative Activity (%)[a]
Arg (wild type)	154	100
Lys	63	43
Ile	16	11
Glu	6	—

[a] The relative specific activities of the wild-type and mutant enzymes parallel the decrease in binding affinity for N_3^-. It is, therefore, likely that the decrease in activity arises mostly from a change in K_m for substrate ($O_2^{-\cdot}$) rather than k_{cat}. It is difficult to obtain values for k_{cat} and K_m by the usual Michaelis–Menten equation (see Section 2.6) since the enzyme is not readily saturated with substrate $O_2^{-\cdot}$.

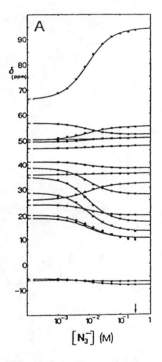

signal	proton	$\delta_{max}(N_3^-)$ppm
A	His-63 Hδ2	94.4
B	His-120 Hδ1	51.4
C	His-48 Hδ1	54.8
D	His-80 Hδ2	49.9
E	His-71 Hδ2	50.1
F	His-80 Hε2	46.9
G	His-48 Hε1	37.9
H	His-120 Hε1	27.2
I	His-80 Hε1	35.8
J'	His-71 Hε1	35.8
J	His-71 Hε2	35.8
K	His-46 Hε2	16.3
L	His-46 Hδ2	12.6
M	His-48 Hδ2	31.9
N	His-120 Hδ2	18.9
O	His-46 Hε1	9.9
P	His-46 Hβ1	9.8

Figure. 8.4 Determination of azide-binding affinities from ^{1}H NMR titration experiments. Illustrative data [adapted from L. Banci et al., *J. Am. Chem. Soc., 110*, 3629–3633 (1988)] are shown for the wild-type enzyme. Binding constants are obtained by fitting to a standard equation [$\delta(L) = \delta_L \cdot [L]/(K + [L]) + \delta \cdot K/(K + [L])$], where δ_L is the chemical shift with saturating ligand, δ is the chemical shift in the absence of ligand, and K is the binding constant. The limiting chemical shifts in the presence of saturating levels of N_3^- are given in the table to the right. Note that the resonances from His-46 (there are two sets since the enzyme is a dimer!) return to the diamagnetic region following displacement of this residue from the paramagnetic copper site. Results for WT and mutant enzymes are given in Table 8.1.

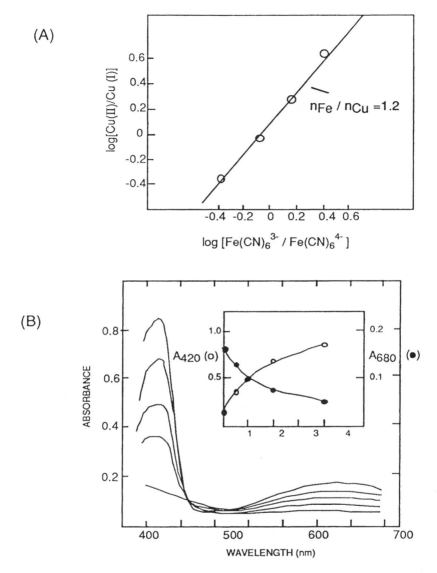

Figure 8.5 Determination of the midpoint reduction potential for Cu(II). A ferrocyanide/ferricyanide redox couple is used and a Nernst plot (A) of the optical titration data (B) is generated by using the 680 nm absorption to determine the relative concentrations of Cu(II) and Cu(I). The 420 nm absorbance values give the concentrations of $Fe(CN)_6^{4-}$ and $Fe(CN)_6^{3-}$. E(Cu) is obtained from the expression $E(Cu) = E(Fe) + (RT/nF)$ $\{\ln[Fe(CN)_6^{4-}]/[Fe(CN)_6^{3-}] - \ln[Cu^{2+}]/[Cu^{+}]\}$, assuming $E(Fe) = 0.441$ V. Within experimental error, $n \sim 1$. [Adapted from J. Fee et al., *Biochemistry, 12,* 4893–4899 (1974).]

where E represents one subunit of the enzyme and $E(H^+)$ is a protonated sub-unit.[4] The oxidation–reduction catalytic cycle represented at the copper center in Figure 8.3 requires that the midpoint reduction potential lies between the limits for superoxide oxidation and reduction $[O_2^-/H_2O_2, E° = 0.98 V; O_2/O_2^-, E° = -0.45 V]$ since the copper ion must be capable of carrying out both reactions. This is indeed the case for all characterized SOD's, including bacterial Fe-SOD and Mn-SOD (e.g., CuZn-SOD, $E° = +0.42V$; Mn-SOD $E° = +0.26$ V). The bacterial SOD's do not show redox-coupled proton uptake, consistent with the lack of bridging histidinate ligands; otherwise, the mechanisms for all SOD's are believed to be related to the preceding model.

8.1.2   Fe- and Mn-SOD

Fe-SOD's and Mn-SOD's are generally dimeric or tetrameric enzymes com-posed of 18–22 kDa subunits and are widespread among microorganisms. The structures of dimeric Fe-SOD's from *Escherichia coli* and *Pseudomonas ovalis* have been established, as has the tetrameric Mn-SOD from *Thermus thermo-philus*. The subunit structures of both Fe-SOD and Mn-SOD are highly ho-mologous, although each is inactive if reconstituted with the incorrect metal ion. Each subunit contains approximately 50% α-helix and three strands of antiparallel β-pleated sheet, and so is structurally distinct from the cuprozinc enzyme. The metal does not appear to serve a structural role, since the structures of Fe-SOD (shown in Figure 8.6) and the apoprotein are similar. The metal ligand set is likely composed of 3 × histidine and 1 × aspartate residues. Note again that the reduction potential of the metal site lies between the poten-tials for the O_2^-/H_2O_2 and O_2/O_2^- couples noted earlier.

8.1.3   Catalases and Peroxidases

Both classes of enzyme are generally termed hydroperoxidases and contain heme b as the active prosthetic center. Catalase is located in peroxisomes (see section 1.14) and the cytosol, and catalyzes the disproportionation of H_2O_2. Two types of peroxidase activity have been identified, according to whether the reducing equivalents derive from a substrate molecule (RH_2) or a natural redox partner. Table 8.2 summarizes the various reactions catalyzed by this family of enzymes.

In spite of the rather disparate appearance of all of these reactions, there are some common themes in the chemistry. The resting state of each enzyme con-tains a ferric heme. In each case, hydrogen peroxide oxidizes the catalytic heme center to form an intermediate species that is generally referred to as *compound I*. An iron-oxo, $Fe(IV)=O$, center is formed, whereas the other electron equiv-alent may come from either the heme ring, to form a π-cation radical, or from a neighboring protein side chain (Table 8.3).

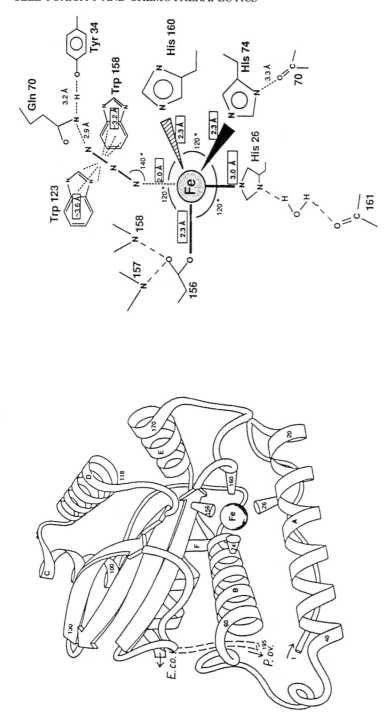

Figure 8.6 Schematic representation of the secondary and tertiary structure of a subunit of Fe-SOD. The conformations of the C-terminae are different for the *E. coli* and *P. ovalis* enzymes. The enzyme is a dimer of two identical subunits, each subunit consisting of ~50 percent α-helix and three strands of antiparallel β-sheet. This stands in sharp contrast to CuZn-SOD (Figure 8.1). Removal of the iron (indicated) does not change the structure of the enzyme. [Reprinted with permission from D. Ringe et al., *Proc. Natl. Acad. Sci. USA*, **80**, 3879–3883 (1983).] The active site interactions of the N_3^- derivative are indicated. [Adapted from B. L. Stoddard et al., *Protein Eng.*, **4**, 113–119 (1990).]

Table 8.2 Summary of Reactions for Catalases and Peroxidases.

Enzyme	Class	Reaction Summary
catalase	disproportionation	$H_2O_2 + H_2O_2 \rightarrow 2H_2O + O_2$
HRP	substrate	$H_2O_2 + RH_2 \rightarrow R + 2H_2O$
		$R'O_2H + RH_2 \rightarrow R + H_2O + R'OH$
CCP	redox partner	$H_2O_2 + \text{cyt-}c_{red} + 2H^+ \rightarrow \text{cyt-}c_{ox} + 2H_2O$
Cl-per	redox partner and substrate	$H_2O_2 + Cl^- \rightarrow ClO^- + H_2O$
		$ClO^- + RH_2 \rightarrow RHCl + HO^-$
Mn-per	redox partner and substrate	$H_2O_2 + 2Mn^{2+} + 2H^+ \rightarrow 2Mn^{3+} + H_2O$
		$2Mn^{3+} + RH_2 + H_2O \rightarrow 2Mn^{2+} + RHOH + 2H^+$

This highly oxidizing center subsequently is reduced by reaction with H_2O_2 (disproportionation), a substrate, or a variety of redox partners (cyt-c_{red}, Mn^{2+}, Cl^-) some of which (Mn^{3+}, ClO^-) may then mediate chemistry on a substrate. In all cases, the oxo center on iron finally ends up in a water molecule, although,

Table 8.3 Variation of Axial Ligand and Radical Sites for Compound I.

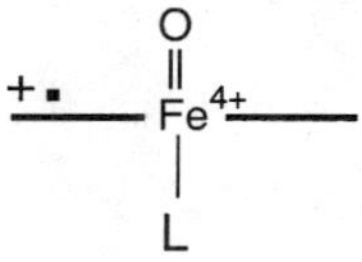

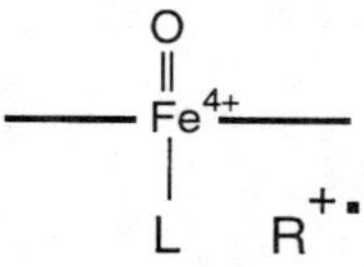

enzyme	L	enzyme	
catalase 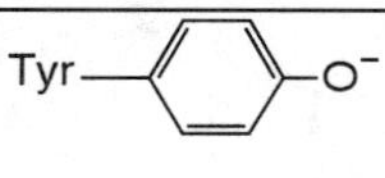 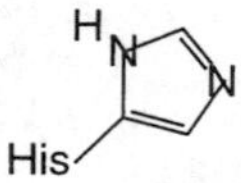	Tyr—⟨benzene⟩—O^-		
		CCP	$L =$ His-imidazole
HRP	His-imidazole		
			$R =$ Trp-indole
Cl-per	Cys - SH		
Mn-per 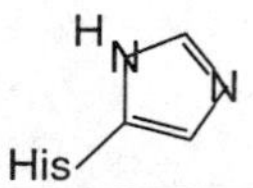	His-imidazole		

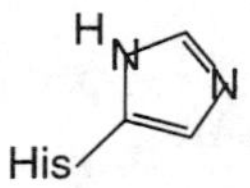

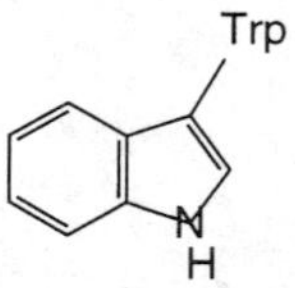

in the case of chloroperoxidase, this is mediated through HOCl. Horse-radish peroxidase (HRP) mediates chemistry directly on a substrate, cytochrome c peroxidase (CCP) is reduced to the resting state by cyt-c_{red}, whereas manganese and chloroperoxidase (Mn-per and Cl-per, respectively) are each reduced to the resting state by a redox mediator which subsequently oxidizes a substrate molecule.

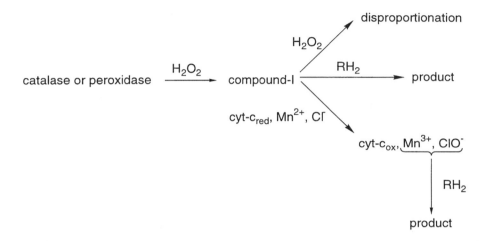

This rich chemistry from a common intermediate results, in part, from variations in the protein side chains and axial ligation to the heme, which tune the chemical reactivity of the heme center (Table 8.3). These differences are reflected in the optical spectra for catalase and HRP and result from a small change in the ordering of the energy levels for each center. The two highest occupied molecular orbitals (HOMO's) of the porphyrin π-system have a_{1u} and a_{2u} symmetry, respectively.[5] The symmetry of the resulting π-cation radical is labeled according to the orbital containing the unpaired electron (Figure 8.7). The relative energies of these orbitals are extremely sensitive to neighboring protein side chains and, especially, the axial ligand to the metal center.

The resting oxidation states of the compound I form of each enzyme can be recovered by two one-electron reductions. The radical site is generally reduced first and the resulting species with a formal ferryl Fe(IV)=O center is called *compound II*. Subsequent one-electron reduction yields the ferric Fe(III) heme. Note that, in each case, the axial ligands help to stabilize the high-valent iron intermediates. In some cases, the intermediate is stabilized by delocalization of charge onto a neighboring protein side chain. This side chain is usually in close proximity (or π-contact) with the proximal heme-binding ligand to facilitate electron transfer. Also, the close connection between this heme ligand and radical site is stabilized by hydrogen bonding to a common Asp residue, as depicted for CCP.

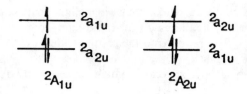

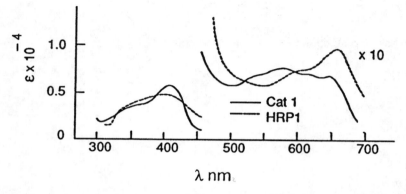

Figure 8.7 Comparison of the optical spectra for the radical ferryl sites for compound-I in catalase and HRP. The differences arise from an inversion of the lower π-energy levels (i.e. the ground state is either $^2A_{1u}$ or $^2A_{2u}$, respectively) as a result of the change in axial ligand (Table 8.3)

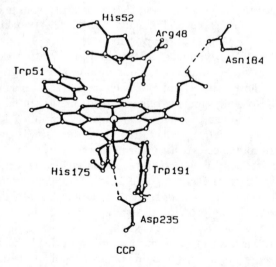

Unlike myoglobin and hemoglobin, where the heme was solvent exposed, the heme in catalase and peroxidase is buried in the protein interior to insulate the

Table 8.4 Reduction Potentials for Globins, Catalase, Peroxidase, and P-450

Protein/Enzyme ($Fe^{3+/2+}$)	$E°$ (mV versus NHE)
Mb	+50
Hb	+170
Catalase	−420
HRP	−170
CCP	−194
Cytochrome P-450$_{CAM}$	−300 (no substrate)
	−173 (+ substrate)

highly oxidizing intermediates. Relative to the oxygen carriers, the reduction potentials of catalase and peroxidases are significantly more negative (Table 8.4), again favoring high-valent iron. These low potentials result from the polar environment created within the heme pocket and the electron donating proximal ligands.

A likely mechanism for peroxidase activity is shown in Figure 8.8. Heterolytic cleavage of the peroxide O–O bond is assisted by the distal histidine and as-

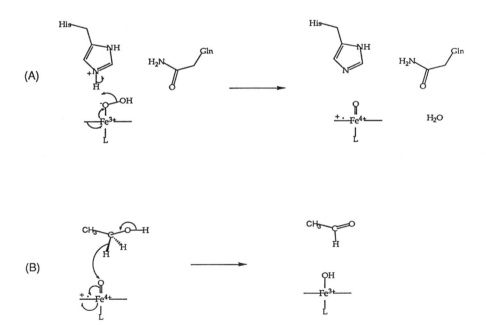

Figure 8.8 (A) General scheme for the heterolytic cleavage of the peroxide bond in catalase and HRP. Note the stabilizing influence of the protein side chains, as described in the text. (B) Possible mechanism for oxidative chemistry by the ferryl intermediate (e.g., the oxidation of ethanol).

paragine, which help to protonate and stabilize the oxygen anion, respectively. Since peroxidase catalyzes the oxidation of many organic and inorganic substrates, most are located in specific cellular compartments where oxidative chemistry is carried out on substrate molecules. However, yeast cytochrome c peroxidase is a soluble enzyme located between the inner and outer mitochondrial membranes where it rapidly oxidizes (10^7–10^8 M^{-1} s^{-1}) reduced cytochrome c.

8.1.4   Cytochrome P-450

Mammalian cytochrome P-450 is a membrane-bound mono-oxygenase that catalyzes the incorporation of one oxygen atom into substrate molecules.[6,7]

$$R_3C-H + O_2 + 2H^+ + 2e^- \rightarrow R_3C-OH + H_2O$$

These substrates are frequently nonpolar steroids, drugs, and pollutants that are trapped in the hydrophobic cell membrane. Hydroxylation lends water solubility to these molecules, allowing removal from the cell. Cytochrome P-450 is widely distributed in mammalian tissues (liver, lung, intestine, kidney) and is also found in bacteria, plants, and yeasts. Unlike the peroxidases, which utilize hydrogen peroxide directly as the initial oxidizing substrate, oxidases and hyroxylases proceed by a series of reductive steps from molecular oxygen.

$$Fe^{3+} + e^- \rightarrow Fe^{2+}$$

$$Fe^{2+} + O_2 \rightarrow Fe^{3+}-O-O^-$$

$$Fe^{3+}-O-O^- + H^+ + e^- \rightarrow Fe^{3+}-^-O-OH$$

The electron transport system for bacterial P-450s (1) differs from the microsomal systems (2) to the extent that a soluble [Fe$_2$S$_2$]-ferredoxin (normally putidaredoxin) acts as an intermediate electron carrier in the former. Both P-450 and the flavoprotein are membrane bound. The most thoroughly characterized enzyme in this class is the bacterial camphor-hydroxylating cytochrome P-450

(1)

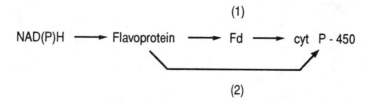

(2)

(designated P-450$_{cam}$) isolated from *Pseudomonas putida*, which assists in the introduction of one atom of dioxygen into a hydrocarbon substrate while the other is released as H$_2$O. NADH provides the two electron-reducing equivalents required for the regio- and stereospecific hydroxylation of camphor to yield the 5-exo alcohol. The active site of the enzyme holds a heme prosthetic center.

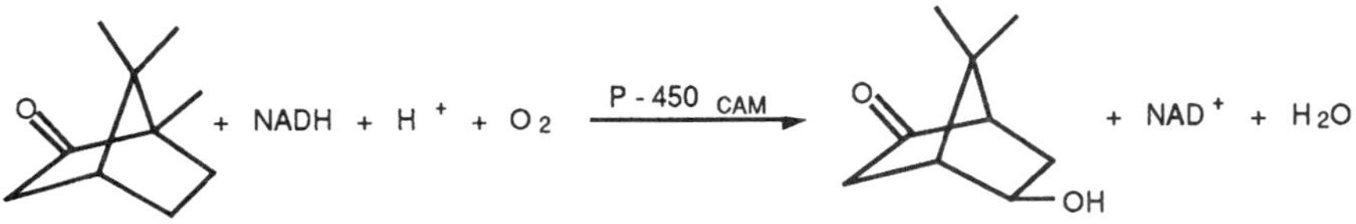

Much of the initial speculation surrounding the axial ligand environment of the heme prosthetic center in this enzyme has been clarified by crystallographic data (Figure 8.9).[9]

Figure 8.10 summarizes the chemical steps in the reaction sequence and shows the coordination state of the low-spin heme ($E° = -300$ mV) in the resting enzyme, with cysteine thiolate and H_2O (or HO^-) as axial ligands, and the changes arising in the course of the reaction. Substrate (RH_2) binds in a pocket close to the heme and forces the displacement of H_2O (or HO^-) from heme iron to yield a high-spin pentacoordinate heme $Fe^{3+} \cdot (RH_2)$ complex ($E° = -170$ mV) that can be more easily reduced to the ferrous–heme substrate complex $Fe^{2+} \cdot (RH_2)$. Displacement of H_2O also opens up the reduced heme to dioxygen. Note that the requirement for substrate binding prior to dioxygen activation prevents formation of reactive oxo-ferryl species in the absence of substrate, since the oxidizing iron center might otherwise attack the protein backbone or neighboring side chains. After oxygen binding to give the oxy-form $Fe^{3+}-O_2^- \cdot (RH_2)$, a second one-electron reduction yields the peroxide adduct $Fe^{3+}-O_2^{2-} \cdot (RH_2)$, which undergoes heterolysis of the O–O bond to produce water and a high-valent iron-oxo center, $Fe^{5+} = O \cdot (RH_2)$. Subsequent two-electron oxidation of substrate yields product and regenerates the ferric enzyme ($Fe^{3+} + RHOH$).

Of the two possible pathways presented in Scheme 1 (a and b), the generally accepted molecular mechanism for oxygen insertion is believed to proceed via radical intermediates (b) and is termed an *oxygen-rebound mechanism*, for obvious reasons.[10]

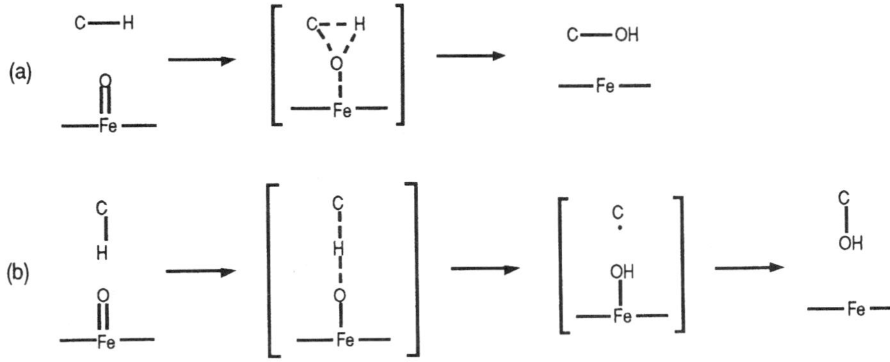

Scheme 1.

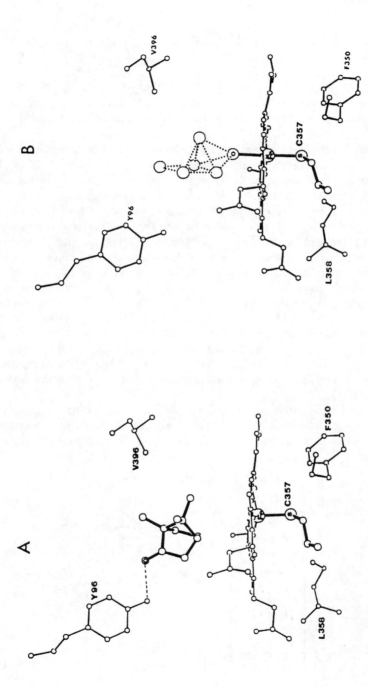

Figure 8.9 Active site of cytochrome P-450$_{CAM}$ in the presence (A) and absence (B) of camphor. Note the breakdown of the water cluster in the active site and release of the axial H_2O/HO^- following substrate binding. [Adapted from T. L. Poulos, *Adv. Inorg. Biochem.*, 7, 1–36 (1988).]

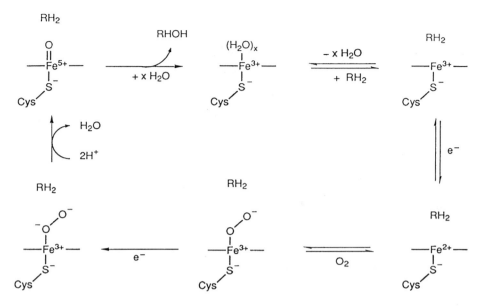

Figure 8.10 Summary of the catalytic cycle for P-450$_{CAM}$ and the postulated structures for intermediates. The mechanistic details of peroxide bond cleavage and oxygen insertion remain the topic of lively debate.

Summary of Section 8.1

1. Two major reactive by-products from oxidative metabolism include superoxide ($O_2^{-\cdot}$) and peroxide (O_2^{2-}).
2. Superoxide dismutase catalyzes the reaction $2O_2^{-\cdot} + 2H^+ \rightarrow O_2 + H_2O_2$. The mammalian enzyme contains a Cu–Zn couple, bridged by a histidinate anion. The copper ion forms the catalytic site.
3. Peroxide is degraded either by catalase or by a number of peroxidases. These contain heme b at the active site. With the exception of cytochrome c peroxidase, which produces an intermediate protein radical, reactions proceed through ferryl porphyrin radical $[PFe(IV) = O]^\cdot$ intermediates.
4. Cytochrome P-450 is normally a membrane-bound heme protein that hydroxylates lipophilic molecules, enhancing their water solubility and ease of removal from the host cell.

Review Question

Cytochrome P450 carries out oxidation chemistry on hydrocarbon substrates, yet it requires the reductant NADH. Provide an explanation for this fact. Also consider the mechanisms of SOD, catalase, and peroxidase, and identify the species that are oxidized and reduced. Account for and detail the flow of electrons.

8.2 Metal Toxicity

Although trace metal ions are vital for normal cell function, elevated concentrations may be deleterious as a result of competitive or adventitious interactions with other metal-binding sites on proteins, ion channels, membranes, polysaccharides, and the numerous other biological ligands that constitute cell metabolism. Cells normally regulate the intracellular concentrations of essential metals by transport mechanisms. However, metal ions that lack a normal biological function may not come under the control of a cell's regulatory machinery. A good example of this are the *heavy metals* from the third transition series and f-block elements that are nonessential for cell function but can bind competitively or adventitiously to many ligands on membrane surfaces, in ion channels, or enzyme/protein active sites. Certain bacteria have acquired resistance mechanisms to such ions (Cd^{2+}, Pb^{2+}, Ag^+, mercurials, arsenicals) that include transmembrane ion pumps to export the ion out of the cell, chemical oxidation or reduction to a more inocuous and/or volatile form, or simple binding or sequestration of the ion by ligands, proteins, or cell membranes (Figure 8.11).

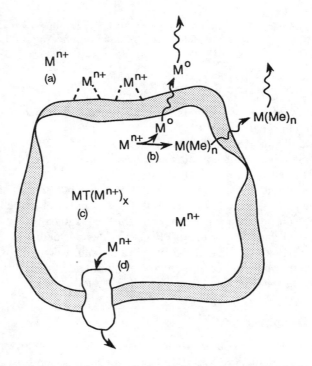

Figure 8.11 Bacterial mechanisms for disposal of heavy metals. (a) Binding to the outer membrane. (b) Chemical reduction and/or methylation to form volatile species. (c) Complexation by ligands or proteins (e.g., metallothionein, MT). (d) Export through an ion channel.

The genes encoding the proteins and enzymes required for specific resistance mechanisms are located in plasmids and transposons within the cell.[11] Transcription of these resistance genes depends on the intracellular concentrations of the toxic metal ion. For specific examples, the reader should review the expression of copper-binding proteins in Chapters 3 and 7 and refer to Chapter 10 for a description of the genetics of mercury resistance.

8.2.1 Specific Detoxification Mechanisms

Metal Binding and Storage

Bacterial cells can accummulate large concentrations of metal ions on their cell walls by binding to the anionic sites of lipid membranes (Table 8.5) and, in particular, to the carboxylate groups of the peptidoglycan layer that forms an important part of the outer cell membrane. If these toxic species are taken into the cell, other mechanisms must come into play.

Yeasts, fungi, and higher plants and animals contain small cysteine-rich proteins called metallothioneins that bind heavy metals. The chemistry of this class of protein was outlined in Sections 3.3 and 7.2. Although metallothioneins are important proteins in copper and zinc homeostasis,[12,13] for other nonessential heavy metals, the function is detoxification. Metals are taken up in the order $Zn(II) < Pb(II) < Cd(II) < Cu(I), Ag(I), Hg(II), Bi(III)$. Heavy metals may also be transported into intracellular compartments for storage. Shortly, we shall see an example of how gold can accumulate in the lysosomes of mammalian tissues.

Membrane Transport

Heavy metal ions may enter a cell by way of transport mechanisms for chemically related species that are essential for cell metabolism. For example, Cd^{2+} uses a Mn^{2+} transport system, whereas Tl^+ and AsO_4^{3-} or AsO_2^- are substrates for K^+ and PO_4^{3-} transport systems, respectively. Resistance mechanisms to these cationic and anionic species involve specific efflux systems that provide for the rapid transport of the toxic ion out of the cell (Figure 8.12).

Table 8.5 Binding of Metal Ions by Isolated Cell Walls and Membranes[a,b]

Organism	Component	Na^+	K^+	Mg^{2+}	Ca^{2+}	Mn^{2+}	Fe^{3+}
B. subtilis	Wall	2.70	1.94	8.22	0.40	0.80	3.58
B. licheniformis	Wall	0.91	0.56	0.40	0.59	0.66	0.76
E. coli-AB264	Peptidoglycan	0.290	0.060	0.035	0.038	0.052	0.010
E. coli-AB264	Outer membrane	0.081	0.025	0.019	0.020	0.012	0.233
E. coli-AB264	Cell envelope	0.042	0.082	2.56	0.035	0.140	0.200

[a] μmol metal (mg dry weight)$^{-1}$.
[b] T. J. Beveridge and R. G. E. Murray, *J. Bacteriol.*, *127*, 1502–1518 (1976).

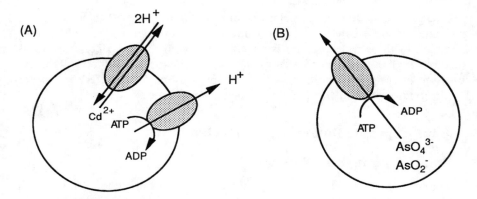

Figure 8.12 Detoxification pathways involving transport mechanisms. (A) Cd^{2+} efflux is controlled by a plasmid-encoded protein complex that defines a Cd^{2+}/H^+ antiport. A transmembrane proton gradient established by ATP hydrolysis actively pumps Cd^{2+} out of the cell. (B) Arsenate and arsenite enter cells by way of a phosphate transport system and are ejected through an anion pump that is driven by ATP hydrolysis.

Chemical Transformation

Methylation of heavy metal ions results in a dramatic decrease in their solubility in aqueous solution and an increase in volatility. It is a common detoxification mechanism used by a number of microbes.

$$M^{2+} \rightarrow CH_3M^+ \rightarrow (CH_3)_2M(g) \qquad M = Hg^{2+}, Sn^{2+}, Pb^{2+}, Tl^+$$

$$[XO_n]^{m-} \rightarrow [XO_{n-1}(OMe)]^{(m-1)-} \rightarrow ... \rightarrow [XO_{n-m}(OMe)_n](g) \qquad X = As$$

Two methylating agents that are typically employed include methyl cobalamin (a source of CH_3^-) for electrophilic cations and S-adenosyl methionine (a source of CH_3^+) for nucleophilic anionic and neutral species.[14] Methylcobalamin methylates a variety of toxic cations, including Hg^{2+}, Sn^{2+}, Pb^{2+}, and Tl^+. Mechanisms for microbial resistance to mercury species are most thoroughly understood. This is the topic of detailed discussion in Chapter 10 and the interested reader may wish to move directly to that section to complete this particular story. Briefly, both mercuric ion [Hg(II)] and organomercurials are reduced to volatile Hg(0). In the case of organomercury species the C–Hg bond is first cleaved by organomercurial lyases and the resulting mercuric ion is then reduced by mercuric reductase. Ultimately, all mercury species are converted to Hg^{2+} prior to reduction to volatile Hg(g).

$$R - Hg - X \xrightarrow{\text{lyase}} RH + Hg^{2+} \xrightarrow[\text{NADPH} \quad \text{NADP}^+]{\text{reductase}} Hg° \qquad (R = \text{alkyl, aryl})$$

Arsenite AsO_2^- is the most toxic arsenic species. Many bacteria oxidize this to the less toxic arsenate (AsO_4^{3-}) anion, which can be readily transported out of the cell; however, reduction and methylation are also commonly used detoxification pathways. Since arsenic species undergo nucleophilic methylation (Figure 8.13), S-adenosyl methionine is used as a source of CH_3^+.[14,15]

Summary of Section 8.2

1. Exogenous metal ions (especially heavy metals) are toxic when they interfere with normal cell metabolism by competitive or adventitious interactions with other metal-binding sites on proteins, ion channels, membranes, polysaccharides, and so on.
2. Resistance mechanisms to extracellular toxins include transmembrane ion pumps to export the ion out of the cell; chemical oxidation, reduction, or methylation to a more innocuous and/or volatile form; or simply binding or sequestration of the ion by ligands, proteins, or cell membranes.

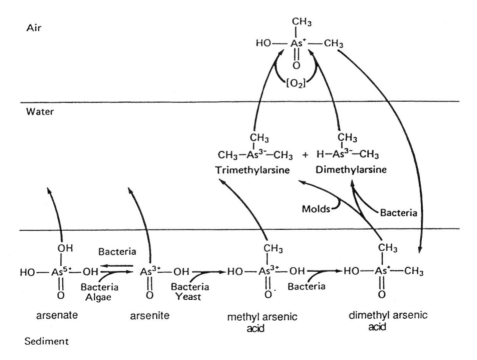

Figure 8.13 Schematic representation of the arsenic cycle showing various stages of reduction and methylation. [Adapted from J. M. Wood, *Science, 183*, 1049–1052 (1974).]

8.3   Coordination Complexes as Therapeutic Agents

8.3.1   Platinum Complexes in Chemotherapy

In 1969 Rosenberg and co-workers reported the anticancer activity of platinum coordination complexes.[16] These workers identified *cis*-dichloro-diammine platinum(II) [Pt(NH$_3$)$_2$Cl$_2$], commonly termed *cisplatin*, as an active species.

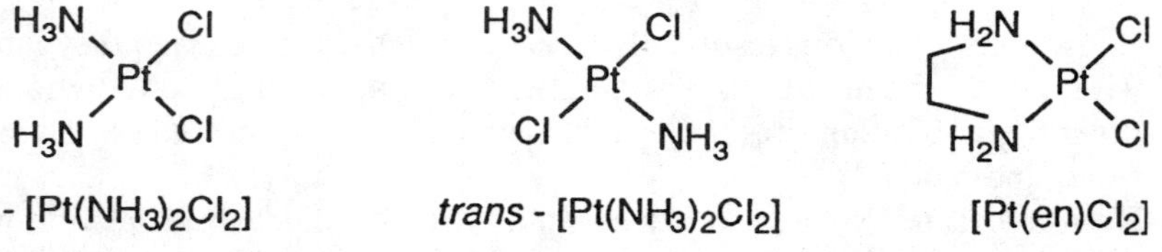

| *cis* - [Pt(NH$_3$)$_2$Cl$_2$] | *trans* - [Pt(NH$_3$)$_2$Cl$_2$] | [Pt(en)Cl$_2$] |

Cisplatin is especially active against testicular, ovarian, and head and neck tumors. The corresponding *trans* configuration was found to be inactive, and a likely reason for this observation will soon be made clear. The reagent exerts its biological influence by binding to DNA, with preferential coordination to guanine bases and the formation of 1,2-intrastrand cross-links in guanine-rich regions (Figure 8.14).[17]

The study of the chemistry of drug molecules under conditions likely to be found in living tissue is extremely important for the understanding of the mechanism of action in vivo. From Scheme 2, it is clear that the aqueous chemistry

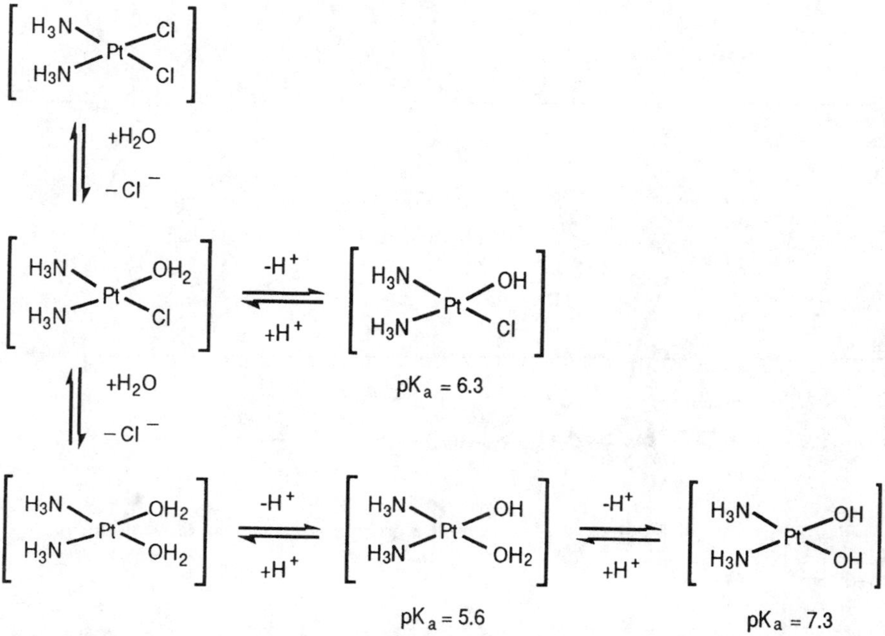

Scheme 2.

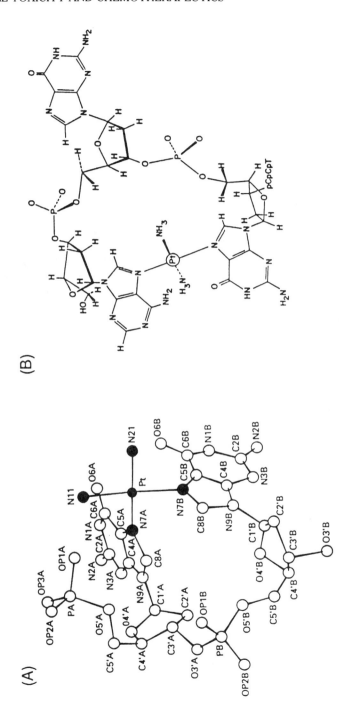

Figure 8.14 (A) The major product from reaction of *cis*-[Pt(NH₃)₂Cl₂] and *d*(pGpG). Pt coordinates to the N-7 atom of each guanine base. The structure may be further stabilized by H-bonding from NH₃ bound to Pt and the O-6 carbonyl oxygen. [Adapted from S. E. Sherman et al., *Science*, *230*, 417 (1985).] (B) In contrast, the *trans* isomer does not bridge adjacent bases. The complex with *d*(ApGpGpCpCpT) is shown. [Adapted from C. A. Lepre et al., *Biochemistry*, *26*, 5651; copyright (1987) American Chemical Society.]

of cisplatin can be extremely complex. At high concentrations of the platinum salt, both hydroxyl-bridged dimeric and trimeric species may form. However, this is unlikely at the low levels typically employed under normal biological conditions. Although cisplatin is unstable to hydrolysis in aqueous solutions, the rather high extracellular concentrations of Cl^- (~160 mM) inhibit this reaction. However, the low intracellular levels of Cl^- (~16 mM) promote rapid hydrolysis, and so the aquo complex is the reactive species that platinates DNA.

$$cis\text{-}[(NH_3)_2PtCl_2] \rightarrow cis\text{-}[(NH_3)_2Pt(H_2O)(HO)]^+ \rightarrow DNA\ binding$$

The solution chemistry may be readily followed by ^{195}Pt NMR (Figure 8.15). Both the *cis* and *trans* isomers form DNA adducts with equal ease; however, the structural consequences are quite distinct (Figure 8.14). Cisplatin forms 1,2-intrastrand cross-links by coordination to N-7 (nitrogen on carbon 7) of two guanine bases,[18] whereas *trans*-Pt forms 1,3- and 1,4-adducts.[19] Binding of cisplatin results in bending of the DNA duplex by ~40° and an unwinding of ~13° for *cis*-GG and *cis*-AG adducts (Figure 8.16). Experimental evidence for bending of the DNA adduct comes from gel electrophoresis studies (see Section 2.8). Bent DNA migrates more slowly than linear DNA, and the degree of bending can be estimated from a calibration plot.[20] The degree of unwinding can also be estimated from gel mobility studies. Both DNA bending and unwinding

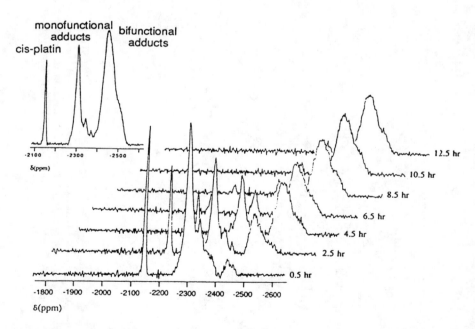

Figure 8.15 ^{195}Pt NMR can be used to monitor the rate of reaction of cisplatin with double-stranded DNA. [Reprinted with permission from D. P. Bancroft et al., *J. Am. Chem. Soc., 112*, 6860–6871; copyright (1990) American Chemical Society.]

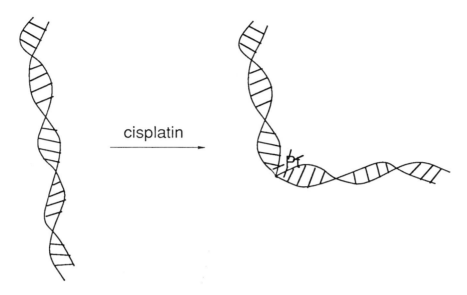

cisplatin

Figure 8.16 Structural influence of cisplatin binding to ds DNA. The DNA is bent 30–40° and is unwound by ~13° at the platination site.

are necessary for recognition of platinated DNA by a damage recognition protein in human cells. The precise mechanism of cisplatin anticancer activity is uncertain; however, several possibilities have been identified that include blocking of DNA replication or binding of damage recognition proteins that inhibit interactions with other cellular components.

8.3.2 Biological Chemistry of Gold Complexes

The treatment of disease and maladies with gold complexes (called *chrysotherapy*) can be traced back to China around 2500 b.c. In 1961 Au(I) complexes were found to halt the progression of rheumatoid arthritis and are currently the only class of drugs clearly implicated with this effect. Since gold lies in the midst of the "coinage" elements, Au(I) ($5d^{10}$) is found to be a soft cation that

GROUP VIII

10	11	12
Ni	Cu	Zn
Pd	Ag	Cd
Pt	Au	Hg

forms stable complexes with polarizable anions and π-acid ligands, CN^- and phosphines in particular.

$$CNO^- < CNS^- \sim Cl^- < Br^- < I^- \leq CN^-, R_3P$$

The ^{197}Au ($I = 3/2$) isotope is 100% abundant and is an effective NMR nucleus. In aqueous solution, Au(I) is unstable and readily disproportionates to Au(0) and Au(III).

$$3Au(I) \rightarrow 2\ Au(0)\ +\ Au(III)$$

Ligands that are known to stabilize Au(I) in water, a prerequisite for in vivo drug delivery, include cyanide (CN$^-$), thiolates (RS$^-$), and thiosulfate (S$_2$O$_3{}^{2-}$). Au(III) is a strong oxidizing agent and is, therefore, highly toxic, and so, when coordinated to thiolate ligands, Au(I) is the most stable state in vivo. This fact has been used in the design of most of the gold compounds that have been approved for clinical use (Figure 8.17). Although few details are known regarding the molecular basis of action of gold pharmaceuticals, two clear facts have emerged. First, gold tends to accumulate in the lysosomes, where it may inhibit the release of hydrolase enzymes that destroy the tissue around joints. Lysosomes loaded with gold are often referred to as aurosomes. Second, serum albumin is the principal carrier in blood serum for many of these drugs. On the basis of EXAFS and chemical studies,[21] Cys-34 has been identified as the binding site in serum albumin.[22]

The chemistry of Au(I) is similar to that of Hg(II), with which it is isoelectronic. Both readily form linear bicoordinate complexes, although coordination numbers of 3 are known and a very few 4-coordinate species have been identified.

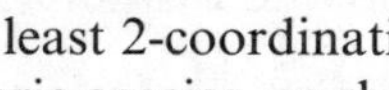

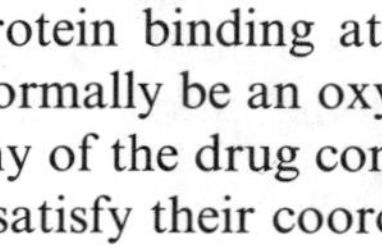

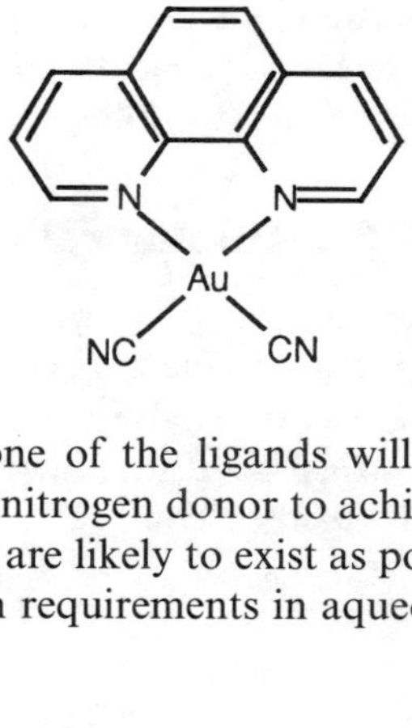

In the particular case of protein binding at least one of the ligands will be thiolate and the other will normally be an oxygen or nitrogen donor to achieve at least 2-coordination. Many of the drug complexes are likely to exist as polymeric species, as shown, to satisfy their coordination requirements in aqueous solution.

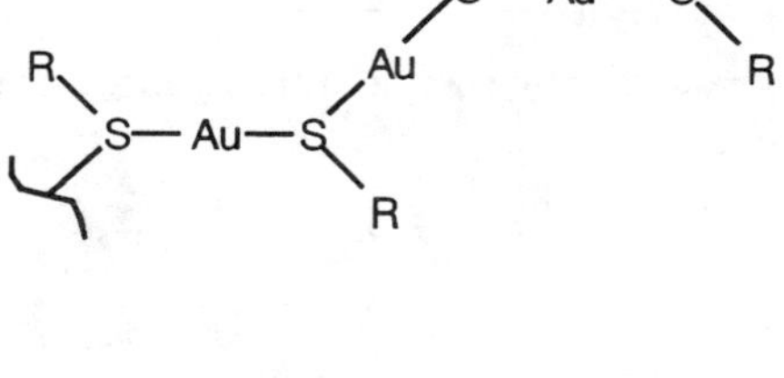

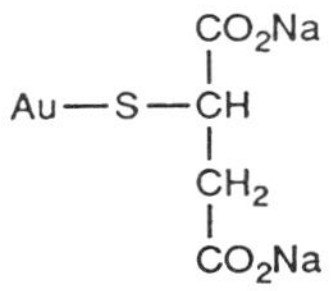

Gold sodium thiomalate
(Myochrisin)

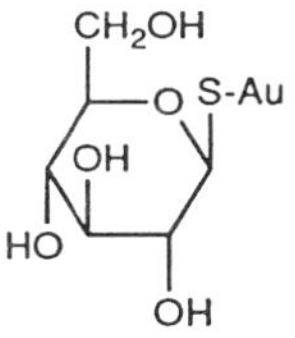

Gold thioglucose
(Solganol)

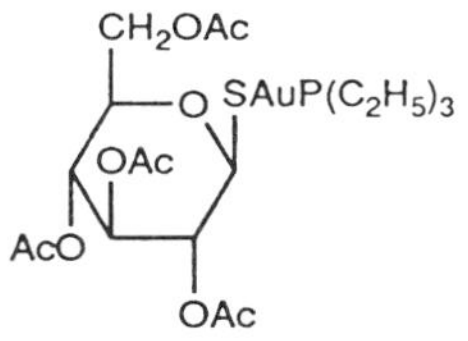

2,3,4,5-Tetra-O-acetyl-1-1-β-D-
thioglucose(triethylphosphine)gold(I)
(Auranofin)

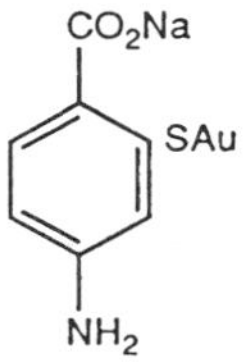

4-Amino-2-mercaptobenzoic acid
(Krysolgan)

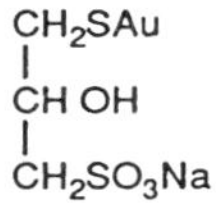

Thiopropanol Sulphonate
(Allocrysine)

Figure 8.17   Structures of five gold complexes commonly used for treatment of rheumatoid arthritis.

8.3.3   Radiopharmaceuticals

A third class of inorganic pharmaceuticals has developed as a direct result of intensive research in the area of nuclear chemistry. Many elements possess one or more radioisotopes that differ from their more stable relatives by possessing an unstable nucleus that may decay by one of three pathways, yielding new particles and/or radiation with sufficient energy to break chemical bonds or ionize atoms to produce reactive radicals (such as HO·). For example,

$$^{232}Th \rightarrow {}^{228}Ra + \alpha$$
$$^{228}Ra \rightarrow {}^{228}Ac + \beta^-$$
$$^{129}Cs \rightarrow {}^{129}Cs + \gamma$$

The α-particle is a helium nucleus ($^4_2$He), the β-particle is an electron that is emitted from a neutron in the nucleus ($^0_{-1}$e), whereas γ-rays are a high energy form of electromagnetic radiation. The use of radioisotopes as pharmaceutical agents is dependent on how well they can be targeted to specific organs. They are commonly used for one of two purposes. As *diagnostics,* they are used to "light-up" particular tissue or organ types in the body, for example, a tumor, the heart, or the thyroid gland, and reflect the pathological state of that tissue. As *therapeutics,* they are used to destroy that tissue. This type of complex is usually targeted toward malignant growths, such as tumors. Because gamma (γ) rays are of very high energy, they do not typically interact efficiently with chemical species, resulting in less chemical damage than the α or β-particles,

Table 8.6 Physicochemical Properties of Radionuclides.

Radionuclide	Emits	Half-Life	Use
^{43}K	β, γ	22 h	imaging (heart)
^{57}Co	γ	271 d	imaging (B$_{12}$)
^{57}Ga	γ	78 h	imaging (tumor)
^{97}Ru	γ	69 h	imaging (tumor, liver)
^{99m}Tc *	γ	6 h	imaging (various)
^{123}I	γ	13 h	imaging (thyroid)
^{131}I	β, γ	8 d	imaging, therapy (thyroid)
^{186}Re	β, γ	89 h	therapy
^{212}Pb	α, β, γ	1 h	therapy

* The m indicates that this isotope is man-made.

and so γ-emitters are typically used in diagnosis, whereas the lower energy α and β-particles, which have a shorter range as a result of rapid reaction with neighboring chemical species, are used in radiotherapy. Table 8.6 summarizes some common radionuclides and their uses.

Several criteria must be met for a chemical complex of a radioisotope to be a viable pharmaceutical agent. It must have a decay time that is not too short so that the relevant complex of the inorganic radionuclide can be synthesized and transported, but short enough (ideally < 8 d) that the intensity of radioemissions is high. Also, for diagnostic purposes, it is useful if the emission levels fall off within a few days so that the treatment can be repeated if necessary. Finally, it must be selectively taken up by organs or tumors. These constraints are quite demanding. As yet, our understanding of how to design ligands that will convey organ or tumor specificity to complexes of metal ions remains very poor. For this reason, there is extensive research into the use of labeled antibodies (proteins produced by the immune response that target specific cell types) that have been tagged with radionuclides to direct these diagnostic and therapeutic agents to where they are needed.

In spite of the "bad press" that often accompanies nuclear chemistry and the undeniable dangers associated with these materials, they have nevertheless proved to be of benign utility when the underlying chemistry and properties of the radionuclides are understood. The majority of radionuclides are metals, and so inorganic radiochemistry is a key area for use of metals in medicine. The ability of the chemist to synthesize new materials is well illustrated by the technetium isotope ^{99m}Tc, a completely man-made radionuclide formed as a result of neutron capture by ^{98}Mo. ^{99m}Tc is almost ideal in terms of its half-life, emissions, and energy range.

$$^{98}\text{Mo} \xrightarrow[]{+\,^{1}\text{n}} \underset{66\ \text{h}}{^{99}\text{Mo}} \xrightarrow[66\ \text{h}]{-\beta^{-}} \underset{}{^{99m}\text{Tc}} \xrightarrow[6\ \text{h}]{-\gamma} ^{99}\text{Tc} \xrightarrow[0.2 \times 10^{6}\,\text{y}]{-\beta^{-}} ^{99}\text{Ru}$$

Molybdenum is usually introduced as the molybdate anion $[^{99}MoO_4]^{2-}$, and, as a result of their distinct chemistries, the pertechnate anion $[^{99m}TcO_4]^-$ is easily separated by ion exchange chromatography. Subsequently this technetium species can be used as a precursor for the synthesis of a variety of complexes.

Summary of Section 8.3

1. Cisplatin $[Pt(NH_3)_2Cl_2]$ is a proven anticancer drug that is especially active against testicular, ovarian, and head and neck tumors.
2. Cisplatin selectively binds to GG and AG sequences, resulting in a local bending (by ~40°) and an unwinding (of ~13°) of the DNA helix. This structural change is most likely a contributor to the activity of the drug. Two speculative mechanisms include (a) direct inhibition of DNA replication by the Pt-bound drug or (b) a damage recognition protein that binds to the bent DNA and inhibits DNA replication.
3. Gold complexes are effective against rheumatoid arthritis. Gold(I) ions show preferential coordination to cysteine. By accumulating in the lysosomes of cells, the metal may inhibit the release of hydrolase enzymes that destroy the tissue around joints.
4. Radionuclides find use as diagnostics for imaging and in radiation therapy. Gamma emitters are preferred for the former and α or β-emitters for the latter. Man-made radionuclides, such as ^{99m}Tc, with useful lifetime and particle emissions, are readily adapted to a variety of uses. Such reagents require careful targeting to select organs or tumors. This can be achieved by labeling antibodies or by careful ligand design.

Review Questions

- Discuss the problems inherent to the transport and selective uptake of medically important metal ion complexes in vivo.
- Technetium(I) can be stabilized by π-acceptor ligands, such as phosphanes, arsanes, and isonitriles. The isonitrile complex $[Tc(CNR)_6]^+$, where R $= -CH_2C(CH_3)_2(OCH_3)$, is commercially sold under the name cardiolyte. This can be taken up by heart muscles and used as an imaging agent. Most likely, it uses transport pathways similar to the cations K^+, Rb^+, and Cs^+. In what ways is this complex well designed for its use? [Deutsch, E. et al., *Science 214,* 85 (1981)]

Notes

1. The interaction of solvent (H_2O) with a paramagnetic (or diamagnetic) metal site can be investigated by nuclear magnetic relaxation dispersion (NMRD) experiments. These measure the longitudinal relaxation time (T_1) of bulk water as a function of the magnetic field strength. At certain field strengths, the frequency of the field correlates with a motional property or exchange reaction of the system under investigation (e.g., a chemical exchange reaction, or an electron

exchange rate). At these frequencies, T_1 changes dramatically and yields information on the chemical process that is correlated with the change in T_1; in this case, solvent exchange.

2. The reasons for this were briefly outlined in Section 2.3 under the heading "NMR of Paramagnetic Molecules."

3. Review Section 2.5 of Chapter 2 for further details of this type of experiment.

4. It should be noted that the rate-limiting step in catalysis is not dependent on [H$^+$]; thus although this scheme is good for keeping track of atoms and electrons, it gives no insight on the mechanistic details.

5. This is a group theoretical notation that reflects the symmetry properties of the orbital set with respect to rotations and reflections about molecular axes and planes.

6. The name comes from the "particulate" nature of the membrane-bound enzyme and the intense absorption band at 450 nm observed in the CO adduct of the reduced form.

7. Another family of oxygenases, the dioxygenases, catalyzes the incorporation of two oxygen atoms into a substrate molecule.

8. Microsomes are vesicles that form from fragments of the endoplasmic reticulum during the homogenization (workup) of tissue.

9. A wealth of spectroscopic data on the native enzyme and ligand adducts has been summarized in recent reviews [e.g., Dawson and Sono, *Chem. Rev., 87,* 1255–1276 (1987)].

10. Refer to J. T. Groves, *Adv. Inorg. Biochem., 1,* 119–145 (1979).

11. Plasmids are circular DNA molecules that can self-replicate independently of the bacterial chromosome. Transposons are DNA elements that can be inserted randomly into plasmids or the bacterial chromosome.

12. Homeostasis refers to the active maintenance of intracellular concentrations of molecules and ions at physiologically appropriate levels.

13. See Sections 3.3 and 7.2.4.

14. See Figure 5.8 for the structure of methyl cobalamin and the following for other methylating agents.

S - methylmethionine   S - adenosylmethionine

15. Other common methylating (CH$_3^+$) agents include *S*-methylmethionine, *S*-methyl-tetrahydrofolate, and coenzyme M (Figure 5.10).

16. Rosenberg has written a wonderful account of the work leading to this discovery that illustrates the elements of serendipity, careful observation, and tireless pursuit by the scientific method that have characterized many important breakthroughs in science (B. Rosenberg, *Interdisciplinary Science Reviews, 3,* 134–147 (1978); also reproduced in *Metals in Biology,* Spiro, T. G. ed., Vol. 1, Chap. 1, pp. 1–29). It is recommended as essential reading for prospective researchers.

17. Evidence includes the inhibitory effect on DNA synthesis and the observation that cells deficient in DNA repair enzymes, which would normally remove Pt-DNA adducts, are more sensitive to the complex than normal parental lines.

18. Guanine–adenine cross-linking may also occur; however, binding to guanine is preferred, possibly as a result of H-bond stabilization to the base carbonyl.

19. The notation 1,2- or 1,3- or 1,4- indicates that the Pt reagent coordinates to two adjacent bases, two bases separated by a single base, and two bases separated by two bases, respectively.

20. The interested reader should refer to S. F. Bellon, J. H. Coleman, S. J. Lippard, *Biochemistry, 30,* 8026–8035 (1991) for further details.

21. The amount of bound gold correlates with the number of free sulfhydryl groups.

22. The chemistry of this carrier protein was briefly outlined in Chapter 3.

Further Reading

Reactive Oxygen Species

Andersson, L. A. and J. H. Dawson. EXAFS spectroscopy of heme-containing oxygenases and peroxidases. *Struct. Bonding 74*, 1–40 (1983).

Dawson, J. H., and M. Sono. Cytochrome P-450 and chloroperoxidase: thiolate-ligated heme enzymes. Spectroscopic determination of their active site structures and mechanistic implications of thiolate ligation, *Chem. Rev., 87*, 1255–1276 (1987).

Fito, I. and M. G. Rossmann. The active center of catalase, *J. Mol. Biol., 87*, 21–37 (1985).

Groves, J. T. Cytochrome P-450 and other heme-containing oxygenases, *Adv. Inorg. Biochem., 1*, 119–145 (1979).

Metalloenzymes involving amino acid residue and related radicals. Metal Ions in Biological Systems, Vol. 30. Sigel H. and A. Sigel, eds., Marcel Dekker, 1994.

Poulos, T. L. Heme Enzyme Crystal Structures, *Adv. Inorg. Biochem., 7*, 1–36 (1988).

Raag, R. and T. L. Poulos. Crystal structures of P-450$_{CAM}$ complexed with camphane, thiocamphor, and adamantane: Factors controlling P-450 substrate hydroxylation. *Biochemistry 30*, 2674–2684 (1991).

Tainer, J. A., E. D. Getzoff, J. S. Richardson, and D. C. Richardson. Structure and Mechanism of Copper/Zinc Superoxide Dismutase, *Nature, 306*, 284–287 (1983).

Metal Toxicity

Metal-Ion Induced Regulation of Gene Expression, *Adv. in Inorg. Biochem., 8*, 1990.

Poole, R. K., and G. M. Gould, Eds. *Metal Microbe Interactions*, Special Publications of the Society for General Microbiology, Vol. 26, IRL Press, 1989.

Thayer, J. S., and F. E. Brinkman. The Biological Methylation of Metals and Metalloids, *Adv. Organometallic Chem., 20*, 313–356 (1984).

Wood, J. M. Arsenic Cycle, *Science, 183*, 1049–1052 (1974).

Inorganic Pharmaceuticals

Bulman, R. The chemistry of chelating agents in medical sciences, *Structure and Bonding 67*, 91–141 (1987).

Clarke, M. J. and L. Podbielski. Medical diagnostic imaging with complexes of technetium-99m *Coordn. Chem. Rev. 78*, 253 (1987).

Deutsch, E. et al. Heart imaging with cation complexes of technetium. *Science 214*, 85 (1981).

Elder, R. C., and M. K. Eidness. Synchotron X-Ray Studies of Metal-Based Drugs and Metabolites, *Chem. Rev., 87*, 1027–1046 (1987).

Howard–Lock, H. E., and C. J. L. Lock. Uses in Therapy, in *Comprehensive Coordination Chemistry* (Eds. G. Wilkinson, R. D. Gaillard, and J. A. McCleverty), Vol. 6, Chap. 62.2, Pergamon, 1987.

Inorganic Chemistry in Biology and Medicine. ACS Symposium Series, Vol. 140. Martell, A. E., ed., 1980.

Lectures in Bioinorganic Chemistry, Nicolini, M. and L. Sindellari, eds., Raven Press, 1991.

Pinkerton, T. C. et al. Bioinorganic chemistry of technetium radiopharmaceuticals. *J. Chem. Educ. 62*, 985 (1985).

Reedijk, J. The mechanism of action of platinum antitumor drugs, *Pure and Appl. Chem., 59*, 181–192 (1987).

Rosenberg, B. Platinum complexes for the treatment of cancer, *Interdisciplinary Science Reviews, 3*, 134–147 (1978).

Sadler, P. J. The Biological Chemistry of Gold, *Structure and Bonding, 29*, 171–214 (1984).

Sherman, S., and S. J. Lippard. Structural aspects of platinum anticancer drug interactions with DNA, *Chem. Rev., 87*, 1153–1181 (1987).

Sundquist, W. E., and S. J. Lippard. The coordination chemistry of platinum anticancer drugs and Related Complexes with DNA, *Coordn. Chem. Rev., 100*, 293–322 (1990).

Problems

1. Saturation kinetics of CuZn-SOD can be observed at low temperatures ($\leq 5°C$). Under these conditions, Michaelis-Menten parameters have been evaluated. At pH 9.5, $k_{cat} \sim 1 \times 10^6 \text{ s}^{-1}$ and $K_M \sim 3.5 \times 10^{-3}$ M. Calculate the average rate at which superoxide binds to SOD.

   The magnitude of k_{cat} was found to decrease in the presence of D_2O and a solvent isotope effect for $(k_{cat})_H/(k_{cat})_D \sim 3.6$ was determined. However, the average rate of superoxide binding was not affected. The magnitude of k_{cat} was found to increase in the presence of the general acid ND_4^+. It was further found that, for the zinc deficient enzyme, the values of k_{cat} and K_M were lowered by only a factor of two or less and that the average rate of superoxide binding was effectively unperturbed.

   Discuss these kinetic data in quantitative detail, with specific reference to the mechanism proposed for SOD in Figure 8.3. Is the data consistent with the mechanism?

[Fee, J. A. and C. Bull, *J. Biol. Chem.* **261**, 13000–13005 (1986)]

2. (a) Suggest a plausible molecular mechanism for the reaction of superoxide with a Mn-containing superoxide dismutase.

   (b) There is an ionizable site close to the Mn ion. Derive the form of the Nernst equation that clearly shows the pH-dependence of E°. At high pH, would you expect the E° value to be more positive or more negative? Explain your reasoning.

   (c) An inactive form of the enzyme has been spectroscopically characterized.

   side-on   end-on

   It has been proposed that this arises from the formation of a side-on bonded metal-peroxo complex, which is known to exhibit weak absorbance bands ($\varepsilon \sim 10^2 - 10^3 \text{ M}^{-1} \text{ cm}^{-1}$). Such reactions are favored by hydrophobic environments that lack protons. Provide an explanation for the optical spectra for the dead-end Mn-SOD complex and two model complexes in the solvents noted (MnEDTA complexes have a vacant coordination site).

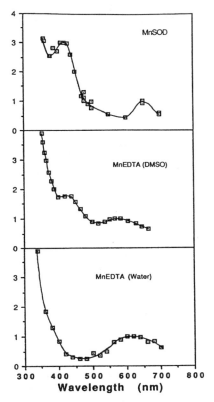

Discuss also the origin and extinction coefficient for the absorbance bands. If the rate of formation and decay of the cyclic peroxo complex for bound superoxide are 650 s^{-1} and 10 s^{-1}, respectively, estimate the equilibrium constant for the reaction,

$$Mn^{2+}.O_2^{-\cdot} \rightleftharpoons Mn^{3+}O_2^{2-}$$

[Bull, C. et al., *J. Am. Chem. Soc. 113*, 4069–4076 (1991), Figure is reprinted with permission.]

3. The axial His ligand to the heme of cytochrome c peroxidase is hydrogen bonded to the buried aspartate, Asp-235. The Asp is believed to modulate the characteristics of His as a heme ligand. Asp-235 also forms a hydrogen bond to the free radical center Trp-191 and helps to align the Trp ring in a close interaction with the axial histidine, most likely serving to facilitate electron transfer between Trp and the heme. Provide an explanation for the following observations.

(a) The mid-point potentials (E_m) for the recombinant native, D235E, D235N, and D235A mutants are -183 mV, -113 mV, -79 mV, and -78 mV, respectively.

(b) The E_m values show a considerable pH dependence. Also, discuss the observed trends for native and mutant CCP. Why do the potentials decrease above pH 6.5 for each protein, but level off below pH 6.5 for the D235N, and D235E mutants?

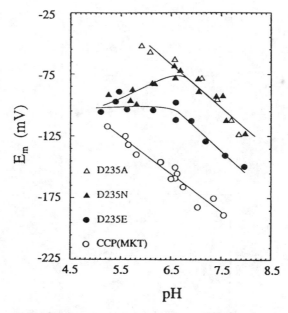

(c) The zero-field splitting parameter varies as follows (given as $2D/k$ in units of Kelvin): 29.6, 26.7, 21.6, and 22.3 for the recombinant native, D235E, D235N, and D235A mutants, respectively.

[Goodin, D. B. and D. E. McRee, *Biochemistry 32,* 3313–3324 (1993), Figure is reprinted with permission.]

4. The stability constants for Hg(glutathione)$_2$ and Hg(cysteine)$_2$ complexes at physiological pH and 25 °C are 9×10^{40} and 2×10^{40}, respectively. Binding of a third ligand to form the tricoordinate complex has been detected by use of ^{13}C NMR measurements. For glutathione, the equilibrium constant for binding of the third ligand is 1.95×10^3.

   (a) Draw complete chemical structures for the bis-coordinate glutathione and cysteine complexes of Hg^{2+}. Show the coordination geometry.

   (b) Glutathione is a major binding site for Hg^{2+} in human erythrocytes (red blood cells). Estimate the fraction of mercuric ion that exists as Hg(glutathione)$_3^-$ if the concentration of Hg^{2+} is 1 mM and the concentration of glutathione is (i) 1 mM, and (ii) 7 mM.

[Cheesman, B. V. et al., *J. Am. Chem. Soc. 110,* 6359–6364 (1988)]

5. Cisplatin shows several prominent side effects that include kidney and gastrointestinal problems and nausea. These arise as a result of the high dosages required and reactivity with a variety of other biomolecules, including the inhibition of enzymes. To counteract these effects, second generation Pt-

drugs have been developed that show similar therapeutic activities, but at lower dosages.

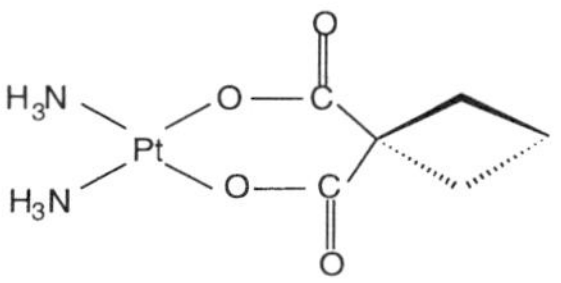

carboplatin

spiroplatin

iproplatin

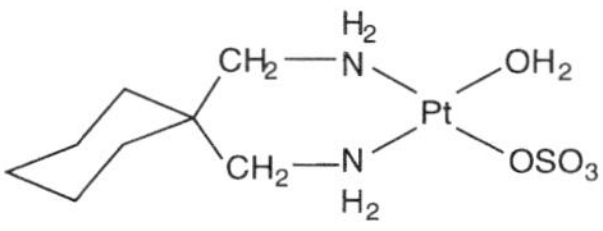

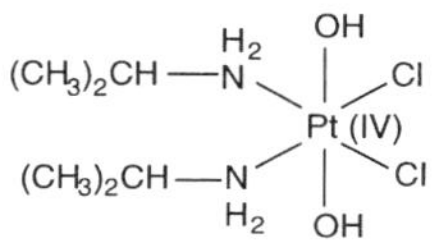

What structural and chemical characteristics of these molecules can be used to rationalize their enhanced therapeutic properties.

[Pasini, A. and F. Zunino, *Angew. Che., Int. Ed. 26,* 615 (1987)]

6. (a) Complexes of ^{99m}Tc (γ; 6h) are particularly suitable as diagnostic agents, whereas those of ^{186}Re (β,γ; 89h) are more commonly used in radiotherapy. Suggest reasons why each isotope is well-chosen for its specific use as a radiopharmaceutical.

   (b) Estimate the fraction of each isotope that would remain after a period of 1 day.

   (c) Determine the identity of the daughter products (a - n) that would result from the following sequence of decays.

$$^{238}\mathrm{U} \xrightarrow{-\alpha} \mathbf{a} \xrightarrow{-\beta} \mathbf{b} \xrightarrow{-\beta} \mathbf{c} \xrightarrow{-\alpha} \mathbf{d} \xrightarrow{-\alpha} \mathbf{e} \xrightarrow{-\alpha} \mathbf{f} \xrightarrow{-\alpha} \mathbf{g} \xrightarrow{-\alpha} \mathbf{h} \xrightarrow{-\beta} \mathbf{i} \xrightarrow{-\beta} \mathbf{j}$$

$$\xrightarrow{-\alpha} \mathbf{k} \xrightarrow{-\beta} \mathbf{l} \xrightarrow{-\beta} \mathbf{m} \xrightarrow{-\beta} \mathbf{n} \xrightarrow{-\alpha} {}^{206}\mathrm{Pb}$$

Metal Complexes as Probes of Structure and Reactivity

Important biochemical problems may occasionally elude the best research efforts, not from the lack of insightful hypotheses to test, but from the absence of an appropriate experimental handle with which to monitor a key parameter (structural or activity) of the system under investigation. With problems that are of interest to inorganic biochemists, this frequently results from the absence of convenient physicochemical tools to probe an experimental variable or the lack of detailed structural information on the biochemical system of interest. Armed with a knowledge of chemistry, however, it is possible to devise new methodologies or strategies to address these points. In this chapter, we shall examine several examples where a detailed understanding of the chemical and spectroscopic characteristics of inorganic compounds has allowed manipulation of metal complexes to provide greater insight on an important structural or mechanistic problem. In some cases, these methods address and provide answers to questions that would not otherwise be tractable.

In the sections that follow, we shall see how metal complexes can be used to provide information concerning binding domains on nucleic acids, probe the coordination sites and chemical reactivities of "spectroscopically silent" ions, such as the alkali and alkaline earth metals, and determine distances between distinct metal cofactors.

9.1 Nucleic Acids

DNA-binding proteins are central to the regulation of genetic events. For example, in Chapter 7 we reviewed several examples of transcriptional activator and repressor proteins, while antibiotics and other complex ligands are also known to bind to specific sites on DNA. Figure 9.1 illustrates how the locations of these binding sites can be determined by *footprinting* experiments. A length of DNA that contains the protein binding sequence is end-labeled with radio-active ^{32}P-phosphate.[1] A hydrolytic enzyme (typically S1 nuclease) or a chemical reagent capable of nicking the deoxyribose-phosphate backbone is added in quantites that are low enough to result in a single nick for each length of ds DNA. On average, considering the entire population of DNA molecules, these nicks will be distributed over the length of the DNA backbone, with the exception of those regions protected by the bound protein or ligand.[2] A population of single-strand fragments is generated that can be separated according to size by gel electrophoresis. Only those fragments possessing a ^{32}P-radiolabel are visualized. Since sequencing gels are capable of single-base resolution, a ladder of bands is identified that map out the entire base sequence. However, the region corresponding to the binding site is protected and a gap is observed. This "footprint" identifies the binding domain. In the absence of protein or ligand, a continuous ladder of fragments is obtained.

9.1.1 Hydroxyl Radical Footprinting

Chemical reagents that nick the DNA backbone find wide utility as probes of DNA conformation and as the catalytic site in synthetic sequence-specific nucleases (see Figure 9.3). Traditionally, the DNA footprinting method has used enzymes called nucleases that cut the ribose-phosphate backbone of DNA by hydrolysis of the phosphodiester linkage. However, the backbone may also be cleaved by use of chemical reagents. As a result of their size these reagents offer the opportunity for greater resolution of the binding site. One of the most common and valuable reagents is also the smallest, the reactive hydroxyl radical, which results in backbone scission by degradation of the ribose ring. The chemistry underlying this method is quite old. In the 1950s, Fenton discovered the reaction between ferrous ion and hydrogen peroxide that now bears his name. Hydroxyl radical (HO$^{\cdot}$) is extremely reactive and readily cleaves the ribose-phosphate chain if formed in its vicinity.

$$Fe^{2+} + H_2O_2 \rightarrow Fe^{3+} + HO^- + HO^{\cdot} \qquad \text{(Fenton reaction)}$$

Normally, the EDTA salt of divalent iron is used, since [Fe(II)(EDTA)]$^{2-}$ does not bind to the negatively charged polynucleotide backbone, and so the structure of the DNA–ligand complex is unperturbed. Rather, the complex produces reactive radicals close to the DNA backbone that are free to diffuse over a limited length of the DNA molecule and probe solvent accessible regions of the backbone. The reagent has proven to be particularly effective for the high-resolution mapping of protein-binding sites on DNA (Figure 9.2).

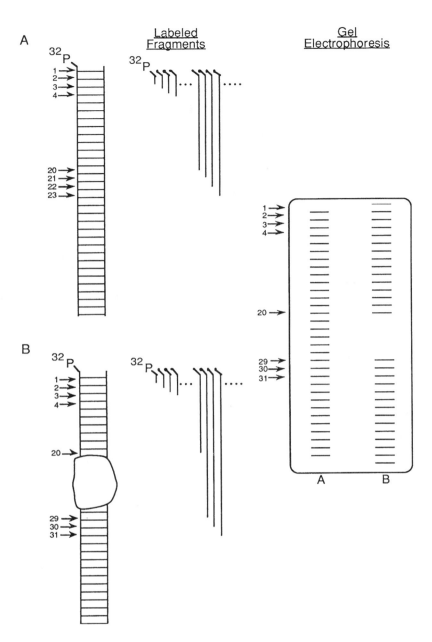

Figure 9.1 Schematic illustration of DNA footprinting. The protein or ligand protects the backbone from attack by a hydrolytic enzyme or chemical reagent. The principles of gel electrophoresis were outlined in Section 2.8. For DNA sequencing gels, a large voltage (2000–3000 V) pulls the DNA fragments through the gel according to their size. Each strand is nicked on average only once at any site that is not protected. By running a sequencing gel with single-base resolution, the footprint of the protein/ligand on the DNA backbone can be visualized. In practice, a separate experiment would be performed after the complementary strand was labeled.

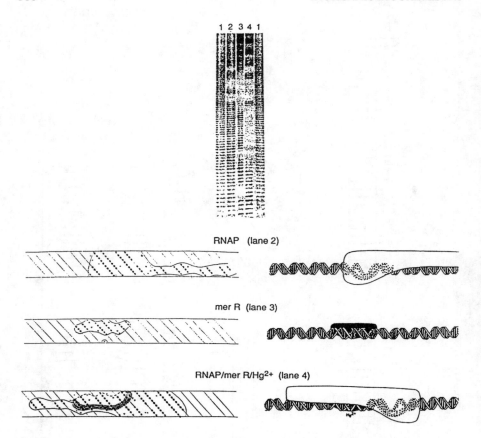

Figure 9.2 Hydroxyl radical footprint showing the interaction of the regulatory protein MerR and RNA polymerase (RNAP) with DNA. The outer lanes show the pattern from free DNA, and the regions protected by protein binding are clearly visible in the inner lanes: lane 1 DNA alone; lane 2 DNA + RNAP; lane 3 DNA + MerR; lane 4 DNA + MerR + RNAP + Hg^{2+}. Structural models deduced from this type of experiment are indicated below the footprinting gel. Regions of the phosphate backbone free from hydroxyl radical nicking are shown on the left and the corresponding protein–DNA complex on the right. For a more complete description of the biochemistry of this system, see Chapter 10. [Adapted from T. V. O'Halloran et al., *Cell, 56,* 119 (1989).]

9.1.2   DNA-Binding Cleavage Reagents

Some other reagents that perform similar chemistry are shown in Figure 9.3. Each of these possesses special structural features that target their use for specific functions. The oxidative chemistry of copper-phenanthroline (Figure 9.3d) is reasonably well developed. Unlike the Fenton chemistry described earlier, where free diffusible HO˙ is generated, the copper reagent forms an intimate complex with the minor groove of the polynucleotide and activates molecular

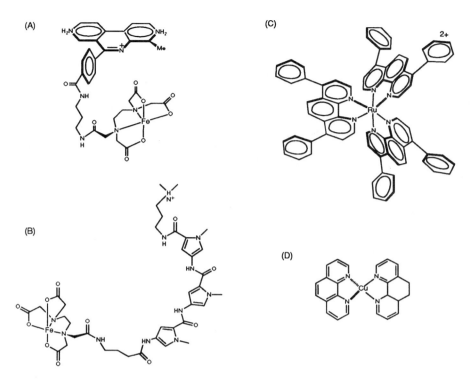

Figure 9.3 A variety of affinity cleavage reagents have been used to study DNA structure and ligand binding, for example, (A) methidiumpropyl-EDTA · Fe(II), (B) distamycin-EDTA·Fe(II), (C) tris(4,7-diphenylphenanthroline)ruthenium(II) (Ru(DIP)], (D) copper phenanthroline. Reagent (A) is general in its activity. In some cases sequence selectivity in the cleavage reaction can be obtained by using an appropriate DNA-binding ligand. For example, by virtue of the distamycin moiety, reagent (B) is specific for A/T tracts of DNA. Reagent (C) is a chiral complex that can exist as Λ or Δ enantiomers. The Δ enantiomer demonstrates high selectivity for right-handed B-DNA. These reagents cleave the DNA backbone in close proximity to their binding domain. Reagent (D) demonstrates DNA conformational specificity (B > A >> Z).

oxygen in situ. Evidence for minor groove binding comes from comparison of the binding footprint for the drug netropsin (minor groove binding) and the restriction enzyme *Eco* RI (major groove binding) to the self-complementary nucleotide 5'-d(CGCGAATTCGCG)-3'. DNA strand scission is blocked in the former case but not the latter. Figure 9.4 summarizes the chemistry of this reagent as it is currently understood. The reagent shows a cleavage pattern that is dependent on secondary structure, insofar as the rates vary in the order B > A >> Z. Both A- and Z-DNA are less able to form stable noncovalent complexes in the minor groove with the correct geometry for cleavage (see Figure 1.21). Z-DNA in particular is completely resistant to attack.

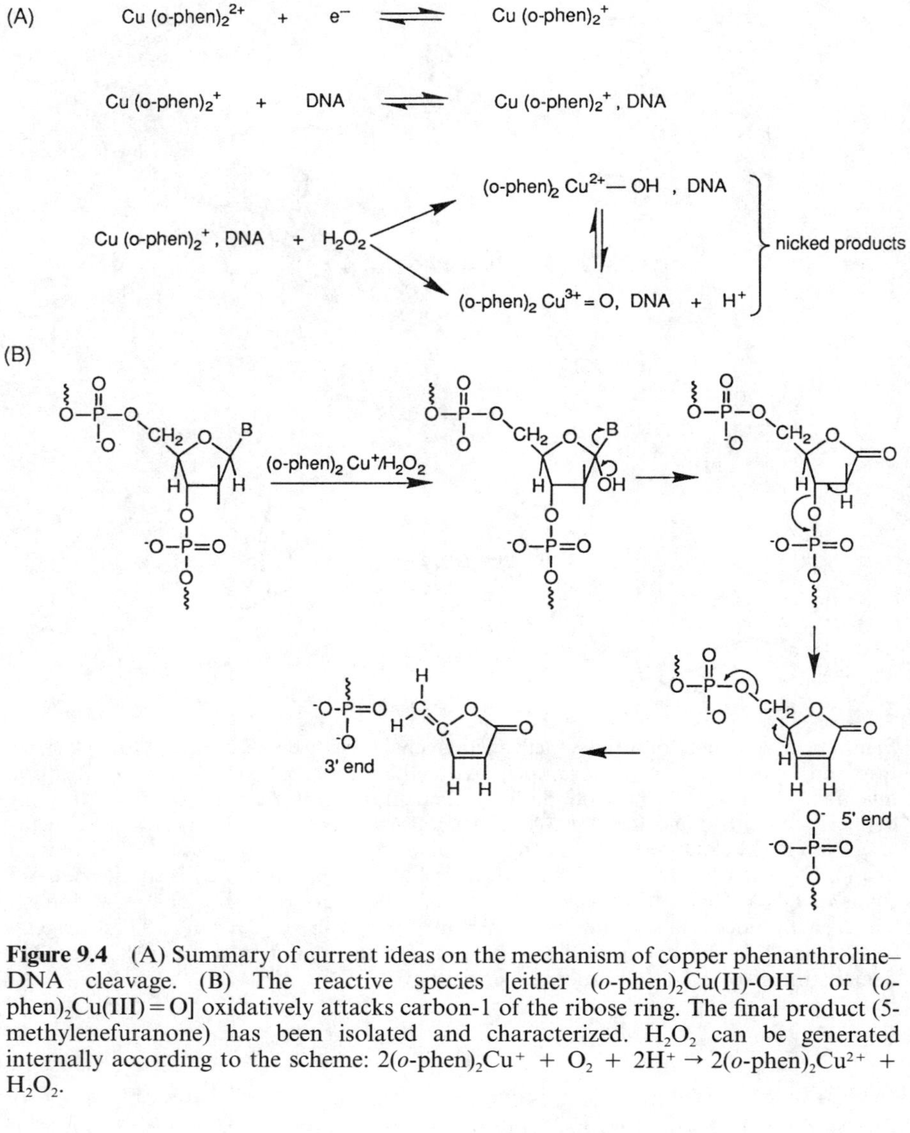

(A)

$$Cu\,(o\text{-phen})_2^{2+} \;+\; e^- \;\rightleftharpoons\; Cu\,(o\text{-phen})_2^+$$

$$Cu\,(o\text{-phen})_2^+ \;+\; DNA \;\rightleftharpoons\; Cu\,(o\text{-phen})_2^+,\,DNA$$

(B)

Figure 9.4 (A) Summary of current ideas on the mechanism of copper phenanthroline–DNA cleavage. (B) The reactive species [either $(o\text{-phen})_2Cu(II)\text{-OH}^-$ or $(o\text{-phen})_2Cu(III)=O$] oxidatively attacks carbon-1 of the ribose ring. The final product (5-methylenefuranone) has been isolated and characterized. H_2O_2 can be generated internally according to the scheme: $2(o\text{-phen})_2Cu^+ + O_2 + 2H^+ \rightarrow 2(o\text{-phen})_2Cu^{2+} + H_2O_2$.

9.1.3 Bleomycin

In Chapter 3 we saw some examples of ion carriers (ionophores) that serve as antibiotics by destroying transmembrane ion gradients, resulting in cell death. There are many other types of natural antibiotics, some of which function by specific cleavage of DNA. In some cases, cleavage may proceed by formation of a covalent linkage to the antibiotic molecule, whereas in others the reaction arises by proton abstraction chemistry from the sugar ring followed by fragmentation (e.g., see Figure 9.4). Figure 9.5 shows the structure and chemistry of one of the most thoroughly studied DNA-cleaving antibiotics. Bleomycin is a glycopeptide antibiotic and a potent anticancer agent.[3] The bithiazole rings and positively charged side chain direct the molecule to DNA. The metal-binding domain coordinates iron ion. In the presence of O_2 and a reducing agent, activated bleomycin attacks the ribose-phosphate backbone and abstracts the C4' proton from the ribose ring. The activated species is the one-electron re-duced product of the bleomycin–Fe^{2+}–O_2 complex. The structural and mecha-

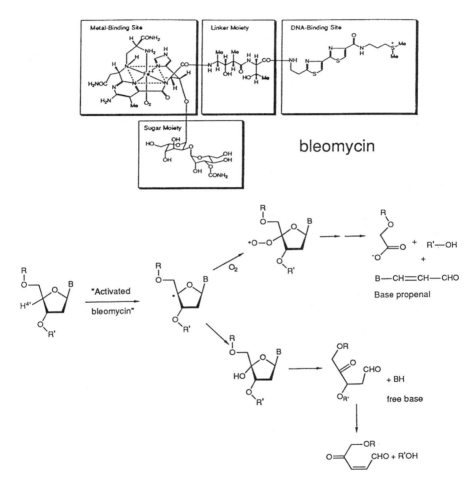

Figure 9.5 Degradative mechanisms for cleavage of the ribose–phosphate backbone after abstraction of the C4'-proton from the ribose ring. Activated bleomycin is produced by a further one-electron reduction of the reduced oxygen-bound complex, bleomycin–Fe^{2+}–O_2. The coordination around the iron center remains ill-defined and the ligand scheme shown is only one of several possibilities.

nistic chemistry of the bleomycin family of antibiotics is discussed more fully in Chapter 10.

Summary of Section 9.1

1. Footprinting is a valuable method for determining the position of DNA-binding sites for proteins or ligands. Either enzymatic or chemical reagents may be used to nick the DNA backbone at accessible sites.
2. The mechanistic chemistry of DNA cleavage reactions (hydrolytic and radical) are reasonably understood for a variety of natural (bleomycin) and synthetic (copper-phenanthroline) reagents.
3. Synthetic reagents have been developed that recognize DNA on the basis of conformation (A, B, Z) or by linking inorganic complexes to sequence selective ligands.

9.2   Metal Substitution: A Probe of Structure and Reactivity

Physicochemical methods have been of great value in probing the structural environments and reactivities of many metal centers in proteins, enzymes, nucleic acids, membranes, and other biological structures. Unfortunately, not all metal ions are amenable to direct investigation. In particular, the most abundant metal ions in biology (Na^+, K^+, Mg^{2+}, Ca^{2+}) and also Zn^{2+} are difficult to study by optical, electrochemical, or magnetic resonance methods. In these cases, it may be possible to substitute the "spectroscopically silent" cation with a transition metal that can then be studied by one or more of the methods described in Chapter 2. Ideally, the biochemistry of the metal-substituted molecule should be similar to that of the native system. That is, the derivative should be *functional*. Several factors enter into the choice of a metal replacement, including (1) size, (2) coordination number and geometry, (3) ligand preference, (4) kinetic and thermodynamic binding parameters, and (5) occurrence of an appropriate physicochemical property of the probe ion that may be readily monitored. Table 9.1 notes some important characteristics of common replacements. The chemistry of the four metal cofactors K^+, Zn^{2+}, Mg^{2+}, and Ca^{2+} have been especially studied by use of substitution methodology.

9.2.1   Potassium(I)

Both Tl^+ and Rb^+ have proved to be effective probe ions for studies of potassium biochemistry. Only Tl^+ is a spectroscopic probe, however, and studies of Rb^+ have focused on comparison of its activity relative to K^+. Section 9.2.4 (including Figure 9.11) gives a specific example of the use of K^+ probes in studies of the enzyme pyruvate kinase.

Table 9.1   Comparison of Some Physicochemical Properties of Na^+, K^+, Mg^{2+}, Ca^{2+} and Zn^{2+} and Their Common Probe Ions

| | Ionic Radius $(\text{Å})^a$ | Coord'n Number | Probe | Probe Ions | | |
				Ionic Radius (Å)	Coord'n Number	Useful Physicochemical Properties
K^+	1.52–1.65	6–8	Rb^+	1.66–1.75	6–8	Similar chemical activity
			Tl^+	1.64–1.75	6–8	^{205}Tl, $I = \frac{1}{2}$, 100% nat. abun.
Mg^{2+}	0.86	6	Mn^{2+}	0.97	6	High-spin d^5, EPR, relaxation agent in NMR, similar chemical activity.
			Li^+	0.73–0.90	4–6	^7Li, $I = \frac{3}{2}$, 92.6% nat. abun.
Ca^{2+}	1.14	6–8	Ln^{3+}	0.93–1.15	$\geqslant 6$	Luminescence, relaxation agent in NMR
Zn^{2+}	0.74–0.88	4–6	Co^{2+}	0.72–0.79	4–6	Paramagnetic probe in NMR, electronic absorption, electrochemistry, similar chemical activity

^{a} Ionic radii may show slight variations in various tables throughout the text. This reflects the variety of sources used and reasonably reflects the errors in such estimates.

9.2.2   Zinc(II). Cobalt as a Chemical and Spectroscopic Probe

Numerous structural studies have shown that zinc metalloproteins typically bind Zn^{2+} in a tetrahedral ligand environment. The metal-free protein (called the apoprotein) can be obtained by dialysis against a ligand that has a high affinity for Zn^{2+} (e.g., EDTA or, 1,10-phenanthroline) and reconstituted with a variety of other metal ions to give a series of metal-substituted derivatives. Divalent cobalt has been widely used as a probe for zinc ion, since it displays

EDTA^{4-}

1,10 - phenanthroline

both similar coordination chemistry and ligand preferences in comparison with Zn^{2+} and exhibits useful magnetic and optical characteristics. In Section 2.2, we saw that there is a large change in extinction coefficient for d–d absorption bands accompanying a change from octahedral to tetrahedral coordination. Remember that d–d bands are symmetry forbidden in octahedral complexes but are formally allowed in a variety of four and five coordinate complexes (Section

2.2.1). The increase in absorption, when Co^{2+}(aq) binds to a protein, allows facile quantitation of the number of metal-binding sites and can be used as an indirect method for determining binding constants of other metal ions by competition experiments. Consider, for example, the following two reactions:

$$E + Zn^{2+} \rightleftharpoons EZn^{2+} \quad (K_{Zn})$$

$$E + Co^{2+} \rightleftharpoons ECo^{2+} \quad (K_{Co})$$

and so for the competitive displacement,

$$ECo^{2+} + Zn^{2+} \rightleftharpoons EZn^{2+} + Co^{2+} \quad (K = K_{Zn}/K_{Co})$$

Occasionally, Zn^{2+} binds too strongly for the determination of K_{Zn} by direct titration studies. Assuming K_{Co} and K can be determined by optical titration experiments, K_{Zn} can be evaluated from such a competition experiment.

Figure 9.6 shows the titration of liver alcohol dehydrogenase with Co^{2+}. The relatively intense d–d absorption bands originate from enzyme-bound Co^{2+}. Since the intensity and λ_{max} of these optical transitions are sensitive to the binding of inhibitors, substrate analogs, and the presence of ionizable functionality in the active site (Figure 9.7), they can be used to evaluate binding constants and pK_a values.

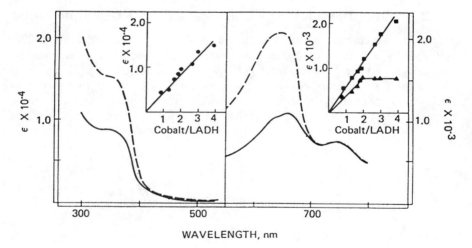

WAVELENGTH, nm

Figure 9.6 Titration of the structural and catalytic zinc sites in liver alcohol dehydrogenase (LADH)Zn_2Zn_2 with Co^{2+}. Since the apoenzyme is unstable, substitution is carried out by equilibrium dialysis against 0.2 M Co^{2+}. The structural pair is selectively replaced at pH 5.9. Both pairs are replaced at pH 5.4. Absorption spectra of (LADH)Co_2Zn_2(——) and (LADH)Co_2Co_2(----) are shown. The insets record absorbances at 340 nm (●), 655 nm (■), and 740 nm (▲). The 740-nm band is responsive to the structural site only. The results accurately quantitate the number of sites per protein and the extinction coefficient is in agreement with the known tetrahedral coordination. [Reprinted in part with permission from A. J. Sytowski and B. L. Vallee, *Biochemistry, 17,* 2850; copyright (1978) American Chemical Society.]

(A)

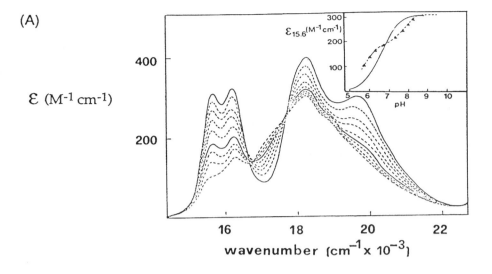

(B)

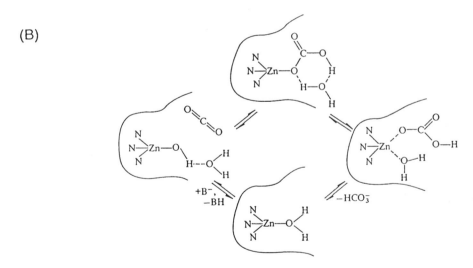

Figure 9.7 (A) Electronic absorption spectra of cobalt carbonic anhydrase as a function of pH. The absorption data at 15,600 cm^{-1} is plotted as an inset and clearly shows the complexity of the pH dependence. The relative intensities indicate tetrahedral or square pyramidal (rather than octahedral) coordination, and the appearance of multiple bands suggests severe distortion of the coordination environment. The pH dependence is indicative of at least two ionizable groups (possibly a bound H$_2$O and a protein side chain). These relate to a putative mechanism shown in (B). [Adapted from I. Bertini et al., *Structure and Bonding, 48,* 45–92 (1982).]

Divalent cobalt is paramagnetic (d^7, $S = 3/2$), however, rapid relaxation of the electrons results in broad EPR signals that are often difficult to resolve and interpret. This rapid electronic relaxation may be put to good use, since Co^{2+} is an excellent paramagnetic shift ion in protein NMR studies (e.g., see Figures 8.2 and 8.4). As a general rule, metal ions that show broad EPR spectra can often be used as paramagnetic probes in NMR studies, and vice versa. Metal ions with slower electronic relaxation rates (e.g., Cu^{2+}, d^9 $S = 1/2$) give sharper EPR features, however, the slow relaxation of the electron spins provides an efficient relaxation mechanism for neighboring nuclei in NMR experiments, and so NMR resonances are often broadened beyond detection. For this reason, paramagnetic metal ions with short relaxation times, such as Co^{2+}, Eu^{3+}, and Yb^{3+}, are often used as shift reagents in NMR studies of proteins.

An ideal probe ion will produce a new macromolecule with indistinguishable chemical properties from the native molecule. In reality, this is seldom, if ever, observed. For example, the cobalt-substituted derivative of carboxypeptidase A is more active than the native zinc enzyme toward peptide substrates, although the activities of both enzymes toward esters is approximately the same (Table 9.2). The retention of significant biological activity is, however, a reasonable criterion for judging an ion to be a useful probe of both structure and reaction chemistry.

9.2.3 Magnesium(II). Manganese as a Chemical and Spectroscopic Probe

Divalent manganese is the most widely used probe of magnesium chemistry. Relative to the other first-row transition metal ions, divalent manganese shows the greatest similarity to the chemical properties of magnesium ion in terms of its ligand preference and geometry, exchange rates, and a propensity for both inner and outer sphere complexation. In many respects, the biological chemistry of Mg^{2+} and Mn^{2+} are closely related, and it is likely that nature's choice of Mg^{2+} over Mn^{2+} reflects the relative abundances of each ion in vivo ($[Mg^{2+}] \sim 10^{-3}$ M; $[Mn^{2+}] \sim 10^{-8}$ M). However, there are certain situations where one or the other ion is specifically required. Examples include the Mn-cluster in the

Table 9.2 Hydrolysis of Amide and Ester Substrates by Metallocarboxypeptidase[a]

Metal	Amide Substrate		Ester Substrate	
	k_{cat} (min^{-1})	10^{-3} K_m^{-1} (M^{-1})	10^{-4} k_{cat} (min^{-1})	K_m^{-1} (M^{-1})
Co^{2+}	6000	1.5	3.9	3300
Zn^{2+}	1200	1.0	3.0	3000
Mn^{2+}	230	2.8	3.6	660
Cd^{2+}	41	1.3	3.4	120

[a] From Auld and Holmquist, *Biochemistry, 13*, 4355 (1974).

oxygen-evolving complex and Mg-chlorin derivatives in the reaction center, both of which are located in the photosynthetic apparatus of plants (see Section 5.3.4). In the oxygen-evolving complex, a redox active metal is required, while the magnesium ion in chlorophyll modulates the redox properties of the ring. The metal ion should not itself be redox active. Generally, redox processes will show high specificity for manganese ion, with acid–base chemistry showing a greater degree of flexibility. Many magnesium-dependent enzymes are also activated by manganese, while manganese transport systems appear to act as backups to magnesium transport.

Manganese(II) cannot be used as an optical probe since d–d transitions from the electronic ground state are spin forbidden and difficult to observe.[4] However, the highly symmetric electronic structure of the ion leads to inefficient spin relaxation, and so the excited spin states are sufficiently long-lived that sharp EPR features may be observed for Mn^{2+}(aq). When bound to an enzyme, there is a significant decrease in the intensity of the EPR signal.[5] This fact provides a simple method for estimating the binding constant to a biological macromolecule, since the concentrations of enzyme-bound and free Mn^{2+} can be readily determined from the intensity of EPR signals (Figure 9.8).

Manganese(II) has also been widely used as a relaxation probe in magnetic resonance studies. The paramagnetic metal ion efficiently relaxes neighboring nuclei, which is reflected by a decrease in the relaxation times T_1 and T_2. In Chapter 2 (Figure 2.15), we saw how the determination of the paramagnetic contribution to T_1 [denoted $(T_1)_p$] provides a good estimate of the distance (r) between Mn^{2+} and specific nuclei on an enzyme or substrate molecule.[6] If two paramagnetic metal ions are in close proximity, the interactions between them may be detected in EPR spectra. Again, there is a $1/r^6$ relationship governing these spin–spin interactions, where

$$\frac{1}{(T_1)_p} \alpha \frac{1}{r^6}$$

9.2.4 Inert Complexes as Probes of Structure and Mechanism

Nucleotide di- and triphosphates exist in cells mainly in the form of magnesium chelates. Consideration of the structures of these bidentate complexes quickly shows that there are several possible isomeric forms (Figure 9.9) that are in rapid equilibrium, only one of which is likely to adopt the proper stereochemistry for reaction. Since magnesium equilibrates rapidly in solution ($k_{ex} \sim 10^5$ s^{-1}), it is not possible to deduce if the enzyme shows a preference for one isomer over another. In the early 1970s, Cleland introduced the use of inert Co(III) and Cr(III) nucleotide complexes as probes of the biological chemistry of ATP-dependent enzymes.[7] By virtue of their electronic configurations (low-spin d^6 and d^3, respectively), trivalent cobalt and chromium compounds are exchange inert (possessing ligand exchange rates $\sim 10^{-11}$ s^{-1}), and so the various isomeric

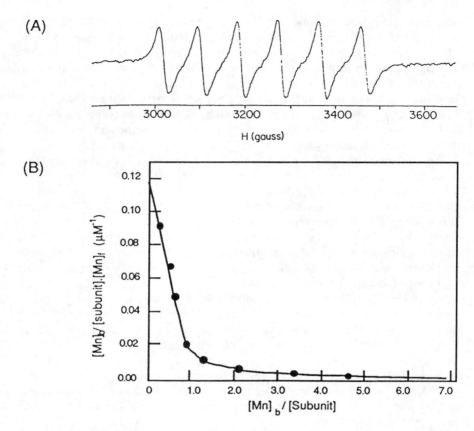

Figure 9.8 (A) The six-line EPR spectrum of Mn^{2+}. In the presence of a metal-binding protein, the intensity is greatly diminished. (B) Titration plot obtained from the addition of Mn^{2+} (aq) to the enzyme mandelate racemase. The concentrations of free and bound Mn^{2+} ($[Mn^{2+}]_f$ and $[Mn^{2+}]_b$, respectively) were established from EPR spectra. Since the manganous ion is essentially EPR silent when bound to the protein, $[Mn^{2+}]_b = [Mn^{2+}]_{total} - [Mn^{2+}]_f$. The Scatchard plot shows that the enzyme binds one Mn^{2+}. [Adapted from E. T. Maggio et al., *Biochemistry, 14*, 1131 (1975).]

forms may be chromatographically separated for use as substrates or inhibitors to test the stereochemical course of enzymatic reaction pathways. Cobalt (III) and chromium (III) are similar in size to magnesium (II) and show a preference for octahedral coordination. Both are amenable to optical spectroscopy, although Co(III) is diamagnetic, whereas Cr(III) is paramagnetic. It is instructive to consider the preparation and use of these complexes.

The Λ and Δ isomers of β,γ-Co(NH$_3$)$_4$ATP can be separated by use of an enzymatic reaction summarized in Figure 9.10.[8] Hexokinase catalyzes the phosphorylation of glucose; however, only one of the two diastereoisomers (Δ or Λ) of β,γ-Co(NH$_3$)$_4$ATP serves as a substrate and the unreacted isomer can be

$$(\Lambda) - \beta, \gamma - MATP^{2-}$$

$$(\Delta) - \beta, \gamma - MATP^{2-}$$

Figure 9.9 Two enantiomeric forms of a generalized β,γ-MATP^{2-} chelate complex of the divalent cation M^{2+}. Other structural forms exist for the α,β-complex, and the mono- and tridentate forms.

chromatographically separated. Note that, since Co(III) is inert, the phosphorylated product is not released (see the schematic below).

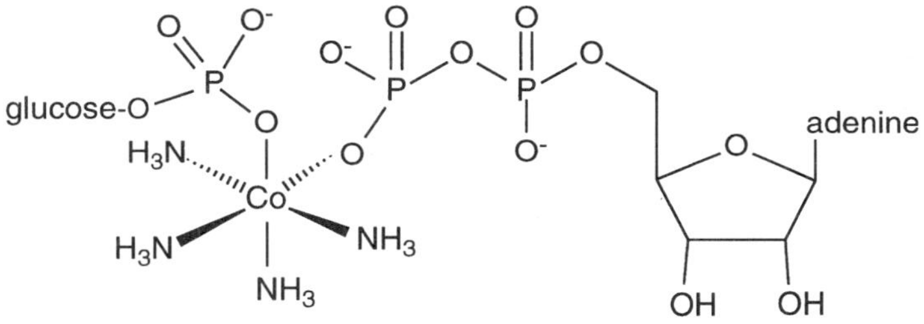

The active isomer [Λ β,γ-Co(NH$_3$)$_4$ATP] could be obtained by reacting the product complex Co(NH$_3$)$_4$(glucose-6-P)-ADP with hexokinase (effectively pushing the above reaction in reverse). After oxidation by HIO$_4$ and treatment with aniline (pH 5) to open the ribose ring, the absolute configurations of the elimination products Δ and Λ Co(NH$_3$)$_4$PP were established by crystallography and characterized by circular dichroism spectroscopy and NMR. In this way, it was firmly established that the enzyme hexokinase used in the reaction scheme in Figure 9.10 was specific for the Λ isomer.

These reagents have been used to study a number of enzymatic reactions that use MgATP as substrates. Application to the biochemistry of pyruvate kinase is particularly instructive, since several other aspects of metal substitution chemistry can be included. Pyruvate kinase is a tetrameric enzyme ($M_r = 237,000$) that catalyzes the reversible phosphorylation of ADP from phosphoenolpyruvate.

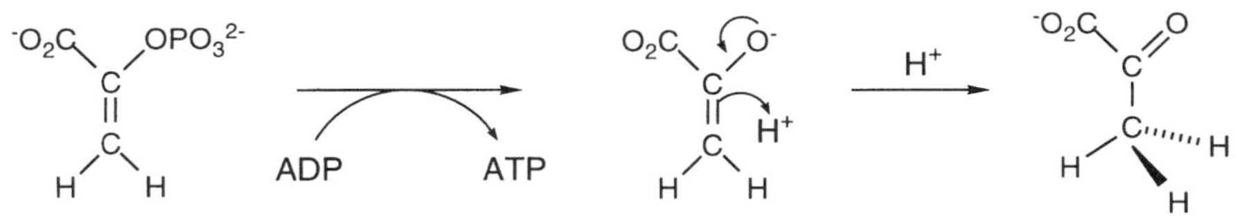

(A)

$$\Lambda, \Delta\text{-Co(NH}_3)_4\text{ATP} + \text{glucose} \xrightarrow{\text{yeast hexokinase}} \Lambda\text{-Co(NH}_3)_4\text{(glucose-6-P)-ADP} + \Delta\text{-Co(NH}_3)_4\text{ATP}$$

$$\Lambda\text{-Co(NH}_3)_4\text{(glucose-6-P)-ADP} \xrightarrow{\text{yeast hexokinase}} \Lambda\text{-Co(NH}_3)_4\text{ATP} + \text{glucose}$$

$$\Lambda\text{-Co(NH}_3)_4\text{ATP} \xrightarrow{\text{IO}_4^- / \text{aniline}} \Lambda\text{-Co(NH}_3)_4\text{PPP}$$

(B)

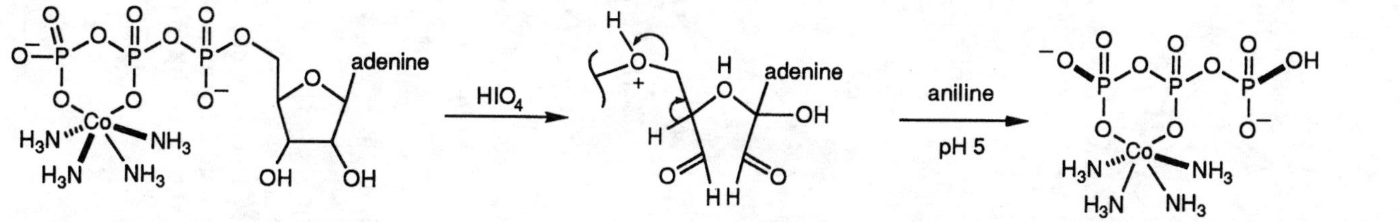

Figure 9.10 (A) Summary of the preparation of the pure Λ-Co(NH$_3$)$_4$ATP isomer from the glucose phosphate (glucose-P) adduct. (B) Periodate oxidation cleaves the ribose–phosphate junction to yield the triphosphate complex Λ-Co(NH$_3$)$_4$PPP. The absolute stereochemistry was confirmed by crystallographic studies of this triphosphate adduct.

Monovalent K^+ and two divalent Mg^{2+} are required for activity. Potassium and one magnesium are enzyme bound, while the remaining Mg^{2+} binds to the nucleotide. Two reactions may be monitored [(9.1) and (9.2)].

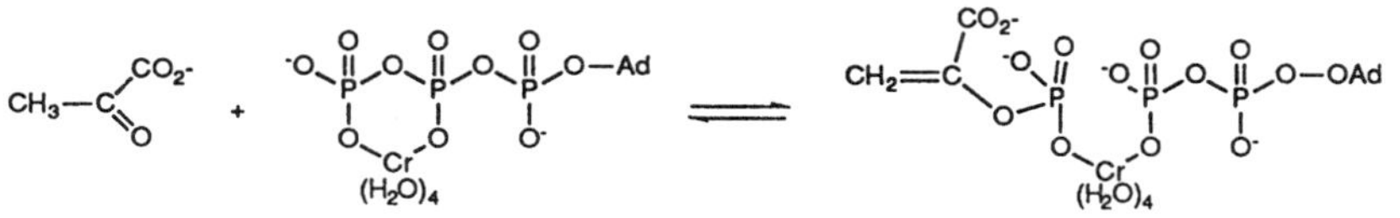

$$(9.1)$$

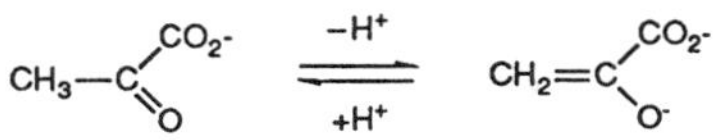

$$(9.2)$$

Only Δ β,γ-CrATP is active in the phosphoryl transfer reaction (9.1). It is also the most active isomer in promoting the enolization (9.2). These results indicate that the Δ isomer of the β,γ–MgATP complex is the active species in phosphoryl transfer to a substrate (S). Since α,β–MgADP is the product of the reaction, Mg^{2+} must migrate from the γ- to the α-phosphate.

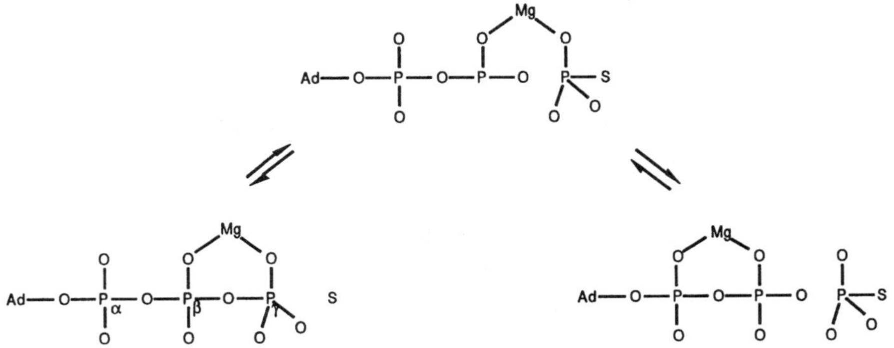

Inasmuch as the inert complex α,β-CrADP cannot undergo phosphorylation by phosphoenolpyruvate [reverse of equation (9.1), but with the chromium bridging the α and β phosphates], magnesium migration must occur before phosphoryl transfer from MgATP in the forward reaction and after phosphoryl transfer to MgADP in the reverse reaction.

A structural model for the enzyme-bound substrate can be established by use of paramagnetic probe metals [where Mn^{2+} substitutes for enzyme-bound Mg^{2+}, and Cr(III)ATP replaces Mg(II)ATP]. Since several metal probes and ligand atoms are NMR active ($^{205}Tl^+$ or $^7Li^+$ for $^{39}K^+$, and ^{31}P), a large number of internuclear distances defining the relative positions of the metal cofactors and substrates may be determined from relaxation measurements. The structural organization of the enzyme–substrate complex that has evolved from these studies is summarized in Figure 9.11.

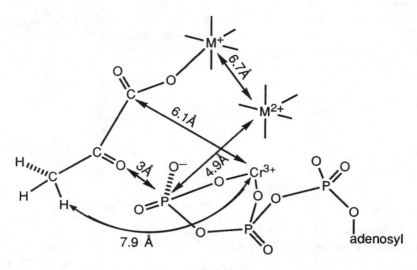

Figure 9.11 Locations of substrates and metal cofactors for pyruvate kinase established by multinuclear NMR. The NMR-active metals M^+ = Tl^+, Li^+ were used in conjunction with relaxation agents $Cr(H_2O)_4ATP$ and Mn^{2+} to establish metal-metal, ^1H-metal, and ^{13}C-metal distances between the metal cations, substrates, and enzyme. This illustration summarizes many experiments from the work of Mildvan and coworkers (see Further Reading section).

9.2.5 Calcium (II). Lanthanide Ions as Chemical and Spectroscopic Probes

Lanthanide ions have proven to be valuable probes of Ca^{2+} biochemistry and, occasionally, Mg^{2+} biochemistry.[9] The chemical characteristics of the lanthanides (Lns) that make this possible are summarized in Table 9.1. Although the ionic radii are similar, the coordination number of the Lns (7–9) tends to be higher than that of Ca^{2+} (6–8), reflecting the higher oxidation state of the former. Both Ca^{2+} and Ln^{3+} are hard metal ions that prefer oxygen ligation. In every calcium-binding protein the ligand environment is made up from O-donors, including carboxylate and hydroxyl residues (Glu, Asp, Thr, Ser), backbone carbonyl and H_2O. Two important differences in the metal–protein interactions of Ca^{2+} and Ln^{3+} are the metal association constants and exchange rates. There is a linear correlation between association constants for Ca^{2+} and Ln^{3+}, although the latter bind ligands by a factor of 10^4 or 10^5 more tightly [e.g., compare $K_a(Ca^{2+}) \sim 10^3$–10^9 M^{-1} and $K_a(Ln^{3+}) \sim 10^7$–10^{14} M^{-1}]. The water exchange rate for Ca^{2+} ($k_{ex} \sim 5 \times 10^9$ s^{-1}) is, however, much larger than that for La^{3+} ($k_{ex} \sim 10^7$ s^{-1}) as a result of the higher charge–radius ratio for Ln^{3+}. When Ln^{3+} replaces Ca^{2+} at a catalytic site, the reaction rates decrease, which partly reflects the lower exchange rates.

Lanthanides have served three major functions in studies of biological molecules: (1) heavy-atom derivatives in X-ray diffraction, (2) luminescent probes,

(3) relaxation and shift reagents in magnetic resonance. The former has been used extensively as a method of solving the phase problem in structure elucidation by diffraction techniques. Our attention will focus on the latter two topics, which find greater use in solution studies.

Luminescence

Lanthanide ions can be promoted to an excited electronic state by absorption of light. The ion might then return to the electronic ground state by either radiative or nonradiative pathways (Section 2.2.4). Nonradiative decay arises through molecular vibrations of the inner sphere ligands and neighboring solvent molecules. For lanthanide ions, the lowest electronic states derive from $4f^n$ configurations (e.g., Er^{3+} $1s^2 2s^2 2p^6 3s^2 3p^6 4s^2 3d^6 4s^2 3d^{10} 4p^6 5s^2 4d^{10} 5p^6 4f^9$).[10] The absorption and emission of light through f–f transitions is formally electric dipole forbidden, and so the extinction coefficients for absorption bands are small (<10 M^{-1} cm^{-1}) and emission lifetimes are long. Nonradiative decay pathways are inefficient, since the $4f$ orbitals are inner orbitals that are shielded from the environment, do not significantly engage in bonding, and so are notstrongly coupled to vibrational decay pathways. Europium(III) and terbium(III) are particularly useful probes, since they luminesce strongly at room temperature when complexed to ligands other than H_2O and have long excited-state lifetimes in the range of 100–3000 μs (Figure 9.12). Both emission and excitation spectra and lifetime measurements can be routinely collected, but laser excitation is required as a result of the weak absorption. Luminescence studies of lanthanide-modified proteins have found application in the determination of the number of water molecules bound to the active-site metal ion and as a probe of aromatic residues and/or other metal ions neighboring the metal cofactor in the active site by determining distances from energy transfer measurements. The physical basis for both of these applications is described later.

Although the coupling of the $4f$ orbitals to the inner-sphere ligands is weak, it is strong enough to give an observable isotope effect when monitoring emission lifetimes in either H_2O or D_2O. By using the difference in decay rates measured in H_2O or D_2O, the number of metal-bound water molecules can be evaluated. The basis for the method is summarized in Figure 9.12. By studying a number of structurally characterized crystalline solids, where the number of water molecules coordinated to Ln(III) was known, a simple relationship between the number of coordinated waters (q) and the difference in reciprocal excited-state lifetimes $[(\tau_{H_2O})^{-1} - (\tau_{D_2O})^{-1} = \Delta\tau^{-1}]$ has been established.

$$q_{Ln} = A_{Ln}[(\tau_{H_2O})^{-1} - (\tau_{D_2O})^{-1}] \tag{9.3}$$

In equation (9.3), A_{Ln} is a characteristic constant ($A_{Eu} = 1.05$, $A_{Tb} = 4.2$ for τ^{-1} in units of ms^{-1}) for each lanthanide. Studies of europium(III) luminescence from thermolysin suggest one water molecule bound to site 1 and three or four each at sites 3 and 4, in agreement with crystallographic data (Figure 9.13).

The use of Ln emission in energy transfer experiments is illustrated in Figure

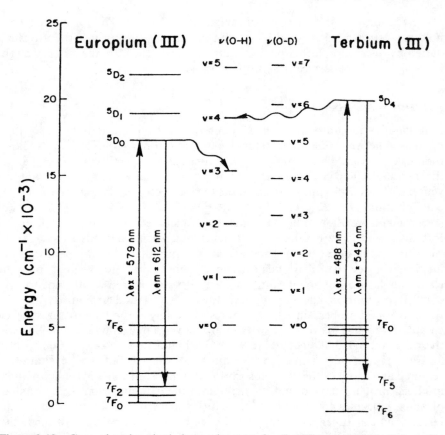

Figure 9.12 Ground and excited electronic states for Eu(III) and Tb(III). Absorption and emission transitions are shown. Radiationless transitions arise through coupling to the O–H (O–D) vibrational states (wavy lines). Coupling is more efficient with H_2O since there is better overlap with lower lying vibrational states. The energy level diagram for Tb(III) has been displaced slightly to align the v = 0 vibrational levels for v(OH) and v(OD). [Reprinted with permission from W. W. Horrocks and D. R. Sudnick, *Acc. Chem. Res., 14,* 384; copyright (1981) American Chemical Society.]

9.14. If the fluorescence emission band of an excited-state ion or chromophore overlaps with the absorption band of another ion or chromophore, energy transfer between these two centers is possible. The basic equations follow.

$$R = R_0[(1 - E_T)/E_T]^{1/6} \tag{9.4}$$

In equation (9.4), R is the intersite distance, R_0 is the distance at which energy transfer is 50 percent efficient, and E_T is the efficiency of depopulation of the excited state by energy transfer. E_T can be determined from the decay lifetime in the presence (τ_T) and absence (τ) of the transfer agent. R_0 (in units of nanometers) can be obtained from equation (9.6),

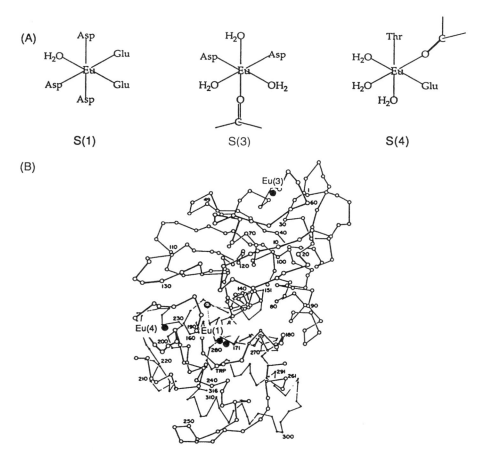

Figure 9.13 (A) Lifetime measurements for Eu(III) centers located at calcium-binding sites S(1), S(3), and S(4) in thermolysin provided the following data. The values for $(\tau_{H_2O})^{-1}$ and $(\tau_{D_2O})^{-1}$ were 1.78 ms^{-1} and 0.61 ms^{-1} for S(1). Sites S(3) and S(4) show comparable decay rates, reflecting similar numbers of H$_2$O ligands. In H$_2$O, biexponential behavior was observed and decay rates of 4.86 or 3.98 ms^{-1} were obtained for $(\tau_{H_2O})^{-1}$ since the decay rates for sites S(3) and S(4) are different. In D$_2$O, a unique rate of 1.00 ms^{-1} was obtained for S(3) and S(4). This gives q_{Eu} = 1.2 for S(1), and q_{Eu} = 3.1 and 4.0 for S(3) and S(4), respectively. The value 4 for S(4) accounts for 3 × H$_2$O and the OH from a threonine residue. [Adapted from A. P. Snyder et al., *Biochemistry 20*, 3334; copyright (1991) American Chemical Society.] (B) This data is in good agreement with crystallographic data on the Eu-derivative of thermolysin. [From B. W. Matthews and L. H. Weaver, *Biochemistry, 13*, 1719 (1974).]

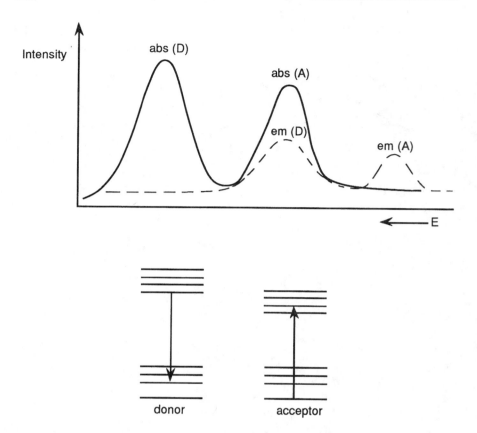

Figure 9.14 Schematic illustration of the energy-transfer experiment. Absorption (———) and emission (– – – –) spectra are shown. The fluorescent emission from the donor (D) can be absorbed by the acceptor (A). The net result is a change from excited → ground-state donor, and ground → excited-state acceptor.

$$E_T = 1 - \tau_T/\tau \tag{9.5}$$

$$R_0 = 9.79 \times 10^2 (J\eta^{-4}\kappa^2\Phi_D)^{1/6} \text{ nm} \tag{9.6}$$

where J is the spectral overlap integral of the absorption and emission spectra of the acceptor and donor, respectively,[11] κ^2 is an orientation factor that is normally given the value 2/3 for an isotropic solution, η is the refractive index,[12] and Φ_D is the fluorescence quantum yield of the energy donor in the absence of acceptor.

Since the thermolysin metal-binding sites $S(1)$ and $S(3)/S(4)$ can each be populated with distinct metals, the energy-transfer studies described in Table 9.3 were made possible. The $S(1) - S(4)$ distance is large, and so energy transfer is dominated by the $S(1) - S(3)$ pair. Thermolysin also contains a Zn^{2+} ion that

Table 9.3 Experimental Determination of Distances (r) Between Sites S(1) and S(4) of Thermolysin for Various Donor–Acceptor Ln(III) Ion Pairs

Donor	Acceptor	τ^{-1}, ms^{-1}	Efficiency E	$J \times 10^{14}$ (cm^6 mol^{-1})[a]	R_0 (Å)[b]	r (Å)[c]
Eu(III)	Pr(III)	1.98	0.101	6.50	8.20	11.8
Tb(III)	Pr(III)	0.78	0.115	2.66	7.77	10.9
Tb(III)	Er(III)	0.80	0.138	3.39	8.09	11.0

Data taken from W. W. Horrocks and D. R. Sudnick, *Acc. Chem. Res., 14*, 384 (1981).
[a] Calculated from the absorption spectra of [Ln(III)(DPA)3]$^{3-}$ complexes.
[b] Calculated using the parameter values $\kappa^2 = \frac{2}{3}$, $\eta^{-4} = 0.294$, $\phi_{Eu} = 0.27$, and $\phi_{Tb} = 0.48$.
[c] r(X-ray) = 11.7 Å.

can be replaced with Co^{2+}. Energy transfer from Tb^{3+} in site 1 and Co^{2+} in the zinc site yields an intersite distance of 13.7 Å, in good agreement with crystallographic data (Figure 9.15).

NMR Probes

A property of many lanthanide ions that we have not yet presented in detail is the paramagnetism that arises from the unpaired f electrons. Two types of probe ions can be identified:

1. *Shift reagents.* These ions are characterized by short electronic relaxation times and have little effect on T_1 and T_2 values for neighboring nuclei, since the oscillating magnetic fields produced by electronic relaxation are not of a frequency suitable for promoting nuclear relaxation. However, the magnetic field generated by the free electron density does contribute to the static field $(H_o)_z$.
2. *Relaxation probes.* These ions are characterized by long electronic relaxation times, resulting in considerable changes in T_1 and T_2 (related to line width), since the oscillating magnetic fields produced by electronic relaxation promote nuclear relaxation but have little effect on chemical shift.

Lanthanide ions that have been extensively used are Gd^{3+}, Yb^{3+}, and Eu^{3+}. The former is a relaxation agent with a long electronic relaxation time ($\tau_s \sim 10^{-9}$ s), while the latter two are shift reagents with short relaxation times ($\tau_s \sim 10^{-13}$ s). Of the transition metals, Mn^{2+} is a relaxation probe, and Ni^{2+} or Co^{2+} are shift probes. (Note that we previously saw the use of the rapid relaxation of Co^{2+} in NMR studies of superoxide dismutase, Section 8.1.1.). Each type of probe ion displays a different distance dependence.

$$\Delta\delta = \text{const}/r^3 \quad \text{(dipolar shift)}$$
$$\Delta(1/T_{1,2}) = \text{const}/r^6 \quad \text{(relaxation)}$$

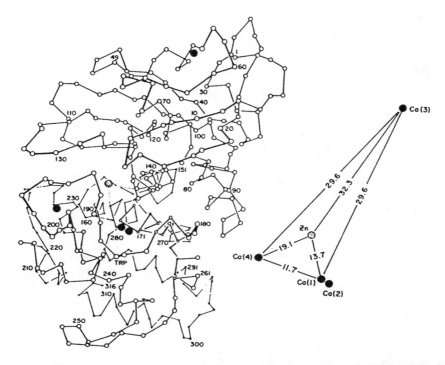

Figure 9.15 Thermolysin contains a Zn^{2+} ion that may be replaced with Co^{2+}. Fluorescence energy transfer from Tb^{3+} to Co^{2+} has an efficiency of 0.90. The spectral overlap integral (J) calculated from the emission spectrum of Tb^{3+} and absorption spectrum of Co^{2+} is 5.96×10^{-16} cm^6 mol^{-1}, and the quantum yield (Φ_D) of the donor is 0.51, yielding a value for R_0 of 19.6 Å and an intersite distance R of 13.7 Å. This shows good agreement with the crystallographically determined Zn^{2+}–Ca^{2+} distances. Adapted from B. W. Matthews et al., *Nature, New Biol. 238,* 41–43 (1972). [Also reprinted with permission from W. W. Horrocks and D. R. Sudnick, *Acc. Chem. Res. 14,* 384; copyright (1981) American Chemical Society.]

A detailed account of the derivation and uses of these equations can be found in the articles by Williams and Lee/Sykes in the Further Reading section. For quantitative work, knowledge of the angular dependence of the metal-nucleus vector is required.

Figure 9.16 illustrates the use of shift and relaxation probes to obtain structural information (metal-nucleus distances) from the Ln^{3+} probe ion. Specific experimental details can be found in the original articles referenced in the figure legend. Note that such studies require prior assignment of ^1H resonances to specific nuclei.

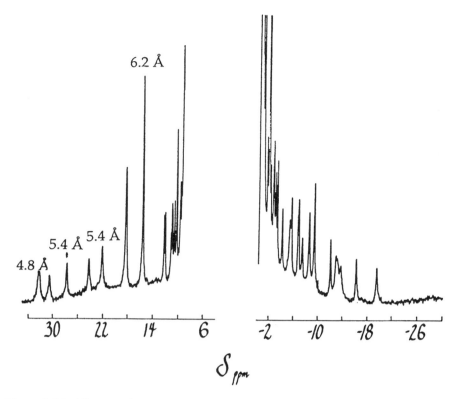

Figure 9.16 The use of Yb^{3+} as a shift probe is illustrated with the calcium-binding protein parvalbumin. Protons on residues coordinated to the metal ion were shifted out of the diamagnetic region. By measuring the relaxation times, the metal–proton distances were determined. [Adapted from L. Lee and B. D. Sykes, *Adv. Inorg. Biochem. 2,* 183 (1980).]

Summary of Section 9.2

1. Many metal ions of interest to inorganic biochemists cannot be readily studied by physical methods. In these cases, substitution by *functional* probe ions is a useful strategy.
2. Common replacements include Co^{2+} for Zn^{2+}, Mn^{2+} for Mg^{2+}, Ln^{3+} for Ca^{2+}. These provide access to optical measurements, magnetic resonance, and luminescence experiments.
3. Comparisons of binding constants and chemical reactivities of probe ions, and the evaluation of enzyme-substrate geometries through distance measurements provide useful insight on the metallobiochemistry of natural metal cofactors.

Notes

1. This reaction is catalyzed by an enzyme called kinase. Kinase assists in the transfer of the γ-phosphoryl unit from ATP to the terminal 5'-hydroxyl of a polynucleotide.

2. Nick implies cleavage of one strand of a double-stranded DNA molecule.
3. Glycopeptides contain both sugar and peptide units. Background reviews of bleomycin chemistry are available [e.g., S. Hecht, *Acc. Chem. Res., 19,* 383–391 (1986); J. Stubbe, J. W. Kozarich, *Chem Rev., 87,* 1107–1136 (1987).]
4. Refer to Section 2.2.1 in Chapter 2.
5. This does not arise from line-broadening due to rapid spin relaxation. Rather, the spectral line intensity is distributed between the normal ($\Delta M = \pm 1, \Delta m = 0$) and forbidden transitions ($\Delta M = \pm 1, \Delta m \neq 0$), where M and m are electron and nuclear spin numbers, respectively. The forbidden transitions are too broad to detect.
6. The paramagnetic contribution is easily determined from the difference in T_1 values with $[(T_1)_{Mn^{2+}}]$ and without $[(T_1)_{no\ Mn^{2+}}]$ added Mn^{2+} $[1/(T_1)_p = 1/(T_1)_{Mn^{2+}} - 1/(T_1)_{no\ Mn^{2+}}]$.
7. See, for example, R. D. Cornelius, W. W. Cleland, *Biochemistry 17,* 3279 (1975); K. D. Danenberg, W. W. Cleland, *Biochemistry 14,* 28 (1975).
8. Co(III) requires a number of N-donor ligands to stabilize the trivalent state. Cr(III) is stable with water as ligands.
9. The generic symbol *Ln* is used to denote the lanthanide metals.
10. Only partially filled orbitals contribute to defining the electron configuration and term symbols (see Section 2.2). For the Ln ions this involves 4*f* electrons.
11. Literally, the area of overlap of the two bands. $J = \int F(\nu)\varepsilon(\nu)\nu^{-4}\,d\nu / \int F(\nu)\,d\nu$, where $F(\nu)$ is the luminescence intensity of the donor, $\varepsilon(\nu)$ is the extinction coefficient of the acceptor in units of $M^{-1}\,cm^{-1}$, and ν is the frequency in cm^{-1}.
12. A value of 1.35 appears to be a good estimate for η, since this is intermediate between the values for water (1.33) and organic molecules composed of first-row elements (1.39).

Further Reading
Magnetic Resonance

Bertini, I., C. Luchinat, and A. Scozzafava. Carbonic anhydrase. An insight into the zinc binding site and into the active cavity through metal substitution, *Structure and Bonding, 48,* 45–92 (1982).

Bertini, I., and C. Luchinat. *NMR of Paramagnetic Molecules in Biological Systems,* Chap. 10, Addison-Wesley, 1986.

Lee, L., and B. D. Sykes. High resolution NMR, *Adv. Inorg. Biochem., 2,* 183 (1980).

Mildvan, A. S., and M. Cohn. Aspects of enzyme mechanism studied by nuclear spin relaxation induced by paramagnetic probes, *Adv. Enzymol. Related Areas Mol. Biol., 33,* 1 (1970).

Luminescence

Horrocks, W. De W., and D. R. Sudnick. Lanthanide ion luminescence probes of the structure of biological macromolecules, *Acc. Chem. Res., 14*, 384 (1981).

Chemical Probes

Barton, J. K. Metals and DNA: Molecular left-handed components. *Science, 233,* 727–734 (1986).

Cleland, W. W., and A. S. Mildvan. Chromium(III) and cobalt(III) nucleotides as biological probes, *Adv. Inorg. Biochem, 1,* 163 (1979).

Mildvan, A. S., and C. M. Grisham. The role of divalent cations in the mechanism of enzyme-catalyzed phosphoryl and nucleotidyl transfer reactions, *Structure and Bonding, 20,* 1–21 (1974).

Pyle, A. M. and J. K. Barton. Probing nucleic acids with transition-metal complexes. *Prog. in Inorg. Chem. 38,* 413–475, Wiley-Interscience (1990).

Sigman, D. S. Nuclease activity of 1,10-phenanthroline copper ion. *Acc. Chem. Res. 19,* 180–186 (1986).

Tullius, T. D., Ed. *Metal-DNA Chemistry*, ACS Symposium Series 402.

Williams, R. J. P. The chemistry of lanthanide ions in solution and in biological systems, *Structure and Bonding, 50,* 79–119 (1982).

Problems

1. Fluorescent calcium indicators are valuable probes of intracellular cation concentrations and the dynamics of ligand exchange. The indicator dye

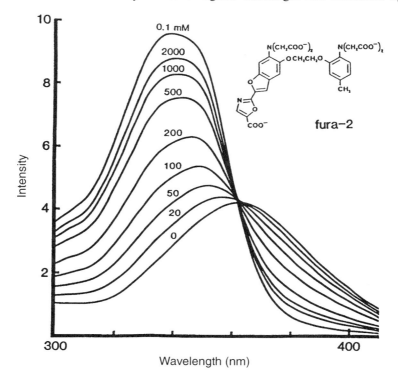

FURA-2 exhibits a change in fluorescence intensity upon Ca^{2+} binding. The association and dissociation rate constants have been estimated as $k_{on} \sim 7.6 \times 10^8$ $M^{-1}s^{-1}$ and $k_{off} \sim 109$ s^{-1} at 22°C and pH 7.4, and 1.5×10^9 $M^{-1}s^{-1}$ and 196 s^{-1}, respectively at 37°C. A modified Mg^{2+} selective ligand (MAG-FURA-2) exhibits $k_{on} \sim 9.9 \times 10^4$ $M^{-1}s^{-1}$ and $k_{off} \sim 493$ s^{-1} at 10°C and pH 7.4, and 5.9×10^5 $M^{-1}s^{-1}$ and 1587 s^{-1}, respectively, at 22°C.

(a) Evaluate the dissociation constants K_d for both metal cations at each temperature. Estimate the activation enthalpy and entropy for association and dissociation and the thermodynamic enthalpy and entropy components for ligand binding.

(b) Is binding driven by enthalpic or entropic factors? Provide a rational explanation.

(c) Contrast the activation energies for association and dissociation and for Ca^{2+} versus Mg^{2+}.

(d) Comment on the importance of the magnitudes of the association and dissociation rate constants if one is planning to use these fluorescent indicators to monitor the dynamics of metal ion uptake or release from biological macromolecules.

[Tsien, R. Y. and M. Poenie, *Trends in Biochemical Sciences 11*, 450–455 (1986), Figure is repinted with permission; Lattenzio, F. A. and D. K. Bartschat, *Biochem. Biophys. Res. Commun. 177*, 184–191 (1991); Martin, S. R. et al., *Eur. J. Biochem. 151*, 543–550 (1985)]

2. Carbonic anhydrase is a zinc enzyme that catalyzes the hydration and dehydration of CO_2 and HCO_3^-, respectively, and is important for the effective removal of the respiratory by-product CO_2 from the blood stream. Replacement of zinc by a paramagnetic cobalt ion provides a spectroscopic handle that allows determination of the distance between the carbon and metal centers. In turn, this can demonstrate whether or not the substrate binds directly to the catalytic metal ion or if the latter only serves to activate a bound water molecule.

The paramagnetic contribution to the longitudinal relaxation rate is defined by

$$\frac{1}{T_{1P}} = \frac{1}{T_1} - \frac{1}{T_1^o}$$

where $1/T_1$ is the measured [^{13}C]-substrate relaxation rate in cobalt carbonic anhydrase solutions and $1/T_1^o$ is the measured rate in a control experiment with the diamagnetic zinc enzyme. The paramagnetic contribution is normalized by a factor f $(= [\text{enzyme}]_o/[\text{substrate}]_o)$. The distance between the relaxing center and the paramagnetic source is defined by the dipolar term of the Solomon-Bloembergen equation

$$\frac{1}{fT_{1P}} = \frac{C}{r^6}$$

where C $(= 4.03 \times 10^5$ $Å^6$ $s^{-1})$ is a product of physical constants that define the electronic and motional state of the protein complex. Under saturating conditions, the following data has been collected.

Enzyme	Substrate	$^{13}C, T_1$ (s)
0.5 mM CoCA	37 mM HCO_3^-	0.33
0.9 mM ZnCA	95 mM HCO_3^-	2.2
0.2 mM CoCA	20 mM CO_2	0.40
none	20 mM CO_2	35

Calculate the $^{13}C–Co$ distance for HCO_3^- and CO_2. Does this support direct binding or not?

[Stein, P. J. et al., *J. Am. Chem. Soc. 99,* 3194–3196 (1977)]

3. A chemical alternative to the use of DNase I for footprinting experiments is the hydroxyl radical (HO·). This is generated from the Fenton reaction below.

$$Fe^{2+}(aq) + H_2O_2 \rightarrow Fe^{3+}(aq) + HO^- + HO·$$

(a) Draw a chemical mechanism for this reaction.

(b) For practical reasons $[Fe(EDTA)]^{2+}$ is used rather than free $Fe^{2+}(aq)$. Why?

Note: EDTA = $(^-OC)_2NHCH_2CH_2N(CO_2^-)_2$.

4. The solution structure of a complex of Zn(II)-bleomycin with d(CGCTAGCG)$_2$ has been probed by 2D-NMR methods (refer to Figure 10.18). As part of this study, sequential connectivities of base and sugar

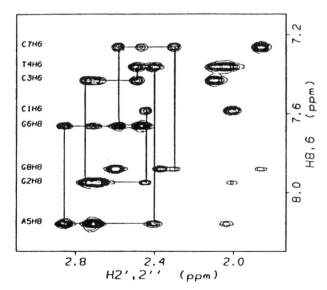

protons were made. The spectrum shown is an expanded NOESY contour plot of the spectral region in which base protons (H6 for T and C, and H8

for A and G) and sugar (H2′,2″) protons are found. The tracing shown outlines the connectivities between the base protons and the 5′-flanking sugar
H2″ protons.

(a) Sketch one complete strand of the duplex, showing the structures of the
    bases and sugars. Indicate the H6,8 and H2′,2″ protons and explain why
    these show through-space connectivities. Use double-headed arrows to
    show these connectivities.

(b) Explain why the cross-peaks in the spectrum show variable intensities.

(c) Trace out the connectivities between the base protons and 5′-flanking
    sugar H2′ protons.

[Manderville, R. A. et al., *J. Am. Chem. Soc.* *116,*10851–2 (1994), Figure is
reprinted with permission].

5. The EPR spectrum of a Co(II)-bleomycin complex in the presence of DNA
   shows a distinct spectrum for the oxygen-free (A) and oxygen-bound (B)
   forms. Given that $I = 7/2$ for cobalt-57 and $I = 1$ for nitrogen-14, provide an
   explanation for the appearance of each spectrum. In particular, comment on
   the fact that $g_\perp > g_\parallel$ for (A) while $g_\parallel > g_\perp$ for (B), and rationalize the hyperfine
   and super-hyperfine splitting patterns observed.

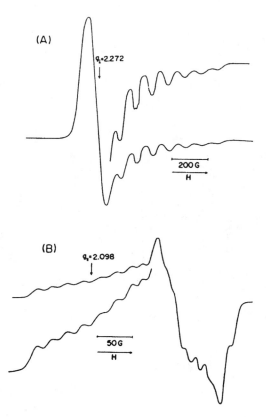

[Sugiura, Y. *J. Am. Chem. Soc. 102,* 5216 (1980), Figure is reprinted with permission]

6. Ferricyanide $(Fe(CN)_6]^{4-}$ quenching of the luminescence for the DNA-bound forms of the two enantiomers of tris(phenanthroline)ruthenium(II) is shown, where the squares denote data for the Δ isomer and the circles correspond to the Λ isomer. Assume that the origin of the quenching is photoexcited electron transfer from the ruthenium reagent.

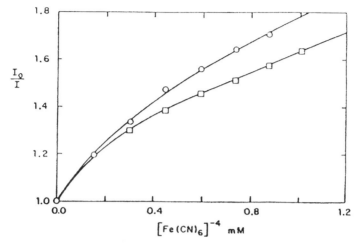

(a) Is the electron-transfer quenching arising from redox chemistry with the bound or free form of the $Ru(phen)_3^{2+}$. Explain.

(b) Why was ferricyanide used in these quenching studies rather than a complex such as ruthenium(III) hexaammine.

(c) Explain why the ratio I_o/I is larger than unity.

(d) Which of the two enantiomers binds more tightly to B-DNA? What would be your prediction for selective binding to Z-DNA?

[Barton, J. K. et al., *J. Am. Chem. Soc. 108,* 2081–2088 (1986), Figure is reprinted with permission.]

CHAPTER

10

Case Studies

In this final chapter we shall discuss three specific problems in some detail. These provide a focal point for the physical, chemical, and biochemical principles developed in earlier sections. Many of the strategies and tactics used are quite general, and so a close study of these topics and others referenced at the end of the chapter will illustrate much of the material we have covered. It should not be thought that the problems selected are completed research. In all three areas there remains much to learn. In this regard, I have tried to indicate the known facts and illustrate the manner in which they are brought together to form a working hypothesis. Such is the progression of science.

10.1 Cytochrome *c* Oxidase

Cytochrome *c* oxidase is the last in a sequence of membrane-bound electron-transfer proteins that make up the mitochondrial respiratory chain (see Figure 5.14). By carefully regulating a sequence of electron-transfer steps that couple a powerful electron donor (NADH) and acceptor (O_2), a large amount of energy (1.14 V) is released in a controlled manner that can be used to establish a trans-membrane proton gradient. This gradient ultimately provides the driving force for ATP synthesis. (Review Mitchell's chemiosmotic theory in Section 5.3.) Cytochrome oxidase is a complex membrane-bound protein (spanning the inner mitochrondrial membrane) that is composed of up to 13 subunits and catalyzes the four-electron reduction of O_2 to H_2O (Figure 10.1). We should bear in mind that the biological *function* of the enzyme is not to reduce O_2 to H_2O (we could

389

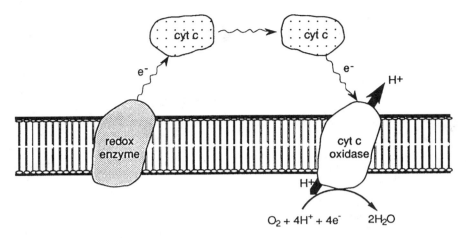

Figure 10.1   Cytochrome oxidase couples the energy released from the reduction of O_2 to vectorial proton transport, generating a transmembrane proton gradient that is used to drive ATP synthesis.

more easily do this by drinking a glass of water!), but to pump protons across a membrane barrier. Electron-reducing equivalents are provided by cytochrome *c*, which binds to cytochrome *c* oxidase on the cytoplasmic side of the membrane. The activity of the enzyme appears to reside with subunits I, II, and III, and so the remaining subunits are presumably required for membrane binding.[1] The chemistry of cytochrome oxidase revolves around four key prosthetic groups (heme a, heme a_3, Cu_A, and Cu_B). These not only catalyze the reduction of O_2, but also regulate the proton pumping action of the enzyme (Figure 10.1). To understand the chemistry of cytochrome oxidase fully, it is necessary to determine not only the details of the catalytic mechanism for the four-electron reduction of O_2 to H_2O, but also to establish the pathway by which this is coupled to proton pumping. A necessary prelude to that discussion is the determination of the structural chemistry of the prosthetic centers and their ligand–electron exchange properties. This has been achieved through a combination of solution studies and x-ray crystallography.

Cytochrome oxidase is a molecular machine that contains a pair of low-potential sites (heme a and Cu_A) and a pair of high-potential sites (heme a_3 and Cu_B). Each part of this machine has been incorporated to serve a specific role. The former serve as an entry port for electrons that come from reduced cytochrome *c* (Figure 10.2); the latter form the catalytic site for O_2 reduction. In the absence of substrate oxygen, the reduction potentials of all four sites are similar, and so the low-potential/high-potential nomenclature appears odd; however, the distinction will be made clear very soon. The low-potential sites are more accessible by the external reducing agent (cytochrome *c*) and the coordination and electronic states of these sites are optimized for efficient electron transfer (a low-spin hexacoordinate heme a, and a "blue" copper center). In contrast,

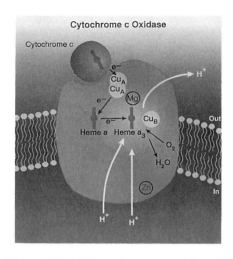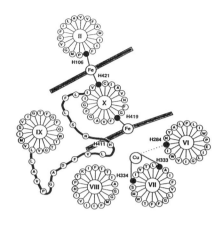

Figure 10.2 Schematic representation of bovine cytochrome oxidase in the inner mitochondrial membrane showing the approximate dispositions of the key prosthetic centers. The cytochrome c binding site and Cu_A site are positioned in subunit II, whereas, heme a, heme a_3 and Cu_B are all interconnected in subunit I by bridging ligands associated with the membrane-spanning helices. The latter illustration it taken from the *E. coli* bo-type ubiquinol oxidase, which also belongs to the structural superfamily of heme-copper respiratory oxidases. [Reprinted with permission from Gennis, R. and S. Ferguson-Miller, *Science 269,* 1069 (1995); Calhoun, M. W. et al., *Biochemistry 32,* 13254–13261 (1993)].

the catalytic sites are designed for substrate binding. The mechanistic details of the O_2 reduction pathway are described later. First, we shall outline some of the experiments that were designed to characterize the coordination chemistry of the copper and heme sites, and then move on to a discussion of their function.

Low-Potential Sites (Heme a and Cu_A)

Figure 10.3 shows that heme a is a six-coordinate, low-spin iron center, ligated by two histidine residues in both the oxidized and reduced states. Cu_A appears to be coordinated by His and Cys ligands. The initial structural model for these sites came not from crystallographic data where we might "see" the ligand atoms, but rather from the interpretation of spectroscopic, physicochemical, and biochemical evidence. In the case of Cu_A, a preliminary model for the coordination state was established by input from a variety of experimental sources that include EXAFS, EPR, electronic absorption, and data from isotopically-labeled amino acid residues. The first suggestion of a mononuclear $(His)_2(Cys)_2Cu^{2+}$ site was later superseded by a dinuclear model that was supported not only by close inspection of the available spectroscopic evidence, but also by results from site-directed mutagenesis experiments and a sequence comparison with the homologous 50-residue peptide that forms a binding site for a

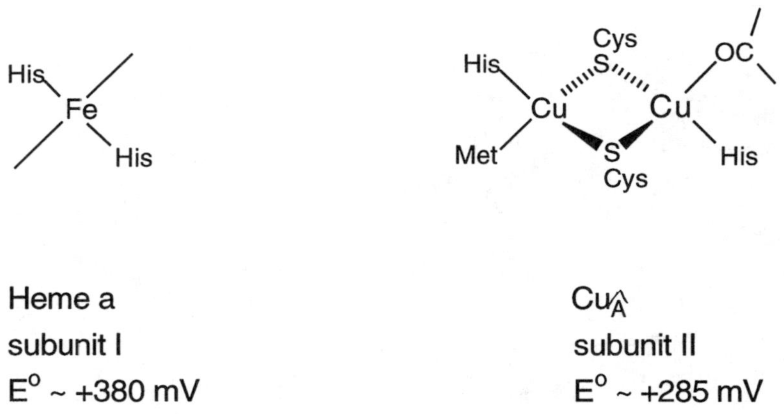

Heme a

subunit I

$E° \sim +380$ mV

Cu_A

subunit II

$E° \sim +285$ mV

Figure 10.3 The low-potential sites: their locations, coordination environments, and physicochemical properties. When heme a_3 and Cu_B are reduced, $E°$(heme a) $\sim +280$ mV.

dinuclear copper center in nitrous oxide (N_2O) reductase. However, an unequivocal demonstration of the nature of the dinuclear Cu_A site required the insight provided by X-ray crystallographic data. Retrospectively one can now see how the available physical data fits well with the dinuclear model illustrated in Figure 10.3. Briefly, EXAFS suggested two N (or O) ligands (remember EXAFS cannot distinguish elements that have similar electron densities) and two S ligand atoms. The more weakly coordinated Met and backbone carbonyl were not so readily identified. By analogy with the type I copper sites, ligation by cysteine thiolate was suggested by the characteristic ligand–metal, charge-transfer band in the electronic absorption spectrum (refer to the discussion in section 5.2 on "blue copper"). In support for this conclusion, EPR studies show a signal from the $S = 1/2$ Cu_A site that displays an unusually small hyperfine splitting typical of Type 1 copper; however, one g-value was found to be lower than 2 (normally Cu^{2+} shows g values greater than 2). This was originally explained by invoking a contribution from an organic radical and, in particular, led to the proposal of a thiyl radical. However, closer inspection of the data at a variety of frequencies and comparison with the EPR signature from the dinuclear copper center in N_2O reductase led to a reformulated hypothesis.[2] The EPR data for Cu_A is, in fact, best interpreted as a weakly coupled binuclear $Cu^{1.5+}$.... $Cu^{1.5+}$ site, which also provides a satisfactory explanation of the fact that highly active cytochrome c oxidase preparations contain a Cu:Fe ratio of 3:2.

More direct evidence for a His/Cys ligand set was provided by studies of [1,3-$^{15}N_2$]-His- and [β,β-$^2H_2$]-Cys-labeled enzymes. These studies with isotopically labeled residues were made possible by the availability of auxotrophic strains of His and Cys mutants of the yeast *Saccharomyces cerevisiae.* An auxotrophic strain cannot synthesize specific vital metabolites, which must be provided in

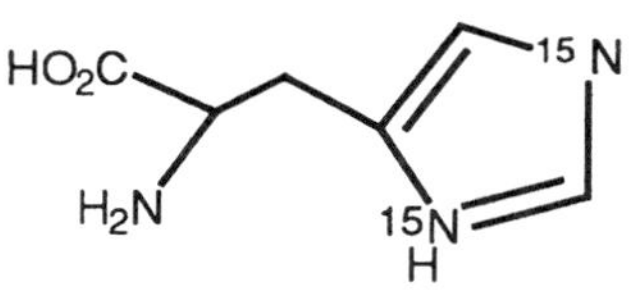

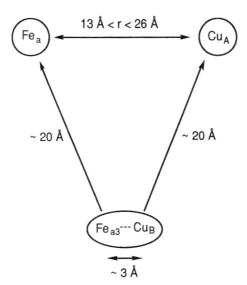

[1,3 – $^{15}N_2$] His

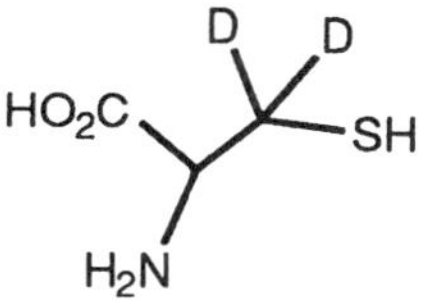

[β, β – $^2H_2$] Cys

the growth medium, and so direct incorporation of labeled amino acids can be readily achieved.[3] EPR and ENDOR studies of oxidase isolated from this organism after growth in a medium enriched with ^{15}N- or ^2H-labeled histidine or cysteine, respectively, demonstrated appropriate coupling features.

The distance between pairs of redox centers is an important parameter, since this regulates electron-transfer rates between them and may have implications for communication between each site either by a bridging ligand or by a cooperative mechanism. The distance between the heme a and Cu_A centers (r) in Figure 10.2 was estimated by EPR (13 Å < r < 26 Å). There are several ways of doing this, since an electron spin may influence one or more of the spectral paramaters (relaxation times, linewidths, g-values) of an adjacent spin in a manner that depends on the distance between them. However, all ultimately depend on the dipolar (through-space) interactions of these spins.[4]

Both heme a and Cu_A show anticooperative interactions in their redox behavior (Figure 10.4). Reduction of heme a makes it more difficult to reduce the Cu_A site (i.e., the formal $E°_{Cu}$ depends on whether heme a is oxidized or reduced). Anticooperativity can arise from either an electrostatic interaction (putting an

(A)

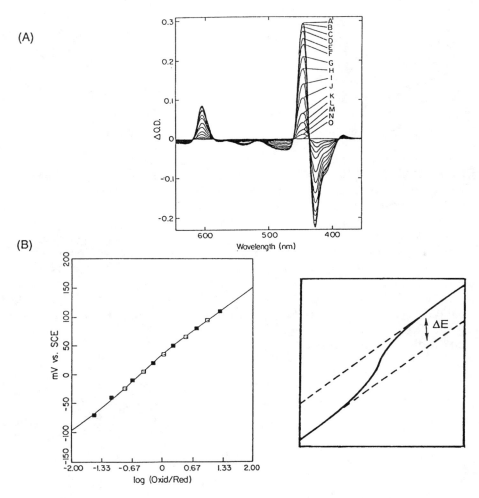

(B)

Figure 10.4 (A) Absorbance difference spectra of the heme a center taken at a variety of potentials (mV): A, -220; B, -100; C, -70; D, -40; E, -25; F, -10; G, 5; H, 20; I, 35; J, 50; K, 65; L, 80; M, 95; N, 110; O, 140. (B) (Left) Nernst plot obtained from this data at ΔA (443 nm). The plot is nonlinear due to the anticooperative interaction of 24 mV between heme a and other centers. (Right) A theoretical plot showing the effect of cooperativity. This may arise from electrostatic or conformational effects. [Reprinted with permission from W. R. Ellis et al., *Biochemistry, 25*, 161–167; copyright (1986) American Chemical Society.]

electron on one center makes the reduction of an adjacent site more difficult) or a structural change. In contrast, a cooperative interaction, where one electron reduction makes the second reduction easier, can arise only from structural changes. This kind of interdependence among redox centers is most often mediated by structural mechanisms and plays a very important role in the regulation of biological redox cycles. Note also that we have previously seen examples of related structural effects in O_2 binding by Hb (Section 4.1) and metal-responsive transcriptional factors (Section 7.2). In the case of cytochrome oxidase, it has been established that reduction of the low-potential Cu_A/heme a sites is accompanied by a structural change that switches on an electron-transfer pathway allowing the electrons stored in the Cu_A/heme a pair to be transferred to the Cu_B/heme a_3 sites. The structural change also opens up the reduced Cu_B/heme a_3 sites to substrate O_2. It is only at this point that the term *high potential* makes any sense. When the powerful oxidant O_2 binds to the reduced Cu_B/heme a_3 pair, the reduction potential of each increases dramatically. We shall see that this change in reduction potential actually provides the driving force for proton pumping.

High-Potential Sites (Heme a_3 and Cu_B)

The heme a_3 and Cu_B sites form the catalytic core of the enzyme and are coupled by a bridging ligand in the oxidized enzyme, but not in the active reduced state (Figure 10.5), and so both heme a_3 and Cu_B possess vacant coordination sites suitable for substrate binding. The close proximity of these sites favors multiple electron delivery to bound O_2. The heme a_3 site is high-spin pentacoordinate and readily binds a variety of ligands (NO, CO, CN^-). Both the reduced (a_3^{2+} $S = 2$, Cu_B^{1+} $S = 0$) and oxidized (a_3^{3+} $S = 5/2$, Cu_B^{2+} $S = 1/2$) states are EPR silent. This is not too surprising for the reduced enzyme inasmuch as Cu^{1+} is diamagnetic, while we saw in Chapter 2 (Section 2.3.2) that integral spins on heme frequently result in EPR silent states, since the allowed transitions frequently lie outside of the normal magnetic field range. As a result of antiferromagnetic coupling between the bridged a_3^{2+}–Cu_B^{2+} sites, we also have an $S = 2$ state ($S = 5/2 - 1/2 = 2$) for the oxidized enzyme.

The coordination environments of heme a_3 and Cu_B have been studied in

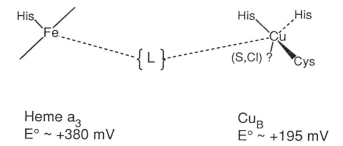

Heme a_3
$E° \sim +380$ mV

Cu_B
$E° \sim +195$ mV

Figure 10.5 The coordination environments and resting potentials of the high potential sites. These centers are located in subunit I. The bridging ligand L (O^{2-}, HO^-, Cl^-, ?) connects the two sites only when both sites are oxidized. The identity of L is uncertain and was not identified in preliminary crystallographic studies. Nevertheless, the magnetic coupling between the metals requires such a ligand.

much the same way as described previously. Since the Cu_A site can be selectively removed, the ligand set around Cu_B can be probed directly by EXAFS. The results are summarized in Figure 10.5. The effective EPR silence of the heme a_3 site in both oxidized and reduced states makes it difficult to probe the axial ligand to this site. This problem can be overcome by use of paramagnetic spin probes that bind to the vacant axial site on reduced heme a_3 and act as reporter groups. For example, nitric oxide (NO˙) is a radical species that binds tightly to reduced heme and gives rise to the characteristic rhombic EPR spectrum shown in Figure 10.6. The unpaired electron is localized on the N-atom of NO˙, and so any nuclear coupling is dominated by this atom, and coupling is largest in the g_z component. Coupling to the g_x and g_y components is often too weak to resolve. A secondary and weaker coupling to a nitrogen ligand bound to the opposite face of the heme is also observed. Although histidine is an obvious candidate, in principle, this nitrogen center could derive from any N-bearing amino acid residue or the peptide backbone. This point was addressed by using the $(1,3-^{15}N_2)$-His-labeled enzyme described earlier, since coupling to ^{15}N can arise only from direct coordination of histidine to heme. Figure 10.6 shows a set of EPR data in which all the combinations of ^{14}N- and ^{15}N-labeled nitrosyl and histidine have been studied. ^{14}N ($I = 1$) gives rise to a three-line coupling pattern, whereas ^{15}N ($I = 1/2$) yields a two-line coupling pattern. The data are, therefore, consistent with an axial His ligand.

Figure 10.2 summarizes the placement of the redox cofactors in cytochrome c oxidase. Both the cytochrome c binding site and the Cu_A site are positioned in subunit II, whereas heme a, heme a_3, and Cu_B are all positioned in subunit I. Twelve membrane-spanning helices had been predicted from analysis of the primary structure of subunit I, while site-directed mutagenesis experiments had identified putative protein ligands for each cofactor. Crystallographic data later confirmed this model, which, in turn, lends itself to a structural interpretation of the negative cooperativity observed between heme a and the heme a_3–Cu_B site, because there is an obvious structural linkage between these two domains.

How the Enzyme Works

We can now proceed with the experiments that have been designed to test the manner in which these various prosthetic groups interact with each other to make the chemistry happen. Figure 10.7 provides a succinct summary of the $O_2 \rightarrow H_2O$ reaction pathway. The reaction proceeds by a number of distinct steps that facilitate ligand binding to the Cu_B–heme a_3 site.

1. Reduction of heme a/Cu_A by cytochrome c produces a change in enzyme conformation. This can be demonstrated in at least two ways. First, there is a change in enzyme fluorescence (from Trp residues). The emission is red-shifted to longer wavelengths (lower energy), since these residues are increasingly solvent exposed, which tends to stabilize and lower the energy of the excited state (Figure 10.8). Second, the rate of ligand binding to heme a_3 is

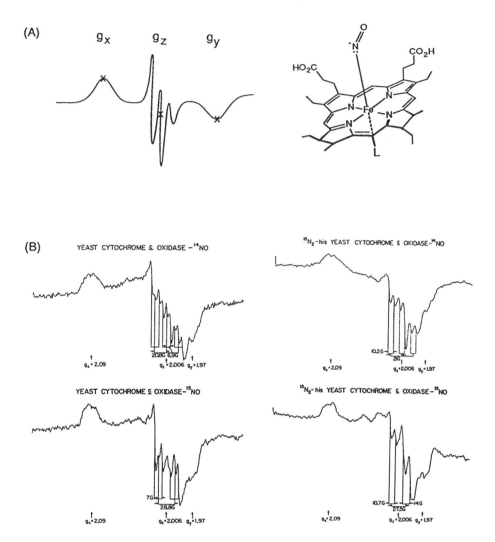

Figure 10.6 (A) Coupling pattern in the ferrous heme–nitrosyl radical EPR spectrum. The central g_z component shows the effects of hyperfine coupling to the ^{14}N nucleus of nitric oxide. If the axial ligand L is histidine, an additional longer range coupling may be observed, as in the following case. (B) EPR data showing all the combinations of ^{14}N- and ^{15}N-nitrosyl and histidine. [From T. H. Stevens et al., *J. Biol. Chem.*, *256*, 1069–1071 (1981).]

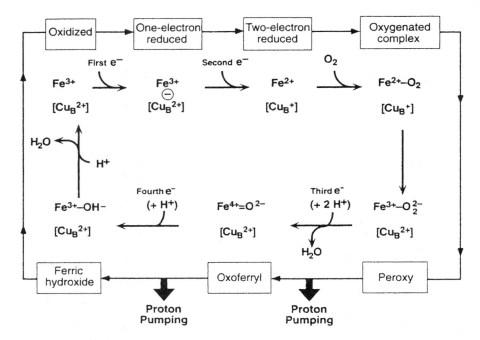

Figure 10.7 Sequence of reactions for bovine cytochrome c oxidase. Starting from the top left corner, dioxygen binds after the heme a_3 and Cu_B centers are reduced. Oxygen may transiently bind to Cu_B, but rapidly moves to heme a_3 to form an oxygenated complex that is spectroscopically similar to $Hb.O_2$. After two electron reduction, a peroxy intermediate is formed. Introduction of an additional electron results in bond cleavage and production of H_2O, and an oxo-ferryl species on heme a_3. Introduction of a fourth electron results in formation of a second molecule of H_2O and regenerates the oxidized dinuclear center. Many of these species show characteristic absorption and resonance Raman signatures that allow the intermediates to be characterized, and the kinetics of formation and decay to be monitored. Proton pumping appears to be coupled to the delivery of the third and fourth electrons, forming the oxo-ferryl and fully oxidized ferric hydroxide centers, respectively. Reaction of fully reduced oxidase with O_2 is rapid and is complete in ~2 ms! [Reproduced with permission from M. W. Calhoun et al., *Trends in Biochem. Sci. 19,* (1994)].

faster since this site becomes more accessible. In summary, reduction of heme a/Cu_A opens up the active site to molecular oxygen.

2. These structural changes also decrease the activation barriers for internal electron transfer (ET) from heme a/Cu_A to the heme a_3–Cu_B site. After reduction, the bridging ligand is released from the heme a_3–Cu_B pair; however, the copper and heme iron remain in close proximity in a geometry that facilitates O_2 binding.

3. Two more electrons enter by way of heme a/Cu_A to complete the four-electron reduction of O_2. Figure 10.7 shows a reasonable model for the reaction pathway.

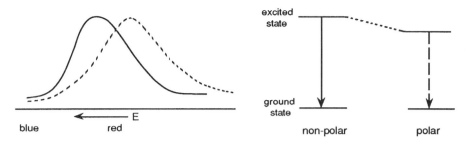

Figure 10.8 A conformational change leads to a red shift in Trp fluorescence since this residue becomes more solvent-exposed. The aqueous environment stabilizes the polar excited state, which results in a lower energy of emission.

Proton Pumping

It is important for us to bear in mind that the function of cytochrome c oxidase is not to reduce O_2 to H_2O, but to pump H^+ across the mitochondrial membrane. Inspection of the reduction potentials for the heme and copper sites in cytochrome oxidase (Figures 10.3 and 10.5) show that the free energy difference between the low- and high-potential sites amounts to only about 100 mV. This is not sufficient to move H^+ against a concentration gradient. To overcome this barrier, the enzyme uses the energetically favorable reduction of O_2 to H_2O $[1/4O_2 + H^+ + e^- \rightarrow 1/2H_2O, \Delta E° = 816$ mV$]$ to drive the flow of protons across the membrane.

$$\Delta E = \left(\frac{RT}{nF}\right) \ln\{[H^+]_{out}/[H^+]_{in}\}$$

To make this work we require an intimate molecular mechanism that couples proton transport to O_2 reduction. This important concept is central to bioenergetics. A reaction may be extremely favorable and might be used to drive a less favorable reaction. However, there must be a mechanism to couple the two reactions together. In the case of cytochrome oxidase, O_2 binds to the heme a_3–Cu_B sites and raises their reduction potentials. This increases the driving force for internal electron transfer. In turn, these electron-transfer steps are coupled to the transmembrane movement of protons. Cytochrome c oxidase is, in fact, a redox-driven proton pump which translocates on average one proton per electron from the inner matrix side of the mitochondrion to the outer cytosol side (Figure 10.7). Translocation occurs vectorially, which means that movement of the electron and proton are distinct—they are not directly coupled. Since the four protons consumed in dioxygen reduction come from the matrix side (Figure 10.1), the effective H^+/e^- stoichiometry is 2. Unfortunately there is, as yet, no good model for the chemical mechanism of proton pumping. It is hoped that the crystallographic data now available, allied with the genetic, chemical, and physical arsenals aligned against this problem will yield an answer in the near future. Certainly it is the ability of researchers to link together diverse pieces of information and to generate working models for further testing that is the key to progress in understanding and requires a firm foundation in basic principles. In this regard such problems are worthy of further study and thought.

10.2 Mercuric Reductase

In an earlier chapter (Section 7.2), we considered the role of metal ions in the transcriptional regulation of proteins and enzymes. The cellular response of mercury-resistant bacteria follows the same general mechanisms and constitutes one of the most thoroughly investigated examples of metal-regulated transcription and cellular detoxification. In this section, we shall briefly consider the regulatory mechanism for transcription of genes involved in mercuric ion resistance and, then, describe the chemistry of mercuric ion reduction catalyzed by mercuric reductase.

Figure 10.9 defines the operon for mercuric ion resistance and Figure 10.10 illustrates the mechanisms that control regulation. The operon contains a number of genes encoding regulatory, transport, and redox proteins. The *mer* R gene encodes a multifunctional regulatory protein that regulates its own expression in the sense that it binds to the *mer* R promoter (P_R) and represses its own synthesis. However, if the intracellular levels of mer R are low, then, the P_R promoter site is left vacant, and so RNA polymerase can bind and the gene product is transcribed and translated to produce more mer R, which then shuts off its own synthesis. Adjacent to the P_R promoter is a second promoter sequence (P_T) that controls expression of the other genes in the *mer* operon. In the absence of mercuric ion, mer R acts as a repressor protein that binds close to the promoter site (P_T) for the gene products T, P, A, D, and B, and inhibits their transcription.[7] When the intracellular concentrations of Hg^{2+} increase, mercuric ion binds to mer R and induces a structural change such that mer R now acts as a transcriptional activator that induces expression of the gene products from the *mer* operon. *Mer* T encodes a transport protein whose role is to trap extracellular Hg^{2+} and release it to the intracellular environment where it

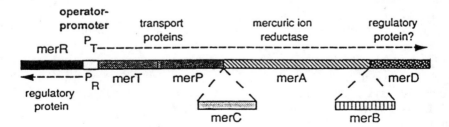

Figure 10.9.   The regulation of bacterial resistance to mercuric ion in gram-negative bacteria. *Mer* A encodes the mercuric ion reductase. [Adapted from J. D. Helman et al., *Adv. Inorg. Biochem.*, *8*, 33–61 (1991).]

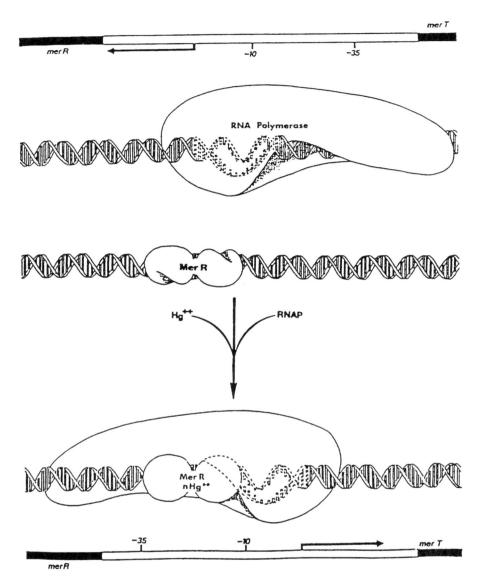

Figure 10.10   In the absence of Hg^{2+}, the Mer R protein negatively regulates its own expression (i.e., it inhibits RNAP binding to the P_R promoter region). Mer R is, therefore, expressed when its solution levels are low. In the presence of Hg^{2+}, the Mer $R-Hg^{2+}$ complex acts as a transcriptional activator that induces RNAP binding to the P_T promoter region. Subsequently, *mer* T and *mer* P (encoding proteins involved in Hg^{2+} and RHgX transport, respectively), *mer* A (encoding mercuric ion reductase), and, in some cases, *mer* B (encoding organomercurial lyase), are expressed. The roles of the mer C and mer D proteins have yet to be firmly established. [Adapted from T. V. O'Halloran et al., *Cell*, *56*, 119 (1989).]

may be reduced by mercuric reductase (encoded by *mer* A). In this way, Hg^{2+} does not build up on the surface of the outer cell membrane where it may react with sulfur-containing residues and impair the function of either membrane receptor or transport proteins. The protein encoded by *mer* P is also involved in trapping extracellular Hg^{2+}. *Mer* B encodes an organomercurial lyase enzyme that cleaves carbon–mercury bonds (see Section 8.2). Since the *mer* B gene is not common to all mercuric ion resistance operons, it is found that some cell systems show resistance to Hg^{2+}(aq) but are sensitive to organomercurials (HgR_2).[8] The function of the protein encoded by *mer* D is currently unknown.

Our attention now focuses on the reaction mechanism of the α_2 dimeric protein mercuric reductase ($M_r \sim 60$ kDa), a flavoprotein that uses NADPH as a two-electron reducing agent in the reaction shown here.[9,10]

$$\text{NADPH} + \text{Hg}^{2+} \rightarrow \text{NADP}^+ + \text{Hg}^\circ + \text{H}^+$$

The flavin unit (FAD) serves as an intermediate electron carrier. The enzyme requires four electron equivalents to generate the active form. These reducing equivalents cleave two redox active disulfide bonds at the active site (Figure 10.11), and further electron equivalents are subsequently used for Hg^{2+} reduc-

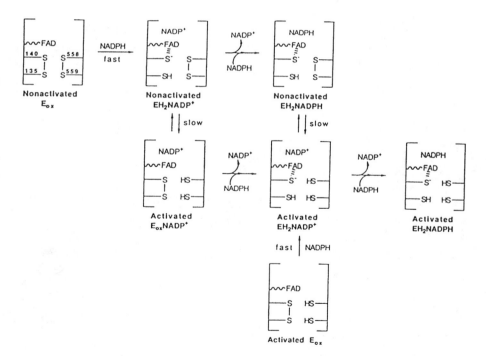

Figure 10.11 Activation of mercuric reductase by NADPH reduction. Note the charge-transfer interaction for the partially reduced enzyme from cysteine to oxidized FAD. [Reprinted with permission from S. Miller et al., *Biochemistry, 28,* 1194–1205; copyright (1989) American Chemical Society.]

tion. A useful spectral response that provides a probe of active-site chemistry is the appearance of a charge-transfer transition after partial reduction of the oxidized enzyme. By comparison with other FAD-containing disulfide reductase enzymes that possess similar spectroscopic characteristics, this can be attributed to a redox active disulfide (cystine) in the active site.[11] One of the cysteines in the partially reduced enzyme lies close enough to the oxidized flavin ring to show charge-transfer behavior from the deprotonated thiolate to the oxidized flavin ring. This charge-transfer band appears at 540 nm as a shoulder on the regular flavin absorbance (Figure 10.12). The function of specific active site Cys residues can be evaluated by site-directed mutagenesis experiments, where each is mutated to structurally related residues (Ser or Ala). In this way, it can be demonstrated that Cys140 gives rise to the charge-transfer (CT) interaction with FAD_{ox}, since the CT absorption disappears if this residue is mutated to serine.

By use of this charge-transfer interaction, the redox chemistry of the active site is open to investigation. For example, by monitoring the absorbance as a function of pH, the pK_a of Cys140 can be deduced, since only the ionized thiolate anion shows the charge-transfer band (Figure 10.12). Also the kinetics of re-

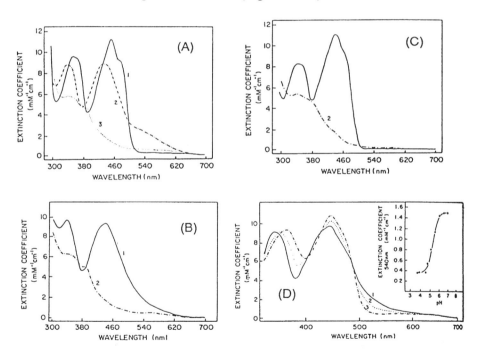

Figure 10.12 Optical spectra for (A) native Cys-135/Cys-140, (B) mutant Ser-135/Cys-140, and (C) mutant Cys-135/Ser-140. Spectra labeled 1–3 are the oxidized and 2- or 4-electron-reduced enzymes, respectively. The weak charge-transfer (CT) band is clearly seen at ~540 nm in (A) and (B), but not in (C) since Cys-140 is required. (D) illustrates how the cysteinate-to-oxidized flavin CT can be used to determine the pK_a of Cys-140. [Adapted from P. G. Schultz et al., *Biochemistry, 24*, 6840–6848 (1985).]

duction of the active-site disulfides can be easily monitored from the appearance or disappearance of the CT band to show that rapid reduction of the Cys135–Cys140 disulfide is followed by a subsequent slower rate-limiting reduction of the Cys558–Cys559 pair (Figure 10.11). The presence of such a redox active disulfide can be further established by quantitating the number of free thiols in the oxidized and reduced enzyme using the reagent 5,5′-dithiobis-(2-nitrobenzoate).[12]

The mechanistic details of Hg^{2+} binding and subsequent reduction chemistry have not yet been clearly elucidated, although it is likely that cysteine binds directly to Hg^{2+}. Typically mercuric ion forms two-coordinate complexes in solution (and very occasionally three-coordinate), and so the individual roles of all four cysteines in binding and catalytic activity remain unclear.[13] Despite these remaining uncertainties, studies on a variety of mutant enzymes and the use of flavin derivatives as cofactors have clearly defined a few chemically reasonable pathways, which we shall now examine.

It has been observed that Cys → Ser or Ala mutants remain active (although greatly reduced from wild-type enzyme). This suggests that the active-site cysteines bind Hg^{2+} and implicates direct reduction by $FADH_2$, rather than two-electron reduction by the thiols Cys_{135} and Cys_{140}.[14] Indeed, there is no known solution chemistry whereby thiols reduce mercuric ion in the following manner:

$$2\ RCH_2SH\ +\ Hg^{2+}\ \rightarrow\ Hg°+\ RCH_2S\text{-}SCH_2R\ +\ 2H^+$$

Direct reduction by $FADH_2$ is, however, well known, and Figure 10.13 shows three possible mechanistic pathways that lead to reduction of mercuric ion. Studies of enzymatic reductions involving hydride transfer by $FADH_2$ have shown that 5-deazaflavin derivatives of mercuric reductase are active. Since 5-deaza-$FADH_2$ cannot reduce Hg^{2+} in free solution, while $FADH_2$ can, the enzymatic mechanism for reduction by $FADH_2$ is unlikely to involve hydride

oxidized 5-deazaflavin

transfer. This leaves outer sphere electron-transfer and addition-elimination pathways as viable candidates. It is of course implicitly assumed that these solution studies are relevant to the enzymatic pathway for mercuric reductase. By studying a mutant form of mercuric reductase (Ala_{135} Cys_{140} Ala_{558} Ala_{559}) which is unable to bind or reduce Hg^{2+}, early reaction intermediates can be detected. NADPH reduces the flavin, and the resulting $FADH^-$ intermediate

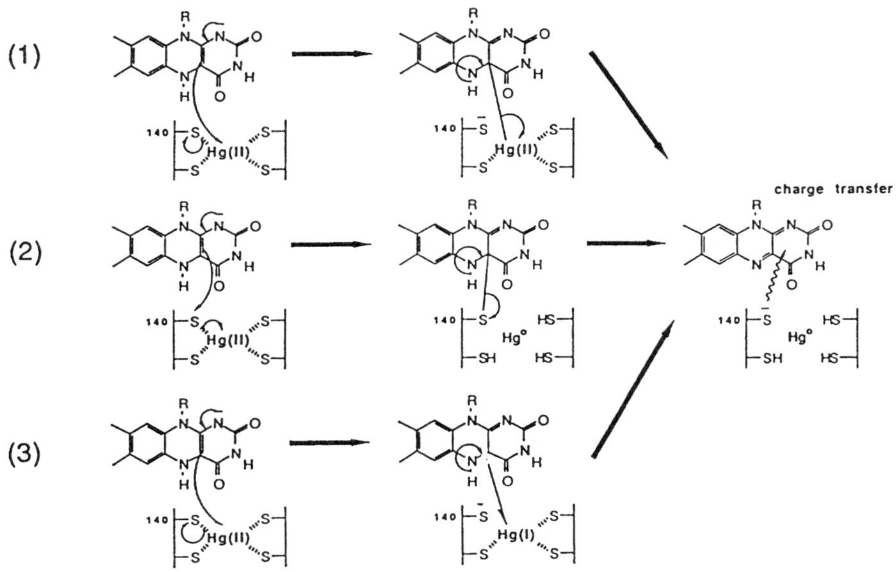

Figure 10.13 Three models for the reduction of Hg^{2+} by mercuric reductase. [Reprinted with permission from S. Miller et al., *Biochemistry, 28*, 1194–1205; copyright (1989) American Chemical Society.]

(A)

(B)

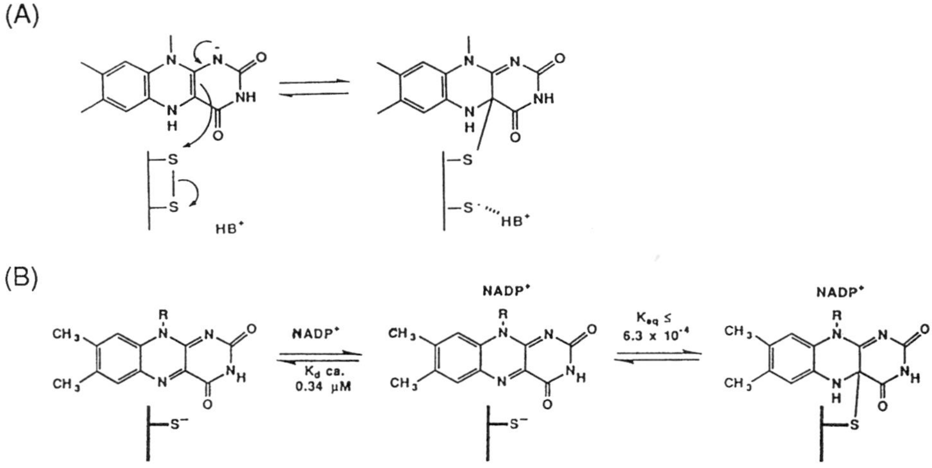

Figure 10.14 (A) FADH$^-$ reduces the disulfide bond in oxidized reductase to form an intermediate species consistent with a flavin C(4a)-Cys 140 adduct. (B) By studying mutant enzymes some early reaction intermediates can be detected. Specifically, NADP$^+$ oxidizes the Ala.Cys-140.Ala.Ala triple mutant to form a cytsteine-flavin adduct (compare with Figure 10-13). [Reprinted with permission from S. Miller *et al., Biochemistry 29*, 2831–2841; copyright (1990) American Chemical Society.]

reacts with disulfide to form a transient adduct (Figure 10.14). Reaction of oxidized mutant enzyme with NADP$^+$ also results in formation of a flavin C(4a)–Cys$_{140}$ thiol adduct. Both of these intermediates are kinetically competent for reduction and turnover of wild-type enzyme.[15] Furthermore, although FADH$_2$ can reduce a variety of redox active metal ions in solution (e.g., Au^{3+}, Ag$^+$, Pt^{2+}, Cu^{2+}), these are not reduced when enzyme bound, which suggests that the inner coordination sphere of the metal ion is an important factor in the enzymatic reaction. The addition-elimination pathway is currently the most promising scheme supported by these data.

10.3   Bleomycin

For our final study, we will explore the chemistry of a glycopeptide antibiotic. This will give us the opportunity to have a closer look at some DNA binding chemistry and the interesting redox chemistry associated with iron-mediated DNA strand cleavage. Some problems based on this section appear at the end of Chapter 8.

The bleomycin family of antibiotics were first isolated in Japan by Umezawa and co-workers in the mid-1960s, and the structure elucidated by a combination of NMR spectroscopy, mass spectrometry, and degradative chemistry. These water soluble substances proved to be effective for the clinical treatment of squamous cell carcinoma's, lymphomas, and testicular cancer. As with many biological molecules under investigation at that time and earlier, it was not initially realized that a metal cofactor might play an essential role in bleomycin (BLM) activity. A large number of BLMs have since been isolated and characterized, and found to differ in the identity of the substituent R which is appended to the bithiazole ring (Figure 10.15). The chemical nature of the side chain R appears to be related to the relative abundance of naturally occurring amines in the culture medium. The molecule is relatively basic, with three ionizable sites in the peptide framework and additional sites on the variable side-chain.

In its optical spectrum, the drug shows π-π* and n-π* bands from the imidazole, 4-aminopyrimidine, and bithiazole heteroaromatic rings. [Try to identify these chromophores in Figure 10.15]. Fluorescence bands observed at 353 and 405 nm have been assigned to the bithiazole and 4-aminopyrimidine rings, respectively, and quenching of the former has proved to be of considerable help in evaluating the binding constants for DNA-BLM interactions (Figure 10.16). In fact, the availability of a wide range of spectroscopic handles from the ligand chromophores and metal center has proved invaluable in elucidating the structural and mechanistic chemistry of DNA-bleomycin complexes. In the following sections, we will consider the structural chemistry of BLM binding to DNA, the chemistry of the metal binding domain and formation of activated BLM, and, finally, the chemistry of DNA strand cleavage. While none of these issues

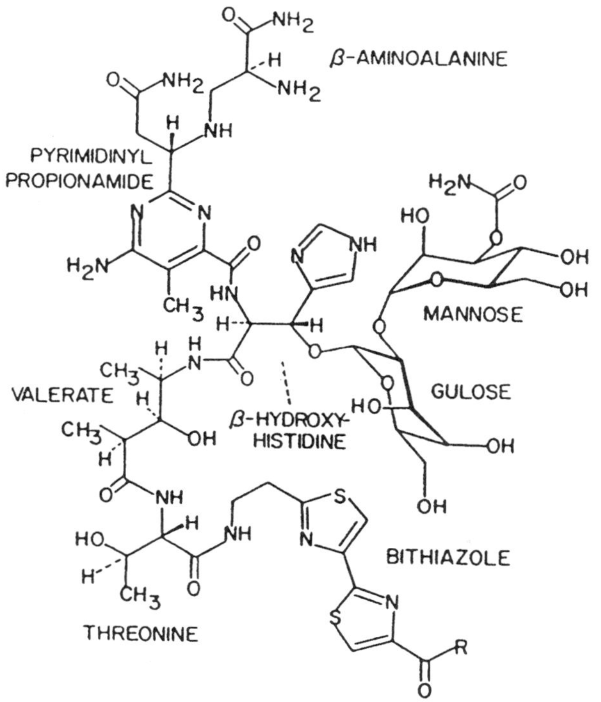

Figure 10.15 General structure of the bleomycin family of antibiotics. Bleomycin-A2 and -B2 have the sidechains $R = NH(CH_2)_3S^+(CH_3)_2$ and $R = NH(CH_2)_4NHC(NH_2)^+ NH_2$, respectively. Adapted from Oppenheimer, N. J. et al., *Proc. Natl. Acad. Sci. USA 76*, 5616–5620 (1979).

have been completely resolved, there is sufficient understanding of the key points to provide insight on the underlying chemistry.

10.3.1 Drug-DNA Complex

Figure 9.5 in the previous chapter illustrates how BLM is composed of four distinct domains. The fluorescence of the bithiazole ring is quenched in the presence of DNA (Figure 10.16), by an energy transfer mechanism analogous to that described in Chapter 9, indicating that this chromophoric unit binds to the DNA. The tripeptide-S domain contains the bithiazole moiety and side chain and appears to account largely for the DNA binding affinity (Figure 10.16 and Table 10.1).

An intercalative model is supported by the observation of helix unwinding and positive supercoiling following binding to DNA. Minor groove binding is suggested from NMR spectral analysis by the absence of a significant resonance shift of the dT(H-6) and dT(CH3-5) protons, located in the major groove, upon

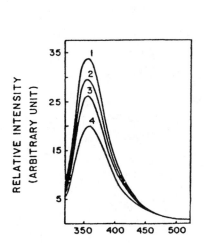
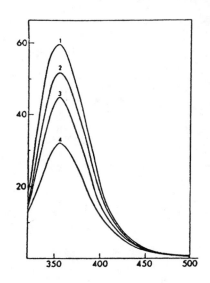

WAVELENGTH (nm)

Figure 10.16   Fluorescence spectra of BLM (left) and tripeptide-S (right) obtained with increasing concentrations of DNA (1 to 4).

Table 10.1   DNA Binding Constants for Tripeptide-S Relative to Intact Bleomycin

R_1	R_2	K (μM)
	$-(CH_2)_3S^+(CH_3)_2$	30
	$-NH(CH_2)_4NHC(NH)(NH_2)$	200
native BLM		0.1

BLM binding to poly(dAdT). Also, there is a significant structural similarity (in terms of size, shape, charge distribution, and hydrogen bonding capability) with the known groove-binding drugs netropsin and Hoechst 33258 (Figure 10.17)

These conclusions are supported by 2-D NMR studies of Zn(II)-BLM and Co(III)-BLM complexes with oligonucleotide duplexes. Figure 10.18 summarizes some intermolecular NOE's identified between Zn(II)-BLM and a d(CGCTAGCG)$_2$ octaoligonucleotide. These results confirm minor groove binding. However, there are differences in the mode of binding of Zn(II)-BLM and Co(III)-BLM which most likely reflect coordination changes at the metal. Although the coordination state for Co(III)-BLM is similar to that for Fe(III) in Figure 10.19 (with O$_2$ replaced by H$_2$O or HO$^-$ or HOO$^-$), in the case of Zn(II)-BLM, the carbamyl amide nitrogen on the mannose sugar, as well as the axial primary amine, bind to the Zn(II) ion and preclude O$_2$ binding to the cofactor.

Such models also place the metal-binding domain in close proximity to the sugar 4'-carbon-hydrogen bond, which is located in the minor groove. This bond is cleaved with a high degree of specificity during oxidative degradation

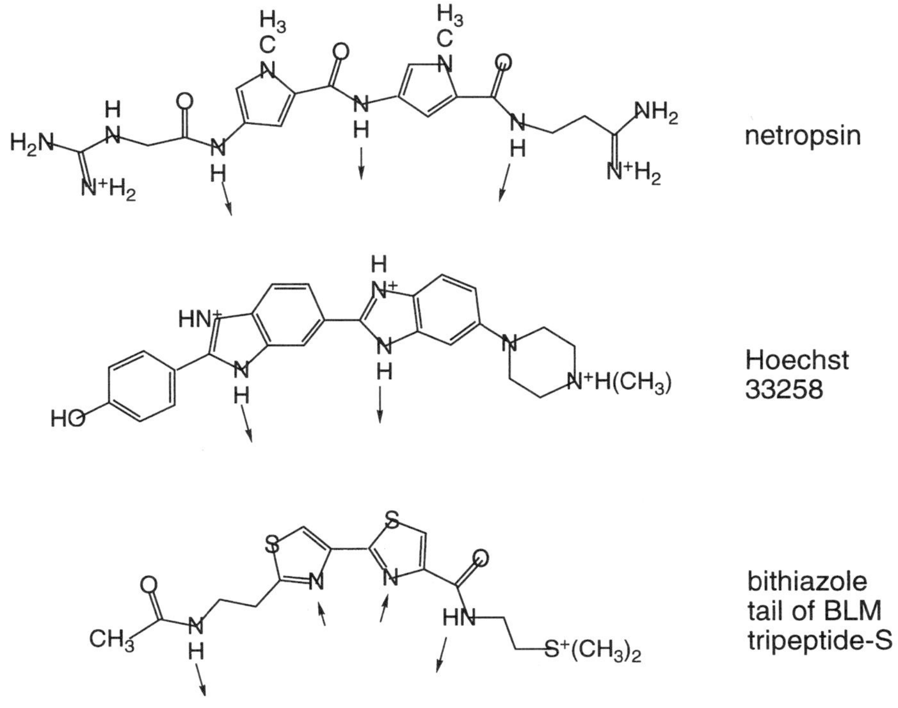

Figure 10.17 Hoechst 33258, netropsin, and tripeptide-S. Note the common crescent shape and pattern of H-bond donors.

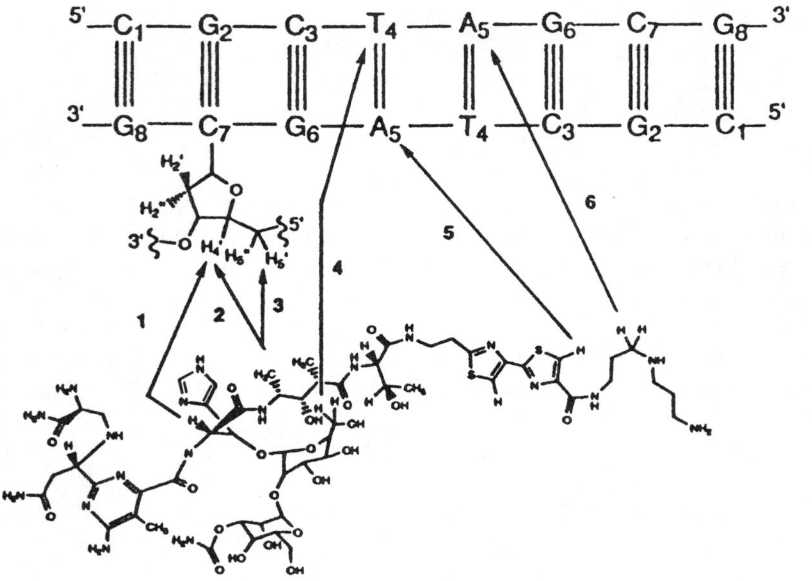

Figure 10.18 Schematic illustration showing six intermolecular NOEs from the drug to base and sugar protons. The assignments of the NOEs, labeled 1 to 6, are as follows: 1, C_7(H4′)-His(Hα); 2, C_7(H4′)-Val(CH$_3$–); 3, C_7(H5′,5″)-Val(CH$_3$–); 4, T_4(H3′)-Gul (H6,H6′); 5, A_5(H2)-Bit(H5); 6, A_5(H2)-Sp(H3). The two methyls of the methyl valerate could not be distinguished. From Manderville et al., *J. Am. Chem. Soc. 116,* 10851 (1994), with permission. See also problem 4 in Chapter 9.

of the deoxyribose ring. We shall now turn to the metal-mediated chemistry underlying this degradative reaction.

10.3.2 Coordination Chemistry

BLM is clearly in possession of a variety of potential binding ligands. There has been considerable confusion as to the identity and coordination arrangement of the complete ligand set, which, in part, arises from the wide variety of metal ions and BLM fragments used in the complexes that have been studied. The metal-binding domain is distinct from the DNA-binding domain (Figure 9.5), and Figure 10.19 shows a reasonable consensus model for the metal ligation sphere of the oxy-complex of iron bleomycin. Similar structures can be assumed for the Co^{3+} derivative, with H$_2$O or HO$^-$ or HOO$^-$ as the axial ligand. For both Fe^{3+} and Co^{3+}, hydroperoxide complexes have been characterized, and the hydroperoxide species is proposed to be a bona fide intermediate for the formation of activated bleomycin.

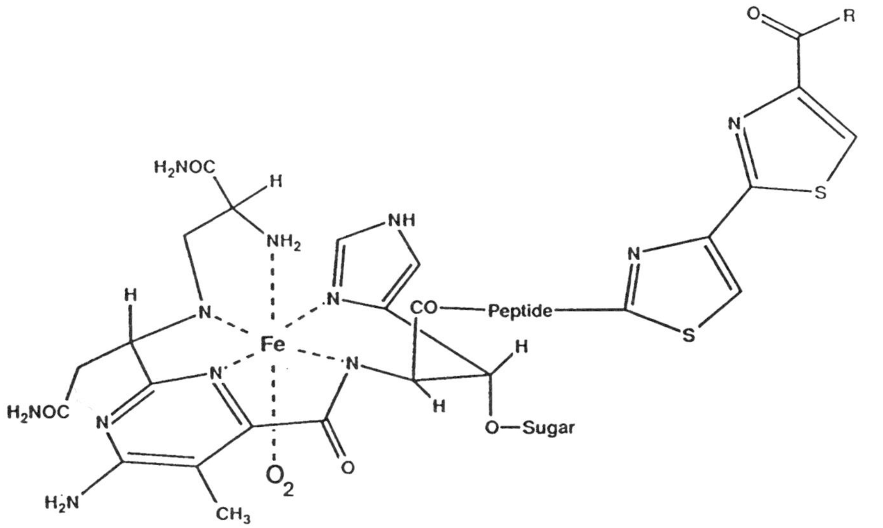

Figure 10.19 Summary of the metal–ligation sphere for BLM. [Reproduced by permission from J. W. Kozarich et al., *Science 245*, 1396–1399 (1989).]

10.3.3 Formation of Activated Bleomycin

It was realized early that the DNA degradation reaction requires both oxidizing and reducing agents. Atmospheric dioxygen can serve as the oxidizing agent. Addition of ferrous ion to anaerobic solutions of BLM produces the pink air-sensitive Fe(II)-BLM complex, although this is only formed at a pH above 6. At lower pH's, some of the ligands are protonated, such as the histidine amide nitrogen, pyrimidine, and histidine ring nitrogens. By use of stopped-flow spectroscopy, two distinct reactions have been identified upon mixing Fe(II)-BLM with O_2: first, a rapid reaction to produce the O_2 adduct with $t_{1/2} \sim 0.2$ s, and, then, a decomposition with $t_{1/2} \sim 6$ s to yield "activated" BLM, which has been formulated as the hydroperoxide derivative, $HO_2^- $-Fe(III)-BLM on the basis of EPR (Figure 10.20) and Mossbauer spectra. Metal ions, such as Mn^{2+}, Ni^{2+}, Cu^{2+} and Zn^{2+}, are incapable of forming activated BLM through reaction with dioxygen, although cobalt ion does form an active species that results in DNA damage after irradiation with UV light.

$$\text{Fe(II)-BLM} \xrightarrow[\text{0.2 s}]{O_2} O_2^- \text{–Fe(II)-BLM} \xrightarrow[\text{6 s}]{e} HO_2^- \text{–Fe(III)-BLM}$$

Formation of this hydroperoxide requires an additional electron. In the absence of external reductants, it is found that only 0.5 mole of O_2 is consumed in the production of "activated" BLM and that the second electron derives from un-reacted Fe(II)BLM. The activated form of BLM can also be generated by ad-

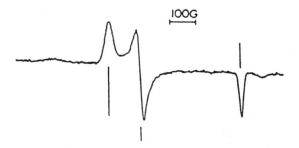

Figure 10.20   EPR spectrum of "activated" BLM with *g*-values of 2.26, 2.17, and 1.94 that are typical of low-spin ferric iron.

dition of H_2O_2 to a solution of Fe(III)-BLM, consistent with the chemistry described earlier. The rate of production of DNA degradation products is comparable to the decay rate of activated BLM, consistent with the view that this form of BLM is responsible for DNA degradation. Now we shall take a closer look at how this degradative chemistry is mediated by the activated iron center.

10.3.4   DNA Degradation

The chemical pathways that lead to strand scission are summarized in Figure 10.21. Two kinds of DNA damage are observed, and both appear to derive from a common 4'-carbon radical intermediate (Figure 10.21). A deuterium isotope effect for rate-limiting cleavage of the 4'-C-H bond by the activated iron center has been determined (Figure 10.22) by use of [4'-D]-labeled dT.

Pathway A (aerobic) requires an additional equivalent of dioxygen and results in formation of base propenal, 3'-phosphoglycolate and 5'-phosphate terminae. Pathway B (anaerobic) results from hydroxylation of the 4'-radical by water, with release of the free nucleic acid base and an alkali-labile abasic site that cleaves at pH 12 with piperidine to yield 3- and 5'-phosphate terminae. The propenal and glycolate products are therefore observed only if additional equivalents of oxygen, beyond those required for formation of activated BLM, are available. As an exercise you should try to draw arrow-pushing mechanisms for these two pathways.

The chemical mechanism for the critical cleavage of the 4'-C-H bond has not yet been defined. However, it is likely that activated BLM undergoes heterolytic oxygen–oxygen bond cleavage to form an iron-oxene center ($Fe^{3+}O \leftrightarrow Fe^{5+} = O$). This is a proposed structure for this transient intermediate state, which has not yet been characterized, although we have already seen such high valent species in Chapter 8 as intermediates in the reactions of Cyt P450 and peroxidases. Such a reactive center can readily abstract the 4'-H to yield the radical intermediate shown in Figures 10.21 and 10.23.

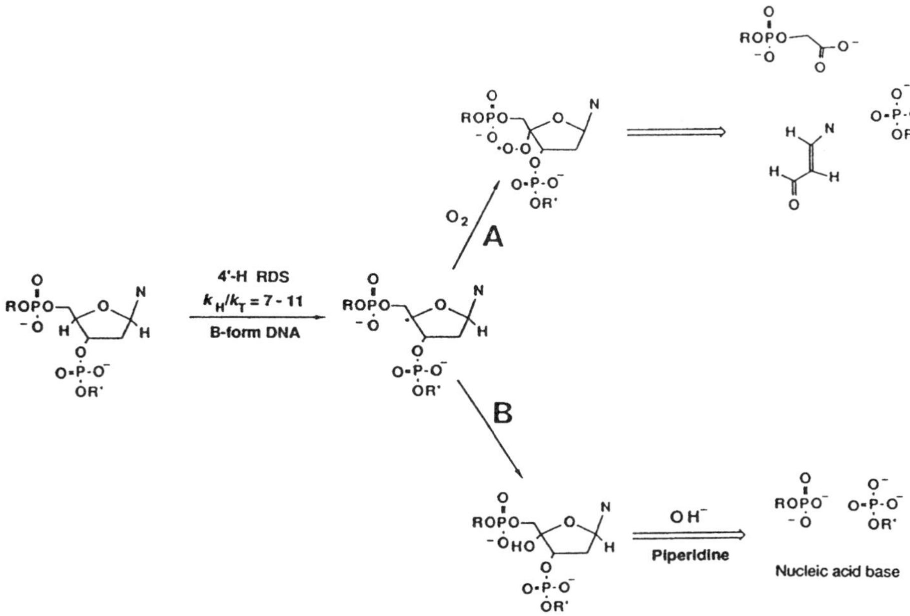

Figure 10.21 Overview of the aerobic (A) and anaerobic (B) pathways and products for cleavage of DNA by activated BLM. Note that either pathway follows from the 4'-carbon radical intermediate. From Kozarich et al., *Science 245,* 1396 (1989), with permission.

Such iron-mediated chemistry is also consistent with the observation that bleomycin does not appear to be bifunctional and that a single molecule of BLM is capable of inducing double-strand breaks, suggesting that activated BLM is regenerated after the first strand cleavage. The chemistry underlying the regeneration is outlined in Figure 10.23. The close proximity of the two breaks further supports two independent backbone scissions from one bound BLM.

10.3.5 Sequence Specificity

BLM cleaves ss-DNA at the pyrimidine (Py) in a G-Py sequence. A similar specificity is observed for the first cleavage in ds-DNA, but is distinct for the second cleavage. In sequences of the type G-Py-Py', a break at Py is usually followed by a break at the purine (Pu) base directly opposite **1**, generating blunt ends. When the sequence is G-Py-Pu, a break at Py is typically followed by a break at the pyrimidine opposite Pu, **2**, yielding a single-base 5'-extension.

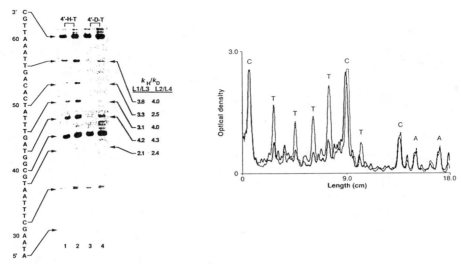

Figure 10.22 An autoradiogram showing the cleavage patterns from the 5'-^{32}P end-labeled oligonucleotide. Samples in lanes 1 and 3 correspond to pathway A in Figure 10.21 and received no additional treatment prior to electrophoresis, whereas the samples in lanes 2 and 4 were treated with piperidine to reveal the products from the anaerobic pathway B. The densitometry scan on the right compares the results from lane 2 (light line) and lane 4 (dark line). The ratio of the peak areas provides a measure of the kinetic isotope effect. The cleavage sites at C and A bases, which were not labeled with D, serve as internal controls. From Kozarich et al., *Science 245,* 1396 (1989), with permission.

These observations indicate that, for both single- and double-strand cleavage, recognition is governed by the first interaction with the single- or double-strand molecule. For the latter, the reactive iron center is already positioned for a second cleavage reaction. Clearly, the chemical and structural reasons underlying these trends have yet to be elucidated. However, much progress has been made in elucidating the key steps that define the reactivity of this interesting class of antibiotics. This knowledge can now be used in the de novo design of DNA-cleaving molecules.

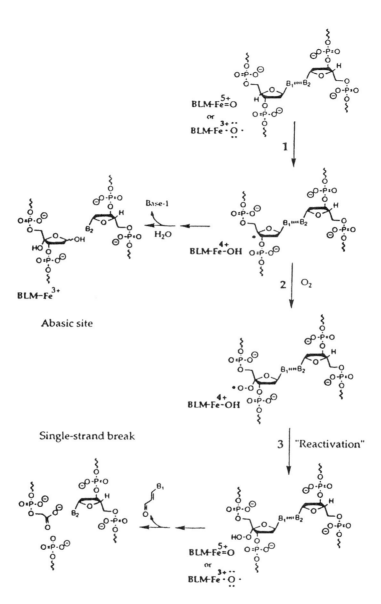

Figure 10.23 Proposed mechanism for metal-mediated oxidative chemistry. Note that the active iron-oxene intermediate is regenerated and can be used in the subsequent cleavage of the second strand for ds-DNA. From Absalon, M. J. et al., *Biochemistry 34*, 2076 (1995), with permission.

10.4   Final Remarks

The examples outlined in this chapter illustrate only a few of a diverse number of problems tackled by inorganic biochemists and highlight the diversity of methods and strategies required to solve them. Each topic remains an active research problem, and the final answers are not yet clear. As one would expect of a young science, few research problems have been solved completely. New challenges and discoveries constantly demand novel approaches and fresh ideas. On this forward-looking note, we end!

Notes

1. Several examples of bacterial cytrochrome c oxidase have also been isolated. These possess two or three subunits, which appears to support the idea of these subunits as the catalytic core.
2. See, for example, P. Kroneck et al., *FEBS Letts,* 268, 274–276 (1990) and *FEBS Letts,* 248, 212–213 (1989).
3. Two common sources of eukaryotic mitochondrial cytochrome oxidase are yeast and beef heart. Clearly, the former is a more convenient organism for labeling studies as a result of the ease of manipulation of yeast cells to generate required mutant strains, and the quantities of labeled material to be used! It is more difficult (and expensive) to carry out labeling with a half-ton cow!
4. For a more detailed discussion, see G. Brudvig et al., *J. Biol. Chem.,* 259, 11001–11009 (1984).
5. Refer to S. I. Chan and P. M. Li, *Biochemistry,* 29, 1–12 (1990) for further details.
6. A residue with a pK_a similar to tyrosine is proposed.
7. Note that the term mer R is used to describe both the gene and the protein. This is also true of the other *mer* gene products of the *mer* operon. Usually, the gene element is written in italics.
8. For example, plasmid R100 and transposon Tn 501 encode mercuric reductase systems but lack *mer* B, and so cells containing these DNA elements are sensitive to organomercurials.
9. The identical subunits are each products of *mer* A expression.
10. The volatility of Hg° allows a facile assay. Air is bubbled through the reaction mixture, and is then carried into an aqua regia (^{c}HNO$_3$/^{c}HCl) bath. This oxidizes and traps the mercury released, which may be quantitated by scintillation counting if the radiosotope ^{203}Hg is used.
11. These are enzymes (such as lipoxygenase) that possess a flavin and a disulfide as the active redox center.
12. DTNB = 5,5'-dithiobis(2-nitrobenzoic acid). The number of thiols can be computed from the enzyme concentration and the concentration of 3-carboxylato-4-nitrothiophenolate (CNT) released following reaction [determined at 412 nm ($\varepsilon = 1.14 \times 10^4$ M^{-1} cm^{-1})].

13. Recent crystallographic studies of *Bacillus* MerA show two active-site Tyr residues that most likely bind the Hg^{2+} center. N. Scheiring et al., *Nature, 352,* 168–172 (1991).

14. Also note that four-electron reduction of the enzyme reduces the redox active disulfides in the active site (Cys_{135}–Cys_{140} and Cys_{558}–Cys_{559}), but the flavin remains oxidized (FAD_{ox}). This partially reduced enzyme can bind but cannot reduce Hg^{2+}. This supports reduction by $FADH_2$.

15. The reaction rates are consistent with those expected by analysis of results from wild-type enzyme.

Further Reading

Cytochrome *c* oxidase

Antholine, W. E. et al. *Eur. J. Biochem. 209,* 875–881 (1992).

Babcock, G. T., and P. M. Callahan. *Biochemistry, 22,* 2314–2319 (1983).

Babcock, G. T. and Wikstrom, M. *Nature 356,* 301–309 (1992).

Blackburn, N. J. et al. *Biochemistry 33,* 10401–10407 (1994).

Blair, D. F. et al. *J. Am. Chem. Soc., 107,* 7389–7399 (1985).

Calhoun, M. W. et al. *Trends in Biochem. Sci. 19,* 325–330 (1994).

Chan, S. I., and P. M. Li. *Biochemistry, 29,* 1–12 (1990).

Ellis, W. R. et al. *Biochemistry, 25,* 161–167 (1986).

Larsen, R. W. et al. *Proc. Natl. Acad. Sci. USA, 89,* 723–727 (1992).

Martin, C. T. et al. *J. Biol. Chem., 260,* 2857–2861 (1985).

Martin, C. T. et al. *J. Biol. Chem., 263,* 8420–8429 (1988).

Morgan, J. E. et al. *Biochemistry, 28,* 6975–6983 (1989).

Shapleigh, J. P. et al. *Proc. Natl. Acad. Sci. USA, 89,* 4786–4790 (1992).

Stevens, T. H. et al. *J. Biol. Chem., 257,* 12106–12133 (1982).

Tsukihara, T. et al. *Science 269,* 1069–1074 (1995).

Witt, S. N., and S. I. Chan. *J. Biol. Chem., 262,* 1446–1448 (1987).

Wong, H. et al. *Biochemistry, 25,* 167–171 (1986).

Woodruff, W. H. et al. *Proc. Natl. Acad. Sci. USA 88,* 2588–2592 (1991).

Mercuric reductase

Cummings, R. T., and C. T. Walsh. *Biochemistry, 31,* 1020–1030 (1992).

Distefano, M. D. et al. *Biochemistry, 28,* 1168–1183 (1989).

Fox, B., and C. T. Walsh. *J. Biol. Chem., 257,* 2498–2503 (1982).

Fox, B., and C. T. Walsh. *Biochemistry, 22,* 4082–4088 (1983).

Miller, S. M. et al. *J. Biol. Chem., 261,* 8081–8084 (1986).

Miller, S. M. et al. *Biochemistry, 28,* 1194–1205 (1989).

Miller, S. M. et al., *Biochemistry, 29,* 2831–2841 (1990).

Moore, M. J. et al., *Acc. Chem. Res., 23,* 301–308 (1990).

Moore, M. J., and C. T. Walsh. *Biochemistry, 28,* 1183–1194 (1989).

O'Halloran, T. V. et al., *Cell, 56,* 119–129 (1989).

O'Halloran, T. V., and C. T. Walsh. *Science, 235,* 211–214 (1987).

Raybuck, S. A. et al. *J. Am. Chem. Soc., 112,* 1983–1989 (1990).

Schultz, P. G. et al., *Biochemistry, 24,* 6840–6848 (1985).

Shewchuk, L. M. et al., *Biochemistry, 28,* 2331–2339 (1989).

Shewchuk, L. M. et al., *Biochemistry, 28,* 2340–2344 (1989).

Shewchuk, L. M. et al., *Biochemistry, 28,* 6140–6145 (1989).

Bleomycin

Absalon, M. J. et al. *Biochemistry 34,* 2076–2086 (1995).

Akkerman, M. A. J. et al. *J. Am. Chem. Soc. 112,* 7462–7474 (1990).

Burger, R. M., J. Peisach and S. B. Horowitz. *J. Biol. Chem. 256*, 11636–11644 (1981).

Dabrowiak, J. C. in *Adv. Inorg. Chem. 4*, Eichorn, G. L. and L. G. Marzilli, eds., Elsevier, 1982, pp. 69–113.

Kozarich, J. W. et al. *Science 245*, 1396–1399 (1989).

Manderville, R. A. et al. *J. Am. Chem. Soc. 116*, 10851–10852 (1994).

Oppenheimer, N. J., L. O. Rodriguez, and S. M. Hecht. *Proc. Natl. Acad Sci. USA 76*, 5616–5620 (1979).

Povirk, L. F., Y-H. Han, and R. J. Steighner. *Biochemistry 28*, 5808–5814 (1989).

Steighner, R. J. and L. F. Povirk. *Proc. Natl. Acad Sci. USA 87*, 8350–8354 (1990).

Stubbe, J. and J. W. Kozarich. *Chem. Rev. 87*, 1107–1136 (1987).

Sugiura, Y. *J. Am. Chem. Soc. 102*, 5208–5215 and 5216–5221 (1980).

Worth, L. et al. *Biochemistry 32*, 2601–2609 (1993).

Xu, R. X. et al. *Biochemistry 33*, 907–916 (1994).

Appendixes

Abbreviation	Name	Structure
DMSO	dimethylsulfoxide	$(CH_3)_2SO$
EDTA	ethylenediamine tetraacetic acid	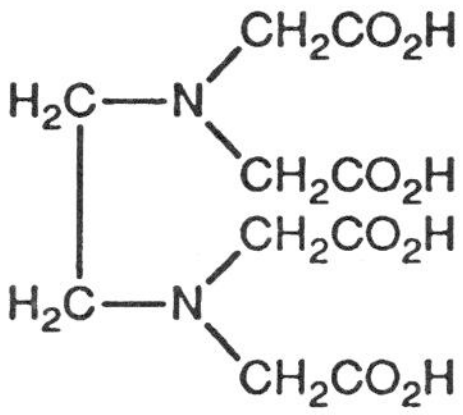
Et	ethyl	CH_3CH_2 -
Me	methyl	CH_3 -
o−phen	1,10−phenanthroline	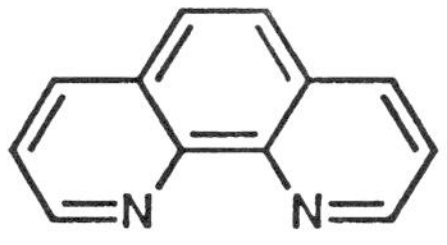
Py	pyridine	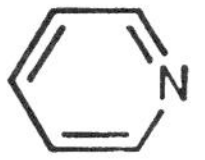
NTA	nitrilotriacetic acid	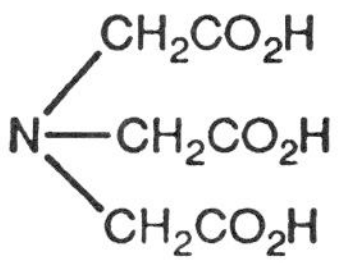
Mes	methylsulfonic acid	CH_3SO_2H

Appendix 2   Glossary of Abbreviations

Abbreviation	Meaning
ADP	Adenosine diphosphate
AMP	Adenosine monophosphate
ATP	Adenosine triphosphate
BLM	Bleomycin
CCP	Cytochrome c peroxidase
CD	Circular dichroism
CTP	Cytidine triphosphate
Cyt	Cytochrome
Da	Dalton
DMSO	Dimethylsulfoxide
DNA	Deoxyribonucleic acid
EDTA	Ethylenediamine N,N,N',N'-tetraacetate
ENDOR	Electron nuclear double resonance
EPR	Electron paramagnetic resonance
ESR	Electron spin resonance
EXAFS	Extended X-ray absorption fine structure
FID	Free induction decay
FT	Fourier transform
HRP	Horseradish peroxidase
Hb	Hemoglobin
HiPIP	High-potential iron–sulfur protein
IE	Ionization energy
Im	Imidazole
IR	Infrared
MCD	Magnetic circular dichroism
MO	Molecular orbital
mol. wt.	Molecular weight
MW	Molecular weight
Mb	Myoglobin
Me	Methyl
NAD^+	Nicotinamide adenine dinucleotide (oxidized form)
NADH	Nicotinamide adenine dinucleotide (reduced form)
$NADP^+$	Nicotinamide adenine dinucleotide phosphate (oxidized form)
NADPH	Nicotinamide adenine dinucleotide phosphate (reduced form)
NMRD	Nuclear magnetic relaxation dispersion
NOE	Nuclear Overhauser effect
NTA	Nitrilotriacetate
NTP	Nucleotide triphosphate
OD	Optical density
Pi	Inorganic phosphate
ppm	Parts per million
Py	Pyridine
rds	Rate determining step
RNA	Ribonucleic acid
SI	International system of units
SOD	Superoxide dismutase
2D	Two-dimensional
ZFS	Zero field splitting

Appendix 3 Ionic Radii for Cations and Anions: The Dependence on Coordination Number (C.N.) and Spin State[a]

Cation	C.N.	Radius (Å)	Cation	C.N.	Radius (Å)
Na^+	6	1.16	Mg^{2+}	6	0.86
K^+	6–8	1.52–1.65	Ca^{2+}	6–8	1.14–1.26
V^{2+}	6	0.93	Co^{2+}	4 (HS)	0.72
Cr^{2+}	6 (HS)	0.94	Co^{2+}	6 (HS)	0.89
Cr^{2+}	6 (LS)	0.87	Co^{2+}	6 (LS)	0.79
Mn^{2+}	6 (HS)	0.97	Co^{3+}	6 (HS)	0.75
$Mn^{2\pm}$	6 (LS)	0.81	Co^{3+}	6 (LS)	0.69
Fe^{2+}	4 (HS)	0.77	Ni^{2+}	4	0.69
Fe^{2+}	6 (HS)	0.92	Ni^{3+}	6 (HS)	0.74
Fe^{2+}	6 (LS)	0.75	Ni^{3+}	6 (LS)	0.70
Fe^{3+}	4 (HS)	0.63	Cu^+	4	0.74
Fe^{3+}	6 (HS)	0.79	Cu^{2+}	4	0.71
Fe^{3+}	6 (LS)	0.69	Cu^{2+}	6	0.87
Zn^{2+}	4	0.74	Mo^{4+}	6	0.79
Zn^{2+}	6	0.88			

Anion	Radius (Å)	Anion	Radius (Å)
F^-	1.17	O^{2-}	1.21–1.28
Cl^-	1.67	S^{2-}	1.70
Br^-	1.82	OH^-	1.18–1.23
I^-	2.06		

[a] The difference between high-spin (HS) and low-spin (LS) configurations is explained in Section 1.6. Data taken from F. A. Cotton and G. Wilkinson, *Advanced Inorganic Chemistry*, 5th ed. Values are derived from crystallographic analyses of simple coordination compounds. Different values from other sources may be quoted in later portions of the text. These inconsistencies reasonably reflect the inherent errors in the experimental data used to derive these values.

Appendix 4 Relative Energies of *d*-Orbitals in Common Coordination Geometries[a]

C.N.	Structure	$d_{x^2-y^2}$	d_{z^2}	d_{xy}	d_{xz}	d_{yz}
2	Linear[b]	−6.28	10.28	−6.28	1.14	1.14
3	Trigonal[c]	5.46	−3.21	5.46	−3.86	−3.86
4	Tetrahedral	−2.67	−2.67	1.78	1.78	1.78
4	Square planar[c]	12.28	−4.28	2.28	−5.14	−5.14
5	Trigonal bipyramidal[d]	−0.82	7.07	−0.82	−2.72	−2.72
5	Square pyramidal[d]	9.14	0.86	−0.86	−4.57	−4.57
6	Octahedral	6.00	6.00	−4.00	−4.00	−4.00

[a] All values are relative to the ligand field-splitting energy for an O_h ligand set ($\Delta_o = 10$).
[b] Bonds lie along *z* axis.
[c] Bonds in the *xy* plane.
[d] Pyramidal base lies in the *xy* plane.

Appendix 5 NMR Properties of Useful Nuclei for the Study of Biochemical Systems

(a) Spin $\frac{1}{2}$ nuclei

Isotope	Natural Abundance	NMR Frequency (MHz)	Relative Receptivity	
			^{1}H	^{13}C
^{1}H	99.985	100.00	1.00	5.67×10^3
^{13}C	1.108	25.14	1.76×10^{-4}	1.00
^{15}N	0.37	10.14	3.85×10^{-6}	2.19×10^{-2}
^{19}F	100	94.09	0.83	4.73×10^3
^{31}P	100	40.48	0.066	3.77×10^2
^{113}Cd	12.26	22.19	1.35×10^{-3}	7.67
^{195}Pt	33.8	21.41	3.39×10^{-3}	19.2
^{205}Tl	70.50	57.633	0.14	7.91×10^2

(b) Quadrupolar nuclei ($I \geq \frac{1}{2}$)

Isotope	Spin	Natural Abundance	Quadrupole moment (10^{28} Q/m^2)	NMR Frequency (MHz)	Relative Receptivity	
					^{1}H	^{13}C
^{2}H	1	0.015	2.8×10^{-3}	15.351	1.45×10^{-6}	8.21×10^{-3}
^{7}Li	3/2	92.58	-4×10^{-2}	38.866	0.272	1.54×10^3
^{11}B	3/2	80.42	-4.1×10^{-2}	32.089	0.133	7.52×10^2
^{14}N	1	99.63	1×10^{-2}	7.228	1.00×10^{-3}	5.69
^{17}O	5/2	0.037	-2.6×10^{-2}	13.461	1.08×10^{-5}	6.11×10^{-2}
^{23}Na	3/2	100	0.10	26.466	9.27×10^{-2}	5.26×10^2
^{25}Mg	5/2	10.13	0.22	6.126	2.72×10^{-4}	1.54
^{35}Cl	3/2	75.53	-0.10	9.809	3.56×10^{-3}	20.2
^{39}K	3/2	93.1	4.9×10^{-2}	4.672	4.75×10^{-4}	2.69
^{43}Ca	7/2	0.145	-4.9×10^{-20}	6.738	8.67×10^{-6}	4.92×10^{-2}
^{51}V	7/2	99.76	-5×10^{-2}	26.336	0.383	2.17×10^3
^{59}Co	7/2	100	0.38	23.61	0.277	1.57×10^3

Appendix 6 Voltage Limits for Various Electrodes in an Aqueous Medium at Neutral pH[a]

Material	Cathodic Limit	Anodic Limit
Pt	-0.9 V	$+0.9$ V
Au[b]	-0.8 V	$+0.9$ V
C (glassy)	-0.8 V	$+0.9$ V
Hg	-1.8 V	$+0.1$ V

[a] Relative to NHE.
[b] Good for cathodic ($+$ve) scans, but anodic ($-$ve) scans are hindered by reactivity toward anions (e.g., Cl$^-$, CN$^-$).

Appendix 7   A Selection of Mediators and Their Potentials vs. NHE

Mediator	$E°$ (mV)	Mediator	$E°$ (mV)
$Ru(phen)_3^{3+}$	+1220	$[Ru(NH_3)_5 \, pyridine]^{3+}$	+260
$Fe(phen)_3^{3+}$	+1107	$Fe(EDTA)^-$	+120
$IrCl_6^{2-}$	+892	$Ru(NH_3)_6^{3+}$	+51
$Mo(CN)_8^{2-}$	+789	Methylene blue	+5
1,1′-Dicarboxylferrocene	+644	2,5-Dihydroxybenzoquine	−60
$Co(oxalate)_3^{3-}$	+570	2-Hydroxy-1,4-naphthoquinone	−139
$W(CN)_8^{3-}$	+510	$Co(en)_3^{3+}$	−216
Ferrocene	+431	Benzylviologen	−352
$Fe(CN)_6^{3-}$	+424	Methylviologen	−446
$Co(phen)_3^{3+}$	+370		

Appendix 8   Commonly Used Materials and Procedures in Protein or Enzyme
Purification

1. **Ion-exchange resins** (separate proteins according to surface charge)
   CM (cation-exchange resin)   binds proteins with pI > 7
   DEAE (anion-exchange resin)   binds proteins with pI < 7
2. **Gel filtration** (separates proteins according to molecular weight)
   a. Pharmacia Sephadex (larger proteins are eluted more rapidly)
       Separate globular proteins in the size range (Da) 1.5×10^3 to 3×10^4 (G-50); 3×10^3 to 7×10^4 (G-75); 4×10^3 to 15×10^4 (G-100); and 5×10^3 to 60×10^4 (G-200), respectively.
   b. Biogels (larger proteins are eluted more rapidly)
       Separate globular proteins in the size range (Da) 3×10^3 to 6×10^4 (P-60); 5×10^3 to 10×10^4 (P-100); 15×10^3 to 15×10^4 (P-150); and 30×10^3 to 200×10^4 (P-200), respectively.

3. **Affinity columns** (separate proteins according to their relative affinities for a defined matrix material)
   a. Hydroxyl apatite. A calcium phosphate mineral material. Proteins or enzymes are eluted with varying concentrations of Ca^{2+} or PO_4^{3-} ions.
   b. Metal affinity columns. Based on the binding of surface residues to metals bound to a column matrix. For example, Cu^{2+} columns bind to histidines on a protein surface. The proteins are eluted by competition with another ligand in the elution buffer (e.g., imidazole).

4. **Ammonium sulfate precipitation** [separates proteins according to their solubility properties in solutions containing variable amounts of $(NH_4)_2SO_4$]

Appendix 9   Codons for Amino Acid Residues[a]

First Position (5' end)	Second Position				Third Position (3' end)
	U/T	C	A	G	
U/T	Phe	Ser	Tyr	Cys	U/T
	Phe	Ser	Tyr	Cys	C
	Leu	Ser	STOP	STOP	A
	Leu	Ser	STOP	Trp	G
C	Leu	Pro	His	Arg	U/T
	Leu	Pro	His	Arg	C
	Leu	Pro	Gln	Arg	A
	Leu	Pro	Gln	Arg	G
A	Ile	Thr	Asn	Ser	U/T
	Ile	Thr	Asn	Ser	C
	Ile	Thr	Lys	Arg	A
	Met	Thr	Lys	Arg	G
G	Val	Ala	Asp	Gly	U/T
	Val	Ala	Asp	Gly	C
	Val	Ala	Glu	Gly	A
	Val	Ala	Glu	Gly	G

[a] The genetic code is composed of 64 triplet codons that are read in the 5' to 3' direction. For example, the codon CAG encodes glutamine (Gln), and GGU encodes glycine (Gly). Codon AUG specifies methionine (Met) but also serves as a start codon to initiate protein synthesis. STOP indicates where the protein sequence ends.

Appendix 10 Physical and Mathematical Constants

Quantity	Symbol	SI	CGS (emu)
Permeability of a vacuum	μ_o	$4\pi \times 10^{-7}$ kg m s^{-2} A^{-2}	1
Speed of light in a vacuum	c	2.9979×10^{8} m s^{-1}	2.9979×10^{10} cm s^{-1}
Elementary charge (absolute value of the electron charge)	e	1.6022×10^{-19} A s	1.6022×10^{-20} coulombs
Planck's constant	h	6.6262×10^{-34} J s	6.6262×10^{-27} erg s
	$\hbar = h/2\pi$	1.0546×10^{-34} J s rad^{-1}	1.0546×10^{-27} erg s rad^{-1}
Avogadro constant	N_A	6.0220×10^{23} mol^{-1}	—
Electron rest mass	m_e	0.9110×10^{-30} kg	0.9110×10^{-27} g
Proton rest mass	m_p	1.6726×10^{-27} kg	1.6726×10^{-24} g
Electron g factor	g_e	2.0023	
Bohr Magneton	μ_B	9.2741×10^{-24} J T^{-1}	9.2741×10^{-21} erg G^{-1}
Electron-to-proton magnetic moments ratio	μ_e/μ_p	658.21	—
Proton magnetogyric ratio	γ_p	2.6752×10^{8} rad s^{-1} T^{-1}	2.6752×10^{4} rad erg^{-1} K^{-1}
Boltzmann constant	k	1.3807×10^{-23} J K^{-1}	1.3807×10^{-16} erg K^{-1}
Hyperfine coupling constant of the hydrogen atom	(A_c/h)	1.4204×10^{9} Hz	8.9247×10^{9} rad s^{-1}
Mathematical constants	e	2.7183	—
	π	3.1416	—

Appendix 11 Conversion Factors

Quantity	To Convert from	to	Multiply by
Length	angstroms (Å)	meters (m)	1×10^{-10}
Pressure	atmospheres (atm)	pascals (Pa)	1.01325×10^5
		(kg m^{-1} s^{-2})	
Mass	atomic mass units (amu)	kilograms (kg)	1.6606×10^{-27}
Energy	calories (cal)	joules	4.1840
	electron volts (eV)	joules	1.6022×10^{-19}
	kilowatt-hours (kWh)	joules	3.6×10^6
Angles	degrees (deg)	radians (rad)	0.017453
Volume	liters (l)	cubic meters (m^3)	1×10^{-3}
Frequency	radians/seconds (rad s^{-1})	cycles/seconds (cps) (Hz)	0.15915
	wave number (cm^{-1})	energy (J)	1.9865×10^{-23}
		frequency (s^{-1})(Hz)	2.9979×10^{10}
		frequency (rad s^{-1})	1.8837×10^{11}
	frequency (s^{-1})(Hz)	energy (J)	6.6262×10^{-34}
	temperature (K)	energy (J)	1.3807×10^{-23}
		wave numbers (cm^{-1})	0.69467

Appendix 12 Some Important Physical Parameters and Their SI Units

Physical Quantity	Symbol	Unit	Fundamental Units
Electric charge (quantity of electricity)	Q	C	A s
Electric current	I	A	A
Electric potential	V	V	kg m^2 s^{-3} A^{-1}
			($=$ J A^{-1} s^{-1})
Energy	E	J	kg m^2 s^{-2}
Frequency	ν	Hz	s^{-1}
Frequency (angular velocity)	ω	rad s^{-1}	rad s^{-1}
Magnetic field strength	H	A m^{-1}	A m^{-1}
Magnetic induction (flux density)	B	T	kg A^{-1} s^{-2}
Magnetic moment	μ	J T^{-1}	A m^2
Magnetic susceptibility	χ	m^3	m^3
Magnetization	M	J T^{-1} m^{-3}	A m^{-1}
Magnetogyric ratio	γ	rad s^{-1} T^{-1}	rad A s kg^{-1}
Molar magnetic susceptibility	χ_m	m^3 mol^{-1}	m^3 mol^{-1}
Power	P	W	kg m^2 s^{-3}
			($=$ J s^{-1})
Viscosity (dynamic)	η	kg m^{-1} s^{-1}	kg m^{-1} s^{-1}

Solutions to Problems Lacking Original Literature References

Chapter 1. Solutions to Problems

1. The sulfinate ligand may coordinate either through the sulfur atom or one of the two oxygen centers. Because the lone pair on the sulfur is stereochemically active, the bound form exists in two enantiomerically distinct forms.

3. Stepwise addition of Br^- to $Cd(H_2O)_6^{2+}$ follows the sequence $Cd(H_2O)_6^{2+}$ → $[Cd(H_2O)_5Br]^+$ → $[Cd(H_2O)_4Br_2]^0$ → $[Cd(H_2O)_3Br_3]^-$ → $[CdBr_4]^{2-}$. The stepwise formation constants initially decrease because the overall positive charge diminishes and lowers the electrostatic attraction for additional Br^-. After addition of the fourth Br^-, tetrahedral coordination becomes the favored geometry because this maximizes the distance between the negatively charged bromide ions. Also the release of water molecules is entropically favorable.

6. (a) $k_{12} = 8.5 \times 10^{12}$ $M^{-1}s^{-1}$. Use the relationships $\Delta G = -RT \ln K_{12} = -nF\Delta E$, and $k_{12} = (k_{11}k_{22}K_{12}f)^{1/2}$ and assume $f = 1$.

    (b) Outer sphere. NH_3 lacks a free lone pair to form a bridge to $Cr(H_2O)_6^{2+}$; consequently an inner sphere pathway would require inner sphere substitution of $Cr(NH_3)_6^{3+}$ by a bridging water ligand from $Cr(aq)^{2+}$. However, the electron transfer rate is more rapid than the ligand exchange rate for $Cr(NH_3)_6^{3+}$. Note that, after reduction, $Cr(NH_3)_6^{2+}$ is labile and the NH_3 ligands will be rapidly exchanged with H_2O.

    (c) The $Cr(NH_3)_6^{2+}$ complex has a labile chromous ion. A water molecule on $Cr(H_2O)_6^{3+}$ can substitute in the inner sphere of the ammine complex, and the bridge facilitates rapid electron transfer. Note that, after oxidation, the inert $Cr(NH_3)_5(H_2O)^{3+}$ product is stable.

7. (a) The self-exchange rates for $IrCl_6^{2-}$ and $Fe(DMP)^{2+}$ are hindered by repulsive electrostatic forces, although the rate constant for the latter is larger as a result of favorable π-π interactions between DMP ligands that facilitate electron transfer and maximize the separation of the two metal centers. The electron exchange rates for the species AmO_2^{2+} and NpO_2^+ are further decreased both by unfavorable repulsive interactions and by in-

creased activation barriers as a result of the larger LFSE's for these two metal ions.

(b) The first cross-reaction is favored by favorable electrostatic attraction between the exchanging species which is not present in the self-exchange process. The self-exchange rates are fairly rapid because the ions are large and the charge density is low, resulting in modest electrostatic interactions. For the second cross-reaction, the ionic species are similar, with comparable electrostatic interactions for both the cross-exchange reactions, and so the calculated and observed rates are similar. The self-exchange rates are significantly lower as a result of the larger electrostatic repulsion resulting from the smaller ionic species

(c) $\Delta E = -31$ mV for $IrCl_6^{2-/3-}$ / $Fe(DMP)^{3+/2+}$ and $\Delta E = +462$ mV for $AmO_2^{2+/+}$ / $NpO_2^{2+/+}$.

11. (a) Refer to Figure 1.16 for the structures of the amino acid side chains.
    (b) Remember to read from the 5' to 3' direction! The sequence is Ile-Phe-Lys-Cys.

Chapter 2.   Solutions to Problems

6. The disproportionation reaction is $3Au^+ \rightarrow Au^{3+} + 2Au^0$, with $\Delta G = -89.7$ kJ mole^{-1}. This follows from the following redox reactions and the standard equation, $\Delta G = -nF\Delta E$.

$$3Au^+ + 3e^- \rightarrow 3Au \qquad \Delta G = -529.7 \text{ kJ mole}^{-1}$$
$$Au \rightarrow Au^{3+} + 3e^- \qquad \Delta G = +440.0 \text{ kJ mole}^{-1}$$

Using the equation $\Delta G = -RT \ln K$ with $\Delta G = -89.7$ kJ mole^{-1}, we can calculate an equilibrium constant of $K = [Au^{3+}]/[Au^+]^3 = 5.3 \times 10^{19}$ M^{-2} for the disproportionation reaction at 298 K, where $[Au^0]$ is neglected because it is a solid with an assumed activity of unity. From the solubility product $K_{sp} = 2 \times 10^{-13}$ for the reaction $Au^+ + Cl^- \rightarrow AuCl(s)$, we can determine $[Au^+] = (2 \times 10^{-13})^{1/2} = 4.5 \times 10^{-7}$, under saturating concentrations of Au^+, assuming $[Au^+] = [Cl^-]$. At this concentration of Au^+ ion, $[Au^{3+}] = 4.7 \times 10^{-4}$ M, and so most of the gold ion in solution will be in the form of Au^{3+} as a result of disproportionation.

9. The reference potential $E^0(H^+/H_2) = 0$ mV is defined for solution conditions of 1 M H^+ at 298 K. At pH 7, $E = E^0 + (RT/nF) \ln [H^+]$ with $[H^+] = 1 \times 10^{-7}$ M.

Chapter 3.   Solutions to Problems

2. (a) See the $Co(en)_3^{3+}$ complexes in section 2.2.2 as an illustration.
   (b) Enantiomers have the same physical properties, and so, the affinity for iron should be the same. Any selectivity in uptake must, therefore, arise as a result of interactions with the siderophore receptor. The binding

pocket will contain chiral amino acids that interact differently with the two enantiomeric iron catec holate complexes, allowing discrimination.
(c) Circular dichroism is one possibility. Can you think of others?

Chapter 4.   Solutions to Problems

1. See section 4.1.1.

Chapter 5.   Solutions to Problems

5. First, the nitrogen N* effects a displacement of sulfite from APS to form the flavin adduct shown and ADP^{2-}. This sulfite complex then rearranges with los of H$^+$ and sulfite, as shown.

Chapter 6.   Solutions to Problems

1. The transport rate is significantly greater than the ligand exchange rate. This implies that the ion is transported as an outer sphere complex with no loss of metal bound water.
2. (a) [ATP]/[ADP] $= 5.9 \times 10^{-5}$, and so the enzyme functions as an ATPase rather than a synthase. Refer to "Energy Transduction in Biological Membranes" study edition, by Cramer & Knaff, Springer-Verlag, 1991.
   (b) At -120 mV, Cl$^-$ ions flow out of the cell. At $+30$ mV, Cl$^-$ ions flow into the cell.
   (c) The free energy cost is 4.88 kcal mole^{-1} for the chemical work and 2.76 kcal mole^{-1} for the electrical work, giving a total of 7.6 kcal mole^{-1}.
3. Biochemical roles for Ca^{2+} ion include structural (bone; crosslinking sugars on membrane surfaces), storage (parvalbumin), catalytic (staphylococcal nuclease), and regulatory (calmodulin). Relevant properties include fast ligand exchange, relatively high charge density, and an expanded and asymmetric coordination sphere. See if you can work out which properties are relevant for the function of the systems listed.
6. K$^+$ is a hard ion and Tl$^+$ is a relatively soft ion, and so Tl$^+$ does not complex

readily with the oxygen rich ionophores that bind K^+. It is clear from its position in the periodic table that Tl^+ is significantly more electron rich than K^+ and also has additional orbitals available for bonding.

8. Look at the melting curves (change in optical absorbance with temperature). Melting of double-strand DNA results in a considerable loss in base-pair stacking with a large increase in absorbance (usually $\geq 30\%$). In the case of single-strand DNA, there is less base stacking and a smaller increase in absorbance.

Chapter 7. Solutions to Problems

3. (a) Assuming that the concentration of P_0 is in excess, $1/K_a = K_d = [P][D]/[DP] \sim [P_0]\{[D_0] - [DP]\}/[DP] = [P_0][D_0]/[DP] - [P_0]$

   (b) In the absence of Cd^{2+}, $K_a = 1.3 \times 10^{10}$ M^{-1}, but is 1.2×10^4 M^{-1} after addition of 0.6 µM Cd^{2+}. The regulatory protein binds weakly in the presence of µM Cd^{2+} and is released from the DNA. The protein functions as a repressor, inhibiting transcription of the storage gene in the absence of significant levels of toxic divalent cadmium.

   (c) Refer to the cited paper.

Chapter 8. Solutions to Problems

6. (a) The ^{99m}Tc isotope decays rapidly with reduced risk to the patient. The high energy γ-radiation is not readily absorbed by tissues and is easily detected. In the contrast, the ^{186}Re isotope emits β-radiation that can interact with and destroy cells over a longer timescale. The slower decay results in an emission of lower intensity, but one that can be continued over a therapeutically reasonable timeframe.

   (b) The franctions remaining after 1 day are 6.3% for ^{99m}Tc and 83% for ^{186}Re. This follows from the relationship % fraction $= 100 \exp(-0.693$ $t/t_{1/2})$. This relationship can be easily derived from equation (1.2). Try to prove it.

   (c) $\mathbf{a} = {}^{234}$Th, $\mathbf{b} = {}^{234}$Pa, $\mathbf{c} = {}^{234}$U, $\mathbf{d} = {}^{230}$Th, $\mathbf{e} = {}^{226}$Ra, $\mathbf{f} = {}^{222}$Rn, $\mathbf{g} = {}^{218}$Po, $\mathbf{h} = {}^{214}$Pb, $\mathbf{i} = {}^{214}$Bi, $\mathbf{j} = {}^{214}$Po, $\mathbf{k} = {}^{210}$Pb, $\mathbf{l} = {}^{210}$Bi, $\mathbf{m} = {}^{210}$Po, $\mathbf{n} = {}^{206}$Pb.

Chapter 9. Solutions to Problems

3. (a)

$$e^- \; HO{-}OH \longrightarrow HO^\cdot + HO^-$$

   (b) The negative charge prevents complex formation with the DNA which would alter the structure and stability of the polynucleotide, inhibit binding of the ligand or protein species that is interacting with the DNA, and interfere with the Fenton chemistry.

Index

Following the page number, the letter f *indicates the entry is a figure;* t, *a table;* n, *a note;* p, *a problem;*
a, *in an appendix.*

431